Architectural
Metals

Architectural Metals

A Guide to Selection, Specification, and Performance

L. William Zahner

JOHN WILEY & SONS, INC.
New York • Chichester • Brisbane • Toronto • Singapore

Library of Congress Cataloging-in-Publication Data:
Zahner, L. William.
    Architectural metals : a guide to selection, specification, and
performance / L. William Zahner.
        p.   cm.
    Includes index.
    ISBN 0-471-04506-3 (alk. paper)
    1. Metals—Specifications.   2. Architectural metal-work.
    3. Decoration and ornament, Architectural.   I. Title.
    TA461.Z34   1995
    624.1'82—dc20   94-40182

Printed in the United States of America

10 9 8 7 6 5 4 3 2 1

To my father, grandfather and great grandfather.

Contents

Foreword

Bill Zahner and I met in 1983 when he invited me to participate in a Sheet Metal Exhibition at the National Building Museum in Washington D.C. I was happy to do so because at that time I was exploring the creative potential of sheet metal and was frustrated because there was so little attention paid to this craft and so little information available.

I designed a Tower of Babel for the exhibit, and since then Bill and I have been talking metal. I use it in almost all my buildings and, thanks to Bill's expertise, I have been able to keep trying and testing new ideas. Apart from being economical, easy to maintain, and quick to construct, it offers endless artistic possibilities.

Metal is sculptural, allowing for free-form structures inconceivable in any other material. It interacts with light and reflected water in a magical way. Changing constantly with the weather, light bounces and shimmers and glides across metal's iridescent surface in a way that is quite poetic. Some metals corrode in really interesting ways. Others maintain their jewel-like sheen over time.

Sheet metals come in different costs and quality. I have clad buildings in the simplest galvanized steel as well as the richer copper and aluminum, and now zinc and titanium. Each has its own beauty and place.

For me, metal is the material of our time. It enables architecture to become sculpture; it also expresses technological possibility as well as the time-honored characteristics of quality and permanence.

I applaud Bill for shedding light on the art of metal in this excellent guide to architectural metals.

FRANK O. GEHRY

Preface

Over the past few years my involvement with different architects and their questions concerning the use of various metals in their designs prompted the idea for a resource to supply such information. The search for alternative uses for metals, the desire to create new finishes or to use old finishes in new and unique ways, required research to distinguish the possible from the problematic.

I have worked around metal and architecture all my life. My father and grandfather operated an architectural metal company started by my great-grandfather. I recall assisting in taking inventory in the metal shop and sorting metal scrap for recycling, as a young man. Metal shops, in those days as well as today, were fastidious recyclers of metal scrap.

At first it was difficult to determine aluminum from coated steel or stainless steel by sight or feel alone. If you had a magnet, you could quickly identify the steel; stainless, on the other hand, was both magnetic and not magnetic, depending on the alloy. Confusion reigned, and the scrap bins were chaos during the summers I worked at my father's shop.

I was shown by the craftsmen how to feel the edge of the metal and determine the softness of aluminum or the hardness of stainless. Among the hazards of working in a metal shop were the constant, endless cuts on your hands that resulted from handling thin sheets, particularly if you were always trying to figure out whether an edge was soft or hard. Stainless steel edges had a lovely razor-blade quality.

The next metals I found perplexing were the coated metals. Here were metals that looked much like the ubiquitous spangled galvanized steel, yet they were softer. On view... an edge I could discern the red copper color. Once I could quickly identify lead-coated copper, I now had to distinguish terne-coated stainless steel, which had a shiny edge and a springy feel.

The more exotic metals, such as the brasses, were more difficult to distinguish. The subtle color differences were often concealed by oxides on the surface. Even for the experienced craftsmen, distinctions between the brasses were difficult. Usually, I would have to polish a portion of the surface to bring out the real color of the brass alloy. A darker color meant commercial bronze or red brass, while a yellow color meant Muntz metal or yellow brass.

Once, while taking inventory in the dark recesses of the old shop, I found a few sheets of metal that had been concealed under dust and old copper downspouts. The sheets were marked "Monel," and only the oldest craftsmen knew what this grayish metal with the green tint was. A magnet would not stick to the metal, so I assumed it must be stainless steel because of the silvery color—but there was a difference. This metal had aged and had a very slight green tint, perhaps caused by the copper downspouts, I thought.

Different trade names, such as Paintgrip, Zincgrip, Chink, Tri-gard, and others entered the inventory—and my vocabulary—each year, expanding the mystery of the seemingly endless list of available metals.

In taking the inventory, I learned the strange units of measure of the metals in relation to one another. The steels and stainless steels were measured by taking a slotted wheel and fitting the edge of a sheet into one of the various slots until it fit tightly. The slots were incremented by what were known as the Brown and Sharpe numbers. The higher the gauge number, the thinner the sheet. When I used the gauge on aluminum, the metal would fit into one of the slots, but if I reported the number (say, 22 gauge aluminum), the craftsmen looked at me as if I had been hanging around the sheet lead too much. Aluminum was measured by the more familiar decimal relationship. Thicknesses were measured in decimal inches by a micrometer.

Copper was another story altogether. Copper was measured in ounces per square foot. The old craftsmen just knew by feel the difference between 16-ounce copper and 20-ounce copper. I had to use a micrometer and then go to a chart on average thicknesses for a particular weight per square foot. Inventory, when I had to take it, was a prolonged, drawn-out affair. But all the questions presented by the various metals, their forms, and their idiosyncrasies required an understanding. Mastering the world of architectural metals seemed to be like traveling across the different countries of Europe. The languages of the various countries have their similarities, but understanding them takes time.

Metal has moved into the limelight as a primary architectural material. In the past, the palette of viable metals was limited not by availability, necessarily, but by general understanding and information available from the various sources of each metal.

There is a greater variety of metal types available to the architect today than at any time in the history of civilization. The ancients knew of just seven metals: gold, silver, copper, iron, tin, lead, and mercury. A century ago, aluminum was considered a precious metal, more precious than gold or silver. Only a very small amount of the metal had ever been isolated. The idea that metals could resist nature (stainless steel, for example) was unheard of. Initially, the properties of stainless steel were only a novelty, good for cutlery. Just a few decades ago titanium was considered to be a brittle, weak material, suitable for creating pigments in paints. Only recently has titanium generated interest in the realm of architecture. Custom finish variations and different chemical coloring techniques for metals are just now finding their way from the world of art into architecture.

It is the aim of this book to provide a tool to the designer and to the artisan working with metal—to uncover some of the mystery and to establish a logic for the various forms and sources of this unique building material. If anything, this book should provide a wealth of information on the various forms and behaviors of the common, and some uncommon, metals used in architecture and building construction.

L. WILLIAM ZAHNER

ACKNOWLEDGMENTS

The author would like to acknowledge the following people:

- Frank O. Gehry, for his many challenging designs that utilize metal surfaces and textures in ways never before imagined. His genius will have a major effect on us all for decades to come.
- My publisher, John Wiley & Sons, Inc., in particular Amanda Miller, for putting up with the madness of a book on metal.
- Bruce Biesman Simons of Frank O. Gehry and Associates.
- Gary Nemchock and the A. Zahner Company employees and staff.

Introduction

Metal
(L, metallum, mine) one of a class of elements, distinguished from a non-metal by physical properties such as ductility, malleability, hardness, conductivity, the ability to form alloys, and some qualities of appearance such as luster.

Like stone and wood, metals are basic, natural building materials. Their varieties of color and texture are as numerous as those of other building materials, and, like the many types of wood and stone, each metal possesses its own unique properties of strength, durability, and appearance. But with metals, we have the ability to enhance and modify these natural properties in a far more effective way.

Like woods, metals have a grain. Cold working with the grain of a metal is different from working against the grain, just as it is with wood. Metals are hard and durable like stone, but, like wood, can be shaped easily into variations of form. Metals have the lightness of wood, but offer a reflectivity and permanence similar to stone and glass.

Unlike stone or wood, however, one metal can be mixed with another to achieve a totally unique metal, with properties and colors altogether different from those of the individual parts. One metal can be coated onto another, producing an entirely new material that can capitalize on the best characteristics of each. Where one is weak, the other gives strength. Where one will corrode, the other protects. No other material known to mankind can offer the vast array of possibilities that metals afford.

Unlike any other building material, metals yield to the entropic nature of the environment, but they can be recovered and reformed. Metals come from the earth, some even from the heavens, and they always seek to return to their earthly, mineral form. Stone, glass, and brick will crumble into dust eventually; wood will decay and dissolve into other organic matter. Metals, however, when they decay or rust or when their usefulness is lost, can be recovered and recast into new forms with enhanced properties equal or superior to those of their previous life.

1

Metals are a resource as old as civilization. They lie in a realm of mystery, shrouded by the supposed alchemy of mixtures and formulations little understood by the people who work with them. Each era introduces a new metal, sometimes made of combinations of other metals, or sometimes with strength and forming characteristics enhanced by new techniques of manufacture.

The concept of alloying has roots in ancient alchemy—the long-held desire to produce a material with entirely new properties by unlocking hidden characteristics of one material through the introduction of another. The ancients first discovered better ways of using copper by adding tin to achieve a metal with a lower melting temperature and improved castability. Then, at about the time of the Romans, adding zinc to copper created a golden material which, to some, would pass for true gold until it tarnished. From these early efforts at alloying mankind discovered that various combinations of metals could develop new materials with different characteristics.

Of all the known elements, metals make up the majority. Within the periodic table of elements (see Figure I-1), there are five families of metals and two families composed of metals and nonmetals. Each family has different atomic structures that dictate the elements' behavior.

THE ALKALI METALS

The family of elements called alkali metals is made up of very reactive metals. Some will spontaneously ignite when exposed to water or to air. None of these metals are used in architecture, except as trace amounts in alloying constituents. There are 6 alkali metals, including sodium, potassium, and lithium.

THE ALKALINE EARTH METALS

There are also 6 alkaline earth metals. Magnesium and beryllium are members of this group. They are not, currently, common architectural metals, except as alloying constituents. Magnesium alloys are used in some bicycle frames and other sporting equipment designed for strength coupled with light weight. Magnesium has a specific gravity of 1.77. Compare this with aluminum, the lightest of the architectural metals at 2.70.

THE FIRST TRANSITION METALS

The family that includes chromium, vanadium, tungsten, and titanium is the group of 14 first transition metals. These metals are known for hardness and strength. Chromium, vanadium, and tungsten, when alloyed with other metals, particularly iron in the form of steel, can develop incredible strength and hardness.

Titanium has entered the realm of architectural metals only recently. This metal, rolled into very thin sheets, is favored as superior roofing material. Its light weight, coupled with superior corrosion resistance, gives titanium an advantage over other materials.

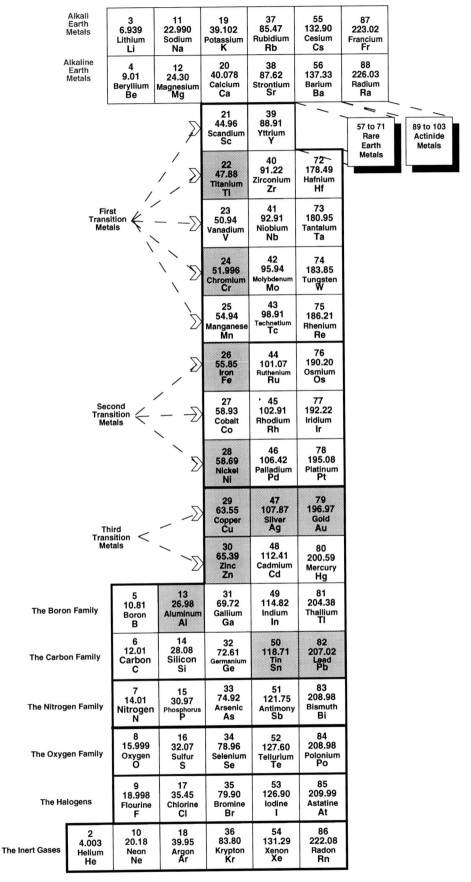

Figure I-1.   Periodic table of the elements.

Chromium provides the characteristic corrosion resistance and hardness necessary to create the stainless steels. In addition, chromium is responsible for the metallic, silver luster of stainless steel known worldwide as a symbol of the technological age.

THE SECOND TRANSITION METALS

Nine different metals make up the family of second transition metals. Iron is the major architectural metal of this group. Nickel, also a member of the family, is alloyed with copper to develop monel, and with iron and chromium to develop the austenitic stainless steels. Both nickel and iron have the characteristic of magnetism. This is not an architectural property, but noteworthy from the perspective of unique attributes: no other material naturally occurs with this characteristic.

THE THIRD TRANSITION METALS

Some of the oldest known metals, gold, silver, copper, zinc, and mercury, belong to the family of third transition metals. This group includes 6 elements; cadmium rounds out the list. All of these play a role in architecture, copper and zinc having the largest part. Gold leaf and, on rare occasions, silver plating offer the designer materials with incredible luster and elegance. Zinc and copper, however, are sought for their superior stability in natural atmospheric conditions.

THE BORON AND CARBON FAMILY

A strange family of elements falls between the metals and nonmetals—the boron and carbon family. Life on earth derives from this family with all its variations of the carbon compounds. The architectural metals used in this family are aluminum, lead, and tin. Lead and tin have been used as architectural or alloying materials since early Roman times. Aluminum, on the other hand, is relatively new to architecture, entering the twentieth century as a realistic building material, then overtaking just about every other metal by the century's end.

THE RARE EARTH METALS

The rare earth metals are a group of 15 elements. These metals are found naturally mixed together. They are not actually rare, except in the laboratory-prepared, isolated state. The rare earths are not architecturally significant metals at this time. In a mixed form, known by the German term *mischmetal*, or mixed metal, these rare earth elements are used with a zinc coating for steel, known as Galfan®.

THE NITROGEN AND OXYGEN GROUP

The nitrogen and oxygen group of 10 elements, containing, of course, nitrogen and oxygen, as well as sulfur and phosphorous, has use only as alloying substances. Like the carbon and boron family, these elements lie on the border between metals and nonmetals. The metallike elements in this family, more familiar to the ancient alchemist than to the architect, are antimony, bismuth, and arsenic.

The basic forms of metals used in architecture, the aluminum, copper alloys, and steels, are produced in completely unique industries with standards of measure and form as different from one another as these metals are from other building materials. The different metals are presented with language describing the nature of their forms as if they were found by different cultures on different planets. Out of the steel industry has come the stainless steels. So the "language" for these materials is the same. The dimensions and available thicknesses, as well as how the jargon is used when describing these ferrous materials, are also common to them.

The aluminum and copper industries are as different from each other as they are from the steels'. Monel, titanium, zinc, for example, address their own industries with a jargon created from a mixture of the others. This leads to confusion for those using the materials. You must translate your understanding from ounces per square foot to decimals or gauge numbers, depending on your familiarity. The following chapters address each metal in its own peculiar jargon, but also refer to measurements of thickness, both inches and millimeters, to provide a basis of comparison.

COLOR

In every design using metal, color is often one of the first considerations. Ever since metals were discovered by mankind, color has played an important role in their use: first, as a means of identification. Copper and gold were the simplest to identify. Next, it was determined that the color of metals change as they weather, or when certain oxidation treatments are applied. Surface coloration of metals was practiced by the Romans to a great extent, and the Japanese developed it into an art form in the sixteenth and seventeenth centuries. Even today, coins are manufactured from different metals as a means of identification. Color has occupied a place in human perception of the worth or artistic possibilities of metals since the beginning of civilization.

Today we still consider the color of metals in design. If the color is produced by a paint coating, the choice of the base metal is simplified by the limitations of the paint applicator. Is galvanized or aluminum the better choice for a base metal? Should one of the proprietary base metals, such as galvalume, galfan, or aluminized steel, be considered? These questions are often limited by availability and whether the paint applicator has the necessary pretreatment facility. The corrosion behavior of the metal may be a consideration, or strength requirements may limit the use of one base metal versus another. The actual color of the metal itself, however, does not enter into the equation.

On the other hand, if the designer wishes to use the natural, metallic tone exhibited by uncoated metals, then a whole new set of variables come to mind. Questions arise regarding how the metal will weather, if exposed to the atmosphere, or how to protect and clean the metal if used indoors? What sort of backing will work with this metal, and will potential corrosion cause the metal to decay prematurely?

METALLIC LUSTER

Metals are unique among the materials known to mankind because of their "metallic" luster. The depth and reflective nature of their surfaces is characteristic of even dull metals. Glass and some stone surfaces are referred to as "metallic" if they exhibit a deep silvery reflection. Air bubbles, rising below the surface of water, possess a metallic appearance as light reflects off this distorted surface through the layer of water.

Metals have a unique behavior at the atomic level. This behavior is responsible for producing their strong reflectivity and associated metallic luster. The high electrical and thermal conductivity, characteristic of all metals is also created by the metals' unique activity at the atomic level.

The nature of metallic luster and electrical and thermal conductivity lies in the outer energy band of the metal atom, known as the Fermi level, in which the electrons have the ability to move about freely. The number of electrons in this level also affects the hardness of metals. For example, lithium, a very soft metal that can be cut with a knife, has only one electron in this level. Chromium, the metal that adds hardness and corrosion resistance to steel, has six electrons in the outer band.

THE PHENOMENA OF REFLECTIVITY

Light behaves as an electromagnetic wave, with different wavelengths corresponding to different colors. Light also behaves as a particle. When light reaches the surface of a metal, it is intensely absorbed, but to a very shallow depth of only a few hundred atoms. Metals are conductors of electricity, and when the electromagnetic wave of light is intensely absorbed, it creates alternating currents on the surface of the metal. Light interacts with the electrons moving freely throughout this thin layer at the surface. The electrons oscillate in the range of visible frequencies and create currents. These currents reemit

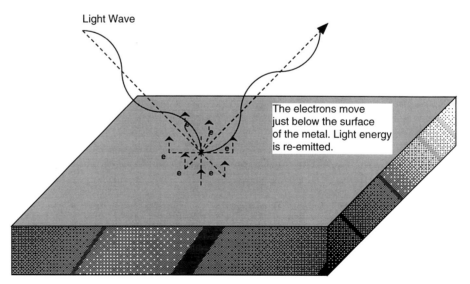

Light Wave

The electrons move just below the surface of the metal. Light energy is re-emitted.

Figure I-2. Phenomena of light absorption and reflection from a metal surface.

the light wave and create the strong reflectivity as the electrons move back to lower energy levels. The more intense the absorption, the more strongly the light is reflected. The more polished a metal surface, the more light is absorbed and the stronger the reflection.

Metals have various colors or tones of color because of the variations in absorption of portions of the light wavelength. Figure I-3 shows a graph of various metals and the portion of the wavelength that is most absorbed and thus reflected.

Silver is reflective over the entire range of light, giving it a white metallic color. The slight dropoff at the violet region of the light spectrum gives silver a very slight yellow cast. This is because silver has reduced absorption at the violet scale and thus reduced reflectivity. Because polished tin and stainless steel also reflect light across the entire spectrum, they make exceptional mirrors without color distortion. Nickel and iron absorb light at various wavelengths, but not at the high levels of silver or stainless steel. Thus these metals have a duller color and lower reflectivity. Polishing the surface helps to increase their absorption. Copper and gold do not absorb as completely at the blue end of the spectrum and thus do not reflect this region. This gives them a red or yellowish cast.

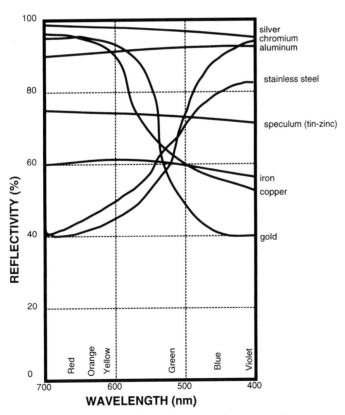

Figure I-3. Graph of wavelength versus reflectivity of various metals.

WEATHERING AND CORROSION RESISTANCE

The atomic level of a metal also plays a role in determining weathering tendencies and corrosion potentials when exposed to other metals and atmospheric corrodents. As metals age, the surfaces roughen with layers of oxides and hydroxides as well as other compounds. These roughened surfaces reduce the absorption of light and thus lower reflectivity.

In regard to corrosion resistance, surface roughness also plays a role. The smoother the surface, the more corrosion resistant. This is because a smooth surface offers little for atmospheric pollutants to grab hold of. A smooth surface also facilitates the natural, self-washing action of rain. Table I-1 gives a comparison of the various architectural metals and how their surfaces change with time.

How, and for how long, a metal will undergo the effects of weathering is, for the most part, related to where the metal is used. What is the local environment? In a regional sense, arid climates tend to preserve metal surfaces better. Moisture, metals' archenemy, is rare and short-lived. The occurrence of salt deposits resulting from the collection of moisture is far less in dry climates than in coastal or industrial environments. In arid environments, the concern is with thermal expansion stresses and erosion from wind-blown sand. Figure I-4 shows the performance of several metals in various exposures.

In coastal regions, the presence of chloride resulting from salt deposits, presents a different situation. Metals used on exterior structures near the coast will be attacked by the corrosive salts found in the ocean. Salts will accumulate in dew and other moisture forming on the metal. The moisture collects on the cooler metal in the early morning, then travels down the surface, collecting particles of salt as it passes. The salt deposits become concentrated at edges, drips, and horizontal ledges. These deposits then go into solution, creating a powerful electrolyte that acts to develop localized polarity on the metal surface. The polarity may be between grains of the metal or between different metals, such as a fastener and the base material, thus allowing intense and rapid corrosion. Weathering of metals used in exterior marine environments occurs especially quickly. The metal tries to protect itself by developing a barrier of oxides. Copper and its alloys develop copper chloride compounds, which are characterized by the common green patina. Copper sulfate, developed in urban and industrial regions, is also green, but copper chloride patina has a slightly bluer tint. Unprotected steel also corrodes rapidly, stainless steel needs the addition of molybdenum to thwart the pitting caused by a coastal exposure.

In urban, industrial environments, corrosion takes a different form. Chemical attack by sulfur dioxide, which combines with moisture to develop sulfuric acid, can deteriorate and stain many metals. Copper develops copper sulfate from the available sulfur, forming the thick green patina. Once the green patina develops, corrosion resistance increases as migration to the base copper is prevented by this impervious layer. Galvanized steel, aluminum, and, to a slight degree, stainless steel, develop corrosion pockets that will eventually deteriorate the metal. Occasional cleaning can slow the effects of pollutants, and proper detailing to fully benefit from the washing effects of rain can also help.

TABLE I-1.

METAL	COLOR	REFLECTIVITY (1 = HIGH, 5 = LOW)	10-YEAR AGING CHARACTERISTICS
Zinc—Natural	Gray-blue tint	2	Dark bluish gray color
Preweathered Zinc	Dark blue-gray	4	Dark bluish gray color
Carbon Steel	Gray-dark blue	4	No change if protected, otherwise red rust
Tin	Gray silver	2	Dark gray tones
Terne	Gray silver	3	Dark gray
Lead	Dark gray	5	Dark gray to gray black
Galvanized	Gray—light tint	3	Gray-white (white rust will develop)
Galvalume	Gray—light tint	3	Gray
Stainless #8	Chrome silver	1	No change
Stainless #2B	Gray, white	2	No change
Stainless #3	Chrome silver	2	No change
Titanium	Gray—medium	3	No change
Clear Anodized Aluminum	Gray—medium	3	Very little color change
Stainless #2D	Gray—medium	3	No change
Stainless—Glass Bead	Gray—medium	2	Some darkening from soot
Stainless—Acid Washed	Gray—medium	4	No change if passivated
Aluminized Steel	Gray-white	2	Light gray, less reflective
Monel	Silver—medium tint	1	Slight brown patina
Mill Finish Aluminum	Silver-white	2	Gray dull, some mottling
Nickel Silver	Silver-yellow tint	2	Gray-green with some mottling
Copper	Reddish pink	2	Green-gray patina
Gilding Metal	Red	2	Green-gray patina
Commercial Bronze	Reddish gold	2	Green-gray patina
Jewelry Bronze	14k gold color	2	Green-gray patina

TABLE I-1. (Continued)

METAL	COLOR	REFLECTIVITY (1 = HIGH, 5 = LOW)	10-YEAR AGING CHARACTERISTICS
Red Brass	Reddish orange	2	Green-gray patina
Yellow Brass	Yellow	2	Green-gray patina, black streaks
Muntz Metal	Golden yellow	1	Green-gray patina, blotchy with streaks
Silicon Bronze	Reddish bronze	3	Green-gray patina
Aluminum Bronze	Pale golden color	3	Gray patina, blotchy
Statuary Finish Brass	Light, medium, dark bronze	4	Unprotected, then spotty, uneven
Blackened Copper	Matte black	5	Brown, almost claylike color, with green patina
Tin-Plated Copper	Reflective, silver-white	1	Gray patina with a green tint

In rural and semirural environments, corrosion moves at a slower rate. Fertilizers used in farming regions can deposit on a metal surface, creating alkaline films. On lead or terne surfaces this effect is evidenced by discoloration, such as white films or even reddish streaks.

The local environment will also affect the corrosion behavior of a metal. Obviously, an interior, controlled atmosphere will expose a metal surface to minimum conditions. However, not all interior exposures are completely safe from the impact of corrosion. If a surface is placed where it is exposed to constant handling, the environmental impact can be severe. Fingerprints and other results of handling may shine parts of the metal while leaving the adjacent parts oxidized. Some metals are etched by the acids deposited from perspiration if they are not cleaned periodically. Glass cleaners, if left on a metal surface in seams or ledges, will oxidize the surface and possibly etch the metal. Stone and tile cleaners, such as builders acid (muriatic acid) and other mild acids, will corrode metal in a very short time. Stainless steels will appear rusty and tarnished from short exposures to such acid fumes used to clean stone floors and walls.

In external applications, metals that are installed in vertical planes or soffits will oxidize at different rates than those installed at slight pitches or flat. Ledges and overhangs will collect moisture and foreign particles. Metals that are self-cleaning through rainfall will be less affected; these areas will require manual cleaning to ensure long-term performance.

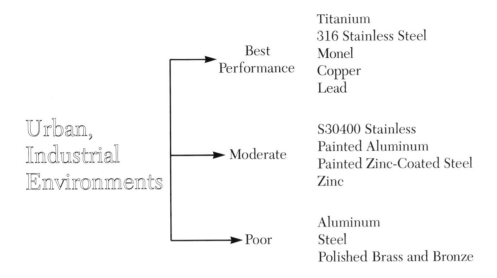

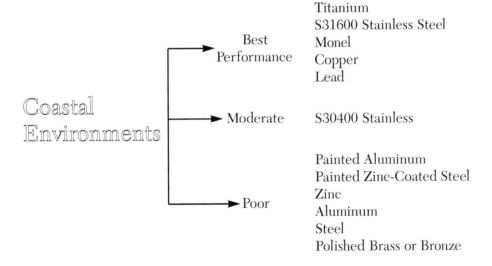

Figure I-4. Various Exposures

Cooling towers expel moisture laden with electrolytes that have fairly acidic pH levels. These electrolytes are generated from chemicals used to clean the pipe and coil assemblies of the heat exchangers and chillers. Deposits from this condensate will collect on nearby metal surfaces, creating streaks and stains. Aluminum, even anodized aluminum, is susceptible to these vapors and deposits. Stainless steel will streak, and copper alloys will darken.

Galvanic relationship is always a consideration when using a metal in combination with other metals. In many architectural situations the relationship is overstated and the corrective measure can sometimes create greater problems. In the following chapters each metal and its galvanic relationship with other metals is discussed. Table I-2 describes the electromotive scale of different metals when they are exposed to a powerful electrolyte, such as seawater. Seawater will take the metal ions into solution more rapidly than fresh rainwater, and the table compares each metal on this basis.

The differential in the electrical charge is more critical than how and where a metal appears on the scale. It does not matter whether one metal has a positive charge and the other a negative charge, or whether two metals have negative charges. It is the overall differential between the two metals that determines the level of polarity, the potential for the two metals to set up opposite charges or poles, that is the determining factor. For instance, if one metal has a -1.03 charge and another metal has a -0.25 charge, the

TABLE I-2. Electromotive Scale of Various Metals in Flowing Seawater

ANODIC POLARITY	VOLTS	
The more active end of the scale—least noble	-1.03	Zinc (galvanized steel)
	-0.79	Aluminum (A30000 series)
	-0.61	Steel or Iron
	-0.61	Cast Iron
	-0.53	S 30400 Stainless Steels (active)
	-0.52	S 41000 Stainless Steels (active)
	-0.36	Copper
	-0.31	Bronze
	-0.29	Brass
	-0.28	Tin
	-0.27	Lead
	-0.25	Monel
	-0.18	S 31600 Stainless Steels (active)
	-0.15	S 41000 Stainless Steels (passive)
	-0.10	Titanium
	-0.08	S 30400 Stainless Steels (passive)
The more noble end of the scale	-0.05	S 31600 Stainless Steels (passive)
	$+1.29$	Gold
CATHODIC POLARITY		

differential is 0.78. The greater this difference, the greater the polarity that will develop, and the more rapid decay of the more active, or in this example, the more negative, metal in relation to the other metal.

Some oxides, such as the thick aluminum oxide developed during anodizing, will inhibit the electrical polarity. The aluminum surface does not become more noble but less reactive. Other oxides, such as iron oxide, offer little protection because of their solubility and porosity; they dissolve and flake away, exposing more metal to corrosive attack. Copper oxides and lead oxides are initially soluble, which explains the green stain seen at the base of many exterior copper applications. Lead oxide is clear. Copper and lead oxide when in solution will quickly attack metals of more negative potentials.

When designing for galvanic corrosion conditions, use coatings or gaskets that do not allow transfer of electrical currents. Be sure the more reactive metal is thoroughly protected. Gaps in the coating, such as a scratch or an exposed edge, can accentuate the effect by concentrating the galvanic attack of the more noble metal to a smaller area of the reactive metal. Galvanic corrosion works as a quantity relation. The ratio of surface area contributable to the noble metal to reactive metal should be low. This is why chrome plating with many small cracks will perform better than chrome plating with only a few cracks. The surface area of the reactive metal exposed to corrosive attack is increased, thus reducing the electrical concentration at any one spot. Basically, the effects of corrosion are spread out and its concentration is restricted.

$$\frac{\text{Noble Metal}}{\text{Reactive Metal}}$$ The ratio of surface areas should be very low.

The flow of moisture from a metal near the cathodic end of the scale to another metal closer to the anodic end should also be prevented. Water will just transfer the metal ions to the surface of the active metal and create cells of galvanic corrosion.

If contact cannot be avoided, use the material mass relationship by increasing the thickness of the active material. If the mass of the more active metal is substantial in relationship to the mass of the cathodic metal, galvanic corrosion will be slowed in most environments.

DENSITY—THE "FEEL" OF METAL

Another concept in the consideration of metals is their basic weight and hardness. Every material has a particular feel to it, a weight and a surface resistance. This feel can be characterized by the density of the material. Materials such as plastic lack the density of metals. Plastics can be painted to match the smooth appearance of metal sheet. They can be formed to shapes similar to those of metals. They can even be given a thin coating of metal to mimic reflectivity or metallic luster. But on touch they feel hollow, soulless. Until recently, you had to beg someone to take scrap plastic, or you buried it to be found by some future generation.

Lead is sort of an aberration among metals. This metal is very dense and heavy, yet it can be formed easily by hand. Lead lacks hardness, and sound, encountering lead, is "swallowed" like a stone falling on soft mud.

TABLE I-3. Comparative Specific Gravity of Metals

METAL	SPECIFIC GRAVITY
Magnesium	1.77
Aluminum	2.70
Titanium	4.51
Chromium	6.92
Zinc	7.14
Tin	7.30
Stainless Steel—Type 410	7.70
Iron/Steel	7.87
Stainless Steel—Type 304	7.90
Muntz Metal	8.39
Cartridge Brass	8.53
Commercial Bronze	8.80
Monel	8.83
Nickel	8.90
Nickel Silver	8.95
Copper	8.96
Silver	10.49
Lead	11.34
Gold	19.32

The different metals have a feel unto themselves. The various specific gravities range from the lightness of titanium and aluminum to the heavy density of lead and gold metals. The specific gravity of an object is a comparative measure of the object's density with that of water. For example, gold has a specific gravity of 19.32. If you took a cubic meter of gold, it would weigh 19.32 times as much as a cubic meter of water. Table I-3 shows the various specific gravities of the architectural metals. Magnesium and silver, not necessarily architectural metals, are indicated for relational comparison.

HARDNESS—ABRASIVE RESISTANCE

Hardness and ductility are other concepts unique to the metals. Hardness is a characteristic that determines a material's wear and abrasive resistance. The ability of a material to resist denting from impact is related to hardness as well as a material's ductility. Various degrees of hardness and ductility can be achieved in architectural metals through different cold working processes and thermal treatments.

Hardness is not a direct correlation to a metal's density. Titanium has a much harder surface than lead, yet lead is almost 2.5 times as dense. Even in the fully annealed state, titanium will not come close to lead in ductility.

Various degrees of hardness can be achieved in many metals by tempering, a heat treatment process used in cold rolled and cold worked metals. As the grain structure of a metal undergoes cold forming, the grains are stretched and altered. The metal takes on a springier characteristic, resisting further cold forming. The surface becomes harder, resisting deformation from contact. A metal can develop additional hardness through smoothing action from cold working, but abrasion resistance is only slightly improved. The yield strength of the metal increases, which is responsible for the increased stiffness. Tempering heats the cold worked metal to temperatures at which the grains begin to dissolve into one another. The stress, built up during the cold working process, relaxes slightly. As it relaxes, the metal becomes more ductile and malleable.

There are series of standard tempers available. These tempers and their availability in a particular alloy vary, depending on the nature of the grains as they recrystallize. The temper designation is actually determined by this grain size, rather than the yield strength of the metal. Table I-4 indicates various hardness comparisons of the different architectural metals under tempers

TABLE I-4. Comparable Hardness of Different Metal Alloys

METAL	ALLOY AND TEMPER	HARDNESS ROCKWELL B-SCALE	YIELD STRENGTH		DUCTILITY DEGREE 1–Very Ductile 5–Stiff
			ksi	MPa	
Aluminum	A93003-H14	20 to 25	21	145	1
Aluminum	A93004-H34	35 to 40	29	200	1
Aluminum	A95005-H34	20 to 25	20	138	1
Aluminum	A96061-T6	60	40	275	4
Copper	1/8 hard (cold roll)	10	28	193	1
Gilding Metal	1/4 hard	32	32	221	1
Commercial Bronze	1/4 hard	42	35	241	2
Jewelry Bronze	1/4 hard	47	37	255	2
Red Brass	1/4 hard	65	49	338	2
Cartridge Brass	1/4 hard	55	40	276	1
Yellow Brass	1/4 hard	55	40	276	2
Muntz Metal	1/8 hard	55	35	241	3
Architectural Bronze	As extruded	65	20	138	4
Phosphor Bronze	1/2 hard	78	55	379	3
Silicon Bronze	1/4 hard	75	35	241	3
Aluminum Bronze	As cast	77	27	186	5
Nickel Silver	1/8 hard	60	35	241	3
Steel—low carbon	Cold rolled	60	25	170	2
Cast Iron	As cast	86	50	344	5
304 Stainless Steel	Temper pass	88	30	207	2
Lead	Sheet lead	5	0.81	5	1
Monel	Temper pass	60	27	172	3
Zinc-Cu, Tn Alloy	Rolled	40	14	97	1
Titanium	Annealed	80	37	255	3

used typically in architectural applications. This table also indicates the degree of ductility, which is a measure of how easily a metal will form. Ductility correlates directly with the yield strength of the metal.

PERFORMANCE

In architecture, the long-term performance of a material is critical. Durability of a material while exposed to the rigors of its environment is a necessary consideration whether that environment is a coastal urban region or the interior lobby of an air-conditioned office building. Metals have a durability unmatched by other building materials. Thicknesses, and thus weight, can be greatly reduced because of this durability. A flat panel used as a spandrel, if stone or glass, will require a thickness in excess of 5 mm, whereas a metal skin may be able to achieve the same flatness at half this thickness.

To achieve long-term performance, specific characteristics of metals must be understood. Metals like to move. They shrink and grow with changes in temperature. Not all metals move in the same way, however. One metal used in a particular application may not perform the same as another metal in a similar application. Wrought forms of metals have different properties, depending on the grain direction. The cold rolling process stretches the grains in the direction of rolling. Such metal will expand more in the direction of the grain than perpendicular to the grain. This characteristic affects forming operations as well. Metals will form differently with the grain than perpendicularly or against the grain. Table I-5 shows the various expansion coefficients of each metal and what total growth can be expected over a 120-inch (3050-mm) fabrication of the metal under normal temperature differentials.

A better way of using this information on the coefficient of thermal expansion is to determine the expected temperature range the metal surface will achieve and compare this with the temperature of actual installation. You first need to determine the maximum temperature (T_{hot}) the metal will possibly achieve. Keep in mind that this temperature can be much greater than the ambient air temperature. Dark metal roofs or blue and black stainless steel walls can absorb heat and develop temperatures twice that of the ambient air temperature. Other conditions, such as angle of exposure, reflectivity, and insulation, will effect the temperature extreme. For dark surfaces, use a temperature at least 50% greater than the maximum ambient temperature expected. Next, determine the lowest temperature ($T_{coldest}$) to which the metal surface will be exposed—typically the coldest expected winter night temperature. Choose a temperature expected at installation ($T_{install}$) and determine the differential between the two extremes. Use the following formulas to determine the maximum expected contraction and the maximum expected expansion from the position at the time of installation. (C_{te} is from Table I-5.)

$$\Delta T_{cold} = T_{install} - T_{coldest}$$
$$\Delta T_{hot} = T_{hot} - T_{install}$$

$$C_{te} \times \Delta T_{hot} \times \text{Length (inches)} = \text{Expected expansion (in inches) for a particular length segment}$$

$$C_{te} \times \Delta T_{cold} \times \text{Length (inches)} = \text{Expected contraction (in inches) for a particular length segment}$$

TABLE I-5. Expansion Coefficients and Expected Expansion of Various Metals at Normal Range Temperatures of 38°C

METAL/ALLOY	COEFFICIENT OF THERMAL EXPANSION μ in./in.°C	EXPECTED EXPANSION (INCHES) OF A 120-INCH SHEET*	EXPECTED EXPANSION ($\mu\mu$) OF A 3-METER SHEET
3003 Alloy Aluminum	23.2	0.11	2.79
5005 Alloy Aluminum	23.8	0.11	2.79
6063 Alloy Aluminum	23.4	0.11	2.79
Copper	16.8	0.08	2.03
Gilding Metal	18.1	0.08	2.03
Commercial Bronze	18.4	0.08	2.03
Jewelry Bronze	18.6	0.08	2.03
Red Brass	18.7	0.09	2.29
Cartridge Brass	19.9	0.09	2.29
Yellow Brass	20.3	0.09	2.29
Muntz Metal	20.8	0.09	2.29
Architectural Bronze	20.9	0.10	2.54
Phosphor Bronze	18.2	0.08	2.03
Silicon Bronze	18.0	0.08	2.03
Aluminum Bronze	16.8	0.08	2.03
Nickel Silver	16.2	0.07	1.78
Iron	11.7	0.05	1.27
Steel	11.7	0.05	1.27
Cast Iron	10.5	0.05	1.27
304 Stainless Steel	16.5	0.08	2.03
Lead	29.3	0.13	3.30
Monel	14.0	0.06	1.52
Tin	23.0	0.10	2.54
Zinc—rolled	32.5	0.15	3.81
Zinc-Cu, Tn Alloy	24.9 with grain	0.11	2.79
Zinc-Cu, Tn Alloy	19.4 across grain	0.09	2.29
Titanium	8.4	0.04	1.02
Gold	14.2	0.05	1.27

*A length of metal, 120 inches (roughly 3 meters), when installed outdoors can experience a temperature differential of as much as 100°F (38°C). When this occurs, the metal will increase in length the amount indicted for that metal. If the metal is to be subjected to a higher temperature range, then you must allow for the additional expansion.

OIL CANNING

Whether a metal will appear flat or not is dependent on the distribution of stresses across the surface. These stresses will change as temperature changes.

The ability of a metal to transfer the effect of these stresses across the sheet, without buckling or distorting out of plane, will determine the level of "oil canning" that will occur. "Oil canning" is a metaphorical term used to describe the tendency of flat surfaces to show variations in reflectivity. For example, a curved mirror will show a stretched and distorted image of a person standing in front of it. So too will a metal surface distort the reflection of light if minor variations in and out of a level plane exist. The appearance of flatness is very much dependent on surface reflectivity. Copper with a dark surface resulting from oxides is less reflective than bright, new copper and thus will not appear with the same degree of surface distortion.

Because oil canning is created by internal stresses within the sheet, control and prevention are difficult. Cold rolled sheet metal forms develop differential stresses across the sheet, as well as through the cross section (Figure I-5). Other operations performed on the sheet metal product can also create these stresses. Welding, forming, and handling operations, as well as improper installation techniques, can influence the effects of oil-canning behavior on thin sheets.

To reduce the possibility of inducing stresses in the sheet during handling and fabrication processes, consider the following recommendations.

- Order stretcher level and architectural quality sheet. Assuming mill tolerances are sufficient or ASTM (American Society for Testing and Materials) criteria will provide the necessary flatness levels is not enough to ensure quality sheet material.
- Make an initial inspection of the sheet, coil, or plate material when it arrives. Often, sheet or plate material will show indications of surface wave prior to shearing or forming operations. Inspect coils for inward or outward bending of the center of the metal, particularly when uncoiling a section of the coil.
- Stresses can be induced in metal by improper handling and storage. Sheet material must be stored flat on smooth, solid supports. When handling the sheet and fabricated form, care should be exercised when lifting to avoid denting the edges or the center of the piece. If the cross section of the sheet is not sufficient, the flat surface will not transfer the load, which will induce a permanent, irreparable buckle in the surface.

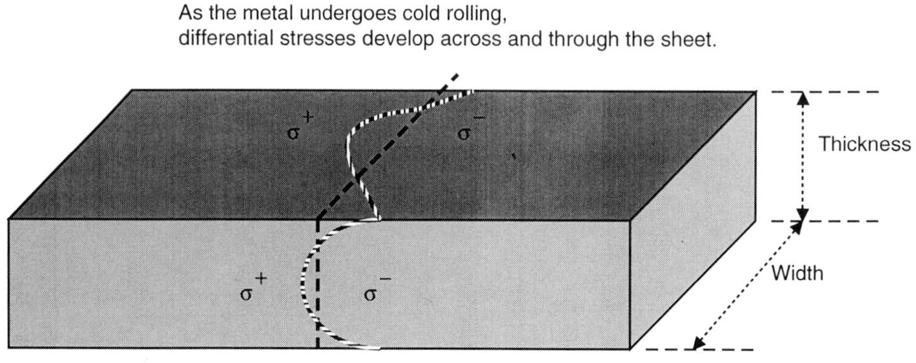

As the metal undergoes cold rolling, differential stresses develop across and through the sheet.

Figure I-5.   Internal stress within a sheet metal cross section.

- Roll-forming operations can induce uneven stresses as the metal is stretched differentially along the edges. Roll-forming operations pass the metal through a series of rolls which gradually shape the edge or surface of the sheet metal. As the metal is brought up or out of the plane, it is stretched slightly in respect to adjacent surfaces. This stretching can induce stress into the flat planes of the metal surface, which will show as oil canning. The stretching can be minimized by using long station roll-forming equipment. Short stations must bring the shape up rapidly, creating more stress than a more gradual shaping of the part.

- Press-brake forming can also induce distortions in the surface. Pan forming the edges of sheets can distort corners of the panel when both sides of the panel are not brought up at the same rate or to the same position. This can also occur if one side is brought up too far in relation to the other side, or if the edge crashes into the brake die as the panel is brought up.

Out-of-level dies and dies with imperfections will distort the flat surface and the edge of folded metal. When the die is not level, or when the outside edges induce more pressure than the inside edges, differential stresses will occur across the length of the part.

Thickening part of a metal sheet relative to its cross section will assist in reducing oil canning. Increasing the gauge will help to a point, then cross-section enhancements are necessary to provide further flatness. A thin sheet with ribs interrupting the flat plane will appear flat. Laminating thin sheets to a thicker support material will also retain flatness as long as the laminated support material is continuous, flat, and unbroken. Inducing a slight convex curve to a thin sheet will also afford a flatter appearance.

There are a number of factors that affect the appearance of oil canning.

- *Reflectivity of surface.* The more reflective the surface, the more apparent surface distortions, even minor imperfections, become. This is also a concern with painted coatings. High-gloss paints, particularly dark colors, will actually accentuate minor surface variations. On some

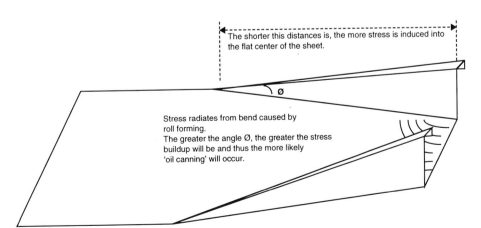

Figure I-6. "Oil canning" stress buildup within a sheet during roll forming.

Figure I-7.   Roll forming stations showing initial shaping of metal.

Figure I-8.   Roll forming stations successive shaping of sheet metal.

metals, copper and zinc in particular, newly installed surfaces are very reflective. Over time they develop surface oxides and lose 60% to 80% of their reflectivity. Initially, oil canning and other distortions may be readily apparent, but soon afterward they tone down, almost disappearing.

- *Angle or orientation of view (position of lighting).* The angle from which a surface is viewed will determine whether minor distortions are visible. On many high-gloss surfaces, you have to move across the surface until the "hot spot" or white light spot over the distortion reflects directly back to the eye. On the outside surface of a building, viewing a wall from one angle versus another, or at different times of day, can make distortions in the surface appear or disappear.

The lighting of surfaces can also affect how a flat surface appears. Slight out-of-plane distortions on a metal surface will be highlighted by light reflecting at acute angles. In addition, if reflective surfaces are not cleaned properly, streaks and surface oils will appear as discolored regions when light reflects off them at acute angles.

COST COMPARISONS BETWEEN THE METALS

Cost is a major consideration for all building construction projects. Metals are often believed to be premium, more costly materials in relation to other options, particularly if initial costs are the sole consideration. There are a number of factors that play a role in establishing the overall cost of using metal, which vary substantially from one metal to the next:

Cost of the base metal
Cost of the metal surface finish
Cost of fabrication and installation
Cost of maintenance over time

Each of these factors influences the resulting final cost of using a metal for a particular application. In construction, there are different, less tangible influences on cost to consider as well. For instance, the knowledge and experience of the metal crafts persons in a particular area will greatly affect cost. If a metal fabrication company is inexperienced, costs could range widely. For example, an inexperienced company will not know how to price the project correctly. Second, a company with little experience may not understand the quality aspects of metal workmanship, nor the need for good detail and joinery technique. No matter how the specification is written, you cannot qualify a subjective statement covering what is meant by "good detail." Experience and understanding of the workmanship requirements should be specified, for example:

"All ornamental work must be performed by experienced craftspersons. Submit at least 5 projects over the last 3 years of equal size and complexity prior to bidding the project."

Pre-qualify the company before it bids. No one would knowingly go to a dentist who was performing his or her first filling installation, especially if the "dentist" used to be a car mechanic.

Consider requiring a visual mockup for quality review and comparison. The mockup should be constructed by the persons actually involved in the manufacture and installation of the final product. The mockup could be an actual section of the finished product or it could be a stand-alone mockup. Include in the requirement that the mockup be presented soon after award and prior to beginning the final installation. Thus, all parties involved in the project will have the same understanding of the quality requirements demanded by an architectural and ornamental metal construct.

THE TANGIBLE COSTS

The Base Metal

The base metal is the material before finishing, forming, painting, or any other post-mill operation. The cost of a base metal is somewhat dependent on the mill state, such as casting or sheet material. Costs are also based on availability and size.[1] Base costs for metals fluctuate with the market pressures of supply and demand. Pricing is established by weight, rather than surface area or piece. For instance, titanium is priced by the pound or kilogram. When rolled into thin sheets, one pound of titanium will cover a large surface area. Even though current prices for one pound of titanium is almost five times that of an equivalent weight of stainless steel, the cost per square foot is only 1.5 to 2 times that of a square foot of stainless steel.

In addition, in considering a base material, its strength and hardness also play a part. A column cover made of aluminum may need to be manufactured from metal of 0.125 inches (3 mm) in thickness to provide the necessary strength. The same column cover manufactured from stainless steel could be 0.080 inches (2 mm) in thickness to provide the necessary strength. Thus, for a particular use, comparing costs of similar thicknesses may not be pertinent to the application.

The comparison in Table I-6 shows the various architectural metals and their relative cost in various forms as well as their relative cost by weight.

The Cost of Finishing

Different finishes impact the cost of the finish product almost as much as the base metal. Paint finishes are relatively constant from one metal to the next. Some metals, however, will not take well to painting without elaborate pretreatments. Many paint applicators are set up for finishing one metal but not another, so the selection may be limited in certain regions.

Other finishes, particularly the mechanical and chemical coloring finishes, add cost to the overall surface area of the metal. Furthermore, because finishes are applied to surface area, equating this to a weight per area of surface is inefficient. For example, pricing a surface finish by using a cost-to-weight relationship makes no sense. This system may be appropriate for

---

[1]The costs are relative to the time of this writing. Some markets may have developed since this time to make the material more cost-effective. New production rates or a change in demand can also affect the price fluctuations of metals.

TABLE I-6. Relative Costs

METAL	COST BY WEIGHT ALONE	COST BY SURFACE AREA— THIN FORM	COST BY SURFACE AREA— THICK FORM	COST BY CAST FORM
Steel	1	1	1	1
Iron	1	1	1	1
Aluminum	2	1	2	2
Lead	2	2	4	4
Stainless Steel	2	2	3	5
Copper	3	2	3	3
Zinc	3	3	5	4
Brass	3	3	3	2
Bronze	3	3	4	2
Tin	4	4	6	—
Nickel Silver	4	4	4	5
Titanium	5	3	6	6
Monel	5	4	6	—
Gold	6	5	—	—

1 is lowest relative cost, 6 is highest relative cost
Dash denotes not applicable.

structural steel, but surface finishing deals with the overall surface area not how much something weighs. Chemical finishes are more expensive than mechanical finishes as a whole. Selective finishing or plating operations are expensive surface treatments. These techniques require special layout and processing controls.

Table I-7 shows the relative cost of the different finishes available on many of the metals. Note that not all finishes are available with all metals. Chemical finishes will have entirely different effects on one metal versus another. Information on each finish and how it relates to a particular metal is discussed in more detail in the chapters to follow.

When considering a finish, ask for samples and be sure to understand the color variation range. Color variations from sheet to sheet or run to run are to be expected. There is no way to control all the variables that affect the end result of the color obtained.

Cost of Fabrication and Installation

Certain metals, because of their softness and ductility, will offer less resistance to fabrication processes than other metals. Other metals may require heat treatment processes, such as annealing to modify the temper, or hot forming operations, before they can be shaped. For most architectural metals, cold working processes can be performed. Some specialized finishes may resist forming. The additions of lead, in some alloys of copper, will assist in cutting and milling operations.

By *fabrication* is meant the common metal-working processes of drilling, bending, piercing, milling, and cutting. More severe forming operations,

TABLE I-7. Relative Costs of Finishing and Coloring

COST	FINISH DESCRIPTION	TYPE	LIMITS
1	Mill finishes	Mechanical	All metals
1	Satin finishes-coil applied	Mechanical	All metals except lead and gold
1	Embossing	Mechanical	All metals
2	Mirror finishes	Mechanical	All metals except lead and gold
2	Electropolish	Chemical	All metals
2	Anodizing	Chemical	Aluminum only
2	Coining	Mechanical	All metals
2	Aluminum coating	Dipping/spraying	Only steels and aluminum
2	Lead and terne	Dipping/plating	Only copper and steels
2	Zinc coating	Dipping/plating	Steels
2	Paint coating by coil	Dipping	All metals
3	Hand-applied satin finishes	Mechanical	All metals
3	Swirl textures	Mechanical	All metals
3	"Angel hair"	Mechanical	All metals
3	"Engine turn"	Mechanical	All metals
3	Glass bead texture	Mechanical	All metals
3	Abrasive blast texture	Mechanical	All metals
3	Buffing and polishing	Mechanical	All metals
4	Interference films	Chemical	Stainless and titanium, steels
4	Forced patination	Chemical	Copper alloys, zinc, aluminum, tin
4	Tin coatings	Dipping/plating	Copper and steel, stainless steel
5	Selective plating	Chemical	All metals
5	Selective finishing	Mechanical	All metals

1 is lowest relative cost, 6 is highest relative cost

such as spinning, deep drawing, and stamping, are often limited by a metal's temper. More technological forming operations, such as vacuum forming, high-velocity forming, bladder-press forming, and stretch forming, are dependent on special alloys or use equipment that is not common to most architectural metal fabricators.

In comparing the different architectural metals from a fabrication standpoint, the actual processes of forming and shaping are considered without regard to handling and finishing. Polished finishes, handling concerns, surface protection, and other secondary, less tangible criteria are not considered. In general, the more polished surfaces require greater care than the mill surfaces, and this care equates to some additional cost. The same is true for the chemical finishes. Some of the chemical processes should be performed on parts that have previously undergone fabrication. Other chemical finishes will allow some level of postfabrication. The handling and control of surface quality should be a part of the particular fabrication facility's quality assurance program. Contamination of the surface by other metals, scratches, and mars are signs of poor understanding and the lack of a thorough quality assurance program.

Installation cost comparisons are similar to those for fabrication. For installation, however, handling is also considered. Metals such as lead and, to a lesser degree, zinc will form with exceptional ease. Handling to the project site, the shifting and bouncing of the fabricated parts while being trucked, can flatten some of the preshaping that the metal underwent during forming operations. Aluminum and brass will scratch easily and must be protected throughout installation. Expensive gold leaf, with its thin, almost diaphanous sheets, can be like butterfly wings in a steady wind. Yet a bronze statue can arrive on a flatbed truck, be hoisted with a crane and plopped down onto a footing, with little regard to weather conditions. Thus the segment of Table I-8 dealing with installation attempts to include some of these factors in comparing installation costs of one metal to another.

TABLE I-8. Relative Costs of Fabrication and Installation

METAL	COST OF FABRICATION	COST OF INSTALLATION
Aluminum	1	2
Lead	1	3
Copper	1	1
Zinc	1	3
Tin	1	2
Steel	1	1
Steel Painted	1	1
Yellow Brass	1	2
Aluminum Painted	2	1
Iron	2	1
Stainless Steel	2	1
Red Brass	2	2
Aluminum-anodized	4	1
Muntz Metal	4	3
Titanium	2	2
Monel	2	2
Nickel Silver	4	3
Bronzes	5	3
Gold Leaf	—	4

1 is lowest relative cost; 6 is highest relative cost

Cost of Maintenance over Time

There are two considerations in evaluating the cost of maintaining a metal. The first consideration asks whether it is desirable to allow the metal surface to age naturally. A metal surface can chemically combine with the oxides, carbonates, and sulfates of the surrounding environment until, in some cases, the end surface appearance is totally different in color and reflectivity than its appearance at installation. However, this final surface is often more inert and protective than any other thin surface known.

The second consideration asks what is entailed in retaining the appearance of the metal surface as originally installed. Interior uses of metals usually require this second criterion, as do many exterior uses. Cleaning and applications of barrier films are necessary to resist changes in color and reflectivity for some metals. Other metals, such as stainless steel, monel, and titanium, offer a relatively stable, unchanging surface appearance, which will require minor cleaning to retain the original color and luster for decades. Table I-9 compares different metals and how they can be expected to respond over time.

The environment in which each metal is intended to be used is the determining factor for the frequency and intensity of any maintenance program. Obviously, interior or protected environments impose the least alterations on a metal's finish; urban, coastal environments and marine environments impose the greatest changes on such a surface. Within interior environments, occasional cleaning is all that is necessary to maintain most metal surfaces. Interior environments do not subject a metal to changing temperatures and humidity. If the anticipated exposure of a metal is handling or fumes from kitchen exhaust, or exposure to abrasive action (for instance, of tables and countertops)

TABLE I-9. Maintenance Costs

METAL	1ST DECADE	NEXT DECADE	BEYOND	COMMENT
Copper—allowed to weather	1	1	1	Stable green patina will develop.
Lead and lead coatings	1	1	1	Stable dark-gray patina will develop.
Monel	1	1	1	Slight dark-gray patina will develop.
Steel-corten	1	1	1	Oxide must develop correctly or corrosion will be severe and rapid.
Titanium	1	1	1	Very stable oxide. Coastal regions have little effect.
Zinc	1	1	1	Gray patina will develop. Potential for corrosion on concealed underside.

TABLE I-9. (Continued)

METAL	1ST DECADE	NEXT DECADE	BEYOND	COMMENT
Stainless steel	1	1	2	Some cleaning necessary. Coastal environments could cause premature staining and discoloration.
Aluminum—Alclad	1	1	2	Most environments should only darken surface.
Gold leaf	1	1	2	Some fraying at edges of leaf allowing attack of base metal.
Aluminum—painted	1	1	2	Life span depends on quality of coating.
Tin coatings	1	2	2	Gray patina will develop. The surface is soft and may erode.
Steel—painted	1	2	3	Depends on quality of coating.
Aluminum—anodized	1	2	3	The surface will collect dirt and require cleaning.
Aluminum	1	2	3	Oxides will develop on surface. Mottling will occur.
Bronzes—lacquered	3	4	4	Lacquer will fail. Surface oxides will develop and require recoating.
Brasses—polished and lacquered	4	4	4	Lacquer will fail in 3 to 5 years. Repolishing and lacquering will be necessary.
Nickel Silver	4	4	4	Lacquer will fail requiring repolishing and lacquering.
Steel	4	4	5	Total replacement is likely unless protected coating is maintained.
Iron	4	4	5	Total replacement is likely unless protecting coating is maintained.

1 is performs well, little maintenance; 5 is performs poorly, maintenance required

additional maintenance may be necessary. In general, however, most interior uses will have a much lower rate of deterioration.

Exterior environments, on the other hand, expose metals to a series of airborne pollutants, ultraviolet radiation, and humidity, all of which will attack metal surfaces and protective coatings. Metals, such as copper (left to weather naturally) and titanium, and to a lesser degree stainless steel and monel, with their impervious oxide films, can resist damaging effects when exposed to some of the most severe environments. Metals exposed to severe environments will need a regular schedule of maintenance to ensure long-term results. Environments of exposure are ranked as follows:

Most Severe	Marine
	Coastal Urban
	Northern Urban Street Level (road salts)
	Urban Industrial
	Coastal
	Urban
	Urban—Arid climate
	Urban—Protected
	Arctic
	Rural
	Interior—Entrances
	Interior
Least Severe	Interior—Protected

Another factor that plays a part in the maintenance of metal surfaces is the nature of the surrounding materials. A lead or copper roof may offer a very low-maintenance surface; however, materials nearby, such as steel handrails or an aluminum skylight frame, will require more frequent maintenance. When these materials deteriorate, staining and streaking will occur on the more inert surfaces of the lead and copper. Maintenance requirements must be accelerated or the surface will be permanently scarred.

Sealant joints will need more frequent maintenance even on the most durable metals. Sealants collect airborne pollutants, which are later deposited on the surrounding metal surfaces. The reflective metal surfaces show dirt stains and loose their sleek appearance if proper cleaning is not performed. Details such as ledges and overhangs should be designed to allow the natural cleaning effect of rain to occur without the deposition of accumulated soot onto visible surfaces of the metal.

In architecture, metals find use both inside and outside a building—as barriers to keep out the environment, at the same time providing a pleasing, esthetic appearance, or as ornamental features, giving a building an artistic relevance. Only the most stable of the metals are used for architectural and ornamental purposes. The following chapters discuss the metals that are commonly used and a few that are less common but offer possibilities for the future.

Aluminum

INTRODUCTION

At one time considered a precious metal, more precious than even gold or silver, aluminum was so rare and hard to come by that Napoleon treated his most honored dinner guests to forks and spoons made from this metal. Second-rate guests had to struggle through their meal using mere gold or silver cutlery.

In 1884 the United States, considering materials to cap off the Washington monument, made a fanfare out of a small cast aluminum pinnacle placed at the top of the monument. At the time, this was the single largest casting of aluminum ever made. It weighed in at 100 ounces.

Why such a commotion over a metal in such common use today—a metal made from ore present in rocks almost everywhere on earth? It is estimated that $^1/_{12}$ th of the earth's crust contains aluminum ore, twice the amount of iron. Aluminum is the third most abundant element found on the earth, exceeded only by oxygen and silicon. Aluminum is an excellent reflector of radiant energy throughout the entire range of wavelengths. Electromagnetic, radio, and radar wavelengths are reflected by thin aluminum housings.

The reason for the early furor lies in the fact that aluminum is not found naturally in its pure state. To purify aluminum to even marginal levels was extremely difficult and, at the time of the Washington Monument construction, very expensive.

Aluminum is a nonmagnetic, nonsparking, white soft metal with a specific gravity of 2.70 (compare this to steel with a specific gravity of 7.87). This is a lightweight element from the strange family of metals and nonmetals—the boron and carbon family. Other architectural metals in this family are the corrosion-resistant lead and tin. Aluminum is second to steel in commercial production. Building construction products such as conduits, window frames, and curtainwall support members make up almost one third of the entire consumption of aluminum.

HISTORY

In 1886 a young American metallurgist, Charles Martin Hall, discovered that metallic aluminum could be produced by dissolving alumina (aluminum oxide ore) in molten cryolite. An electrical current was applied to the solution to draw the oxygen out. In the same year a Frenchman, Paul T. Heroult, discovered the identical process. When both men applied for patents, they discovered each other. The process is known today as the Hall-Heroult process, signifying their joint discovery.

Once a more cost-effective refining process was developed, aluminum came into architectural use. Its initial uses were for castings for monuments and statues; One of the first ornamental castings was the aforementioned pinnacle of the Washington Monument.

The first project to use the wrought form of the metal was the Church of St. Gioacchino in Rome in 1897. The church had the first aluminum roof ever manufactured. In the same year in Canada, cornices made of aluminum decorated the Canada Life Building and the Canada National Railways Building in Montreal. Beyond the expensive tip of the Washington Monument, the first architectural uses of aluminum in the United States were the sand-cast spire of the German Evangelical Church in Pittsburgh in 1927 and the cast aluminum spandrels decorating New York's Rockefeller Center in 1931 and the Empire State Building in 1929. The Empire State Building, designed by architects Shreve, Lamb, and Harmond, featured the first major architectural use of the metal. Its tower segment was fabricated from aluminum, and the spandrel panels below the windows were manufactured from cast aluminum.

It was not until after the second World War that architectural uses of the material accelerated. With the lack of war demand on the material, the enhanced production methods could be transferred from military to commercial uses. The Mormon Tabernacle in Salt Lake City, Utah, was one of the first large structures at this time to utilize a beautiful aluminum roof. The roof was completed in 1947 and to this date is still functioning adequately as well as maintaining a lustrous appearance. The 1950s brought aluminum to the architectural forefront as new decorative and functional processes were developed. Today virtually every new building uses aluminum. Curtainwall extrusions, window framing mullions, storefront supports, spandrel panels, siding and roofing panels, louvers and many other common architectural products are made from this versatile material. Aluminum can be economically manufactured into complex shapes by extruding, or shaped with relative ease from plate and sheet forms. The metal can be anodized, adding decorative color treatments that use the natural corrosive behavior of the metal. Aluminum also receives paint well, with the proper pretreatment of the surface to remove the protective oxide film.

Aluminum develops a natural oxide film upon exposure to the atmosphere. This film, one millionth of an inch thick, will renew itself if scratched. The film becomes impervious to attack on the base metal even in corrosive atmospheres. Runoff moisture from an aluminum surface does not stain adjacent materials, as happens with other weathering materials such as copper and corten steel.

Left to weather naturally, aluminum will turn to a light gray tone; sometimes, depending on the atmospheric pollution, dark gray mottling will be intermixed on the surface. Aluminum, unlike stainless steel, will not sustain a polished surface when exposed to the atmosphere. Oxides form on the surface, dulling the finish. Frequent cleaning and polishing are required to sustain a luster. In seaside regions unprotected aluminum will pit extensively.

Exterior aluminum is typically provided coated or anodized. Coatings can be thin layers of a more corrosion-resistant aluminum alloy, paints, or ceramic coverings. Anodized finishes enhance the natural oxide film and can impart colors to the aluminum surface through chemical reactions with the alloying components or by the addition of colored dies or metal salts into the pores.

Anodized surfaces used on exteriors have a satin texture. Mirror surfaces, produced by electrochemical means, will not hold the reflective tones without constant maintenance when they are used in an exterior environment or undergo extensive handling. Aluminum will receive paint finishes readily after it has gone through pretreatment. Baked-on finishes such as KYNAR 500® and urethanes are excellent coatings for exterior aluminum. Paint scratches, sheared edges, and fastener penetrations quickly develop the protective aluminum oxide coating. When the proper pretreatment techniques are used, painted aluminum surfaces offer durable and long-term color solutions to the designer.

PRODUCTION AND PROCESSING

Aluminum ore can be found in most rocks and soil. However, commercial quantities are obtained from bauxite. The mineral bauxite, named for Les Baux, the town in southern France, contains 50% to 60% alumina. Alumina is aluminum oxide, the compound used in the refining process to produce aluminum. Major sources for high-grade bauxite are Australia, Guinea, Brazil, and Jamaica. Other sources of lower-grade ore are Surinam, British Guiana, Russia, Greece, Hungary, Croatia in the former Yugoslavia, and Indonesia. In the United States, Arkansas contains the major domestic supply. Bauxite is usually found near the surface and is mined through strip-mining techniques.

In the refinement process, bauxite is first ground and dried for transport to the refinery. The ground ore is mixed with soda ash, crushed lime, and water to form a caustic soda mixture. Aluminum oxide is dissolved into the caustic soda solution to form sodium aluminate. The solution is further purified, refined, and then calcinated at 2000°F (1093°C) to form a relatively pure aluminum oxide cake. For every pound of aluminum refined, 0.36 pounds of waste sludge is produced.

The cake is poured into a carbon-lined vessel containing cryolite, (sodium aluminum fluoride), which dissolves the aluminum oxide. An electric current is passed through the solution, and oxygen is separated from the aluminum oxide. The aluminum, 99% to 99.5% pure, is removed and cast into pigs weighing 50 pounds each. The pigs are used to develop the alloys of aluminum which, in turn, are alloyed with other metals to form the multitude of products manufactured from the metal.

ENVIRONMENTAL CONCERNS

Aluminum refinement requires tremendous amounts of electricity, some 20,000 kilowatt hours per ton of aluminum refined. Most small towns use less electricity per year than aluminum refinement uses per day.

Aluminum recycling has become a substantial secondary business. Reducing the ravages on the environment caused by mining, recycling bypasses the large ore refining costs. Recovery of scrap and discarded aluminum products uses less than 4% of the energy required in the refining process and reduces the vast quantities of waste sludge byproduct. Recycling of aluminum also reduces the need for additional mining and the destruction of the natural landscape caused by mining. Metal recovered from scrap accounted for 30% of the entire world supply in 1991. It is hoped that the secondary metal will play an even larger role in the future of aluminum production.

ALLOYS

Commercially pure aluminum has a tensile strength of 13,000 psi (90 MPa). This is a very soft material, but the most corrosion-resistant form of the aluminum alloys. The temper can be enhanced to improve strength, but only to a slight degree. The high-purity forms of aluminum are used where strength is not a factor and when corrosion resistance is desired.

Very minute amounts of other elements will have major effects on the strength and forming characteristics of aluminum. Additions of certain metals to the alloy mix impart varying degrees of strength and hardness as well as the ability to enhance these structural qualities. Adding copper and zinc to aluminum allows the alloyed metal to be heat-treated to improve strength. In addition, adding silicon and magnesium in combination will also allow the alloyed metal to be heat-treated, whereas magnesium or silicon alone will not produce an alloy capable of heat treatment for structural enhancement.

Some of the heat-treated alloys of aluminum can achieve incredible improvement in yield strength to levels in excess of 80,000 psi (550 MPa). When compared to the high-purity forms of aluminum, this increase is remarkable for a metal. Additional properties of the various alloys of aluminum are also of interest. Some alloys will age-harden as they rest at room temperature. Known as "natural aging," this characteristic can affect subsequent forming operations on materials that have changed over time while stored.

Additionally, anodizing treatments on aluminum can be alloy specific. Processes performed on one alloy of aluminum may have a different effect on another alloy. Cold forming operations performed on one alloy of aluminum may not work on another alloy. Temper of the wrought form has a lot to do with this, but some alloys allow for cold forming operations to occur with ease while others may split or crack under the same operation.

The alloy combinations available in aluminum are numerous. There are five major alloying elements commonly added to aluminum in various degrees and combinations, which impart specific characteristics important to the architectural user: copper, zinc, silicon, manganese, and magnesium.

Alloy Designation System

The Aluminum Association has devised a method of alloy identification based on the major alloying element constituents. Nine basic groups are established for wrought (sheet, plate) aluminum alloys, the first seven of which are common to the building industry. The other two are for special alloys devised for particular uses not pertinent to architecture.

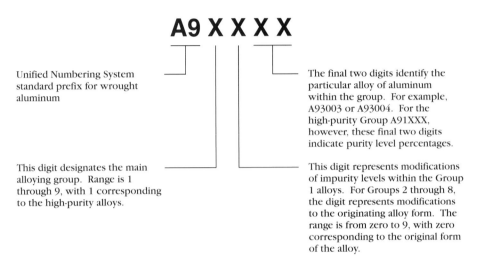

Unified Numbering System standard prefix for wrought aluminum

The final two digits identify the particular alloy of aluminum within the group. For example, A93003 or A93004. For the high-purity Group A91XXX, however, these final two digits indicate purity level percentages.

This digit designates the main alloying group. Range is 1 through 9, with 1 corresponding to the high-purity alloys.

This digit represents modifications of impurity levels within the Group 1 alloys. For Groups 2 through 8, the digit represents modifications to the originating alloy form. The range is from zero to 9, with zero corresponding to the original form of the alloy.

Table 1-1 provides a basic description of what distinguishes these nine groups.

Temper

Further classification of aluminum alloys includes the temper designation. Temper is the amount of stress or hardness an alloy exhibits. Some tempers form readily around a small-diameter mandrel, others will crack. The temper corresponds to the stiffness or hardness imparted to the alloy as it becomes stronger. Temper correlates to a particular alloy's overall strength.

The proper specification of aluminum should include the temper separated by a dash from the alloy designation. Temper is indicated by the letter "H" for the non-heat-treatable alloys such as the A91000, A93000, A94000, or A95000 alloy groups. A "T" designation is used to indicate the tempers for the heat-treatable alloys, followed by a series of numerical digits representing the various degree of temper induced into the metal.

For instance, the alloy A93003-H14 is a common architectural metal used for general sheet metal flashing and fascia work. The H14 addition designates a half-hard temper produced by cold working the metal at room temperature through metal rolls. Rolling in between pressure rolls puts a slight strain on the metal, stretching the grains in the direction of the rolls. This imparts strength and hardness to a controlled degree, in this case, half-hard temper. Another example is the alloy designated as A96061-T6. The T6

TABLE 1-1. The Nine Alloy Groups of Aluminum

ALLOY GROUP	MAJOR ALLOYING ELEMENT	CHARACTERISTICS
A91000	99% pure	Excellent corrosion resistance Soft and ductile, low-strength characteristics High thermal and electrical conductivity
A92000	Copper	Heat treatable to improve strength Poor corrosion-resistance tendencies High strength obtainable
A93000	Manganese	Non-heat treatable Good ductility, medium strength Good corrosion resistance
A94000	Silicon	Non-heat treatable Silicon reduces the melting point Good corrosion resistance
A95000	Magnesium	Non-heat treatable Fair ductility, medium strength Good strength and corrosion resistance
A96000	Silicon and Magnesium	Heat treatable Medium to high strength Good ductility and corrosion resistance
A97000	Zinc	Heat treatable to arrive at very high strength Fair corrosion resistance
A98000	Various Elements	Not considered an architectural alloy
A99000	—	Currently an unused series

addition designates a temper and hardness created from solution heat-treatment processes. A solution heat-treatment process requires the metal to be heated at a controlled elevated temperature for a period of time. This temperature must be sufficient to cause one or more elemental constituents to enter into solid solution. The metal is then rapidly quenched to capture the element in solid solution. See Table 1-2 for the various temper designations used on aluminum.

In the commercially pure form, alloy group A91000 is very soft and ductile. Adding certain elements increases the strength, but additional strength is obtained from cold working or heat-treating. Cold-working strain hardens the aluminum, making wrought materials less malleable. Other plastic deformation, such as deep drawing and stamping, also strain harden the aluminum. An example of work-hardening aluminum can be demonstrated by manually bending a strip of metal back and forth. Eventually, the edge where the bending occurs work-hardens and becomes brittle. Soon a crack will develop.

Further strengthening and hardening can be obtained by age hardening. Age hardening, particularly for the alloys containing silicon, magnesium, and zinc, occurs at room temperature when the metal is at rest. This is known as "natural aging." Heating for a short period of time at a high temperature will stabilize the properties, further strengthening the alloy. This process is known as "artificial aging," or precipitation hardening.

TABLE 1-2. Temper Designations for Aluminum Alloys

TEMPER	CATEGORY	DESCRIPTION
F		This is an "as fabricated" temper. Should not be used in architectural specifications because no special controls are established.
O		Annealed (wrought alloys only). Softest temper. This is used when maximum ductility is required.
H	H a-b-c	Strain hardened (wrought alloys only). This is to describe non-heat-treatable tempers. Hardness and strength obtained by cold working. The *a* digit stands for degree and type of strain hardening. The *b* digit represents the degree of temper. The *c* digit is a custom clarification. (optional)
	H12	Strain hardened. Quarter hard temper (2 designation).
	H14	Strain hardened. Half hard temper (4 designation).
	H16	Strain hardened. Three-quarter hard temper (6 designation).
	H18	Strain hardened. Full hard temper (8 designation).
	H22	Strain hardened, then partially annealed. Quarter hard temper (2 designation).
	H24	Strain hardened, then partially annealed. Half hard temper (4 designation).
	H26	Strain hardened, then partially annealed. Three-quarter hard temper (6 designation).
	H28	Strain hardened, then partially annealed. Full hard temper (8 designation).
	H31	Strain hardened, then stablized to a reduced, full annealed state.
	H32	Strain hardened, then stablized. Quarter hard temper (2 designation).
	H34	Strain hardened, then stablized. Half hard temper (4 designation).
	H36	Strain hardened, then stablized. Three-quarter hard temper (6 designation).
	H38	Strain hardened, then stablized. Full hard temper (8 designation).
	Hx11	Strain hardened to a slight degree over O temper, but less than Hx1 classification.
	H112	Strain hardening developed from working at elevated temperatures. Extrusions can develop this temper.
	H311	Strain hardened to a degree less than H31 temper.
	H321	Strain hardened to a degree less than H32 temper.
	H323 to H343	Products that have been specially fabricated to reduce the occurrence of stress-corrosion cracking.

TABLE 1-2. (Continued)

TEMPER	CATEGORY	DESCRIPTION
W		Represents solution heat-treated tempers. The temper is unstable and changes with time.
T		Heat-treated alloys. The T designation is always followed by one or more numerals. There are ten major classifications representing stable tempers.
	T1	Cooled from an elevated temperature and naturally aged.
	T2	Cooled from an elevated temperature shaping process, followed by cold working processes such as flattening or straightening, and then naturally aged.
	T3	Solution heat-treated then cold worked, followed by natural aging.
	T4	Solution heat-treated and naturally aged.
	T5	Cooled from high temperature and artificially aged.
	T6	Solution heat-treated, then artifically aged.
	T7	Solution heat-treated, then stablized.
	T8	Solution heat-treated, followed by cold working, and artificial aging.
	T9	Solution heat-treated, artifically aged, then cold worked.
	T10	Cooled from an elevated temperature, cold worked, then artificially aged.

When variations to one of the ten major T temper classifications are necessary, additional digits can be added. The following classifications have been created to represent those tempers which have been stress relieved.

	Tx51	Stress relieved by stretching after cooling from an elevated temperature shaping operation such as hot rolling or extruding.
	Tx510	Stress relieved plate, bar, or rod which has undergone cold finishing operations with no additional straightening after stretching.
	Tx511	Stress relieved plate, bar, or rod which has undergone cold finishing operations with some minor degree of straightening after straightening.
	Tx52	Stress relieved by compressing after solution heat treatment or cooling down from elevated temperature forming operations.
	Tx54	Stress relieved by a combination of compression and stretching such as die stamping hot, then die stamping cold.
	T42	Solution heat-treated tempers created from the O and F tempers, then naturally aged.
	T62	Solution heat-treated tempers created from the O and F tempers, then artificially aged.

Common Architectural Wrought Alloys of Aluminum

There are innumerable aluminum alloy variations available. The variations are needed to develop specific strength and corrosion characteristics. Some alloys are designed for strength and formability, others for colors obtainable in the anodizing processes. As used in architecture, the alloys and their corresponding characteristics are listed in the following pages. This is not considered to be a complete list of possible aluminum alloys, but it covers the major ones. The initial alloy number is presented in the Unified Numbering

TABLE 1-3. Architectural Wrought Alloys

ALUMINUM ASSOCIATION ALLOY NUMBER	UNS ALLOYING DESIGNATION	TYPICAL USES
1100	A91100	Sheet metal flashings, Alclad coating on other metals
3003	A93003	Sheet metal flashings, siding, column covers—aluminum plate
Alclad 3003	Alclad A93003	Sheet metal flashings, siding, column covers—aluminum plate
3004	A93004	Sheet metal flashings, siding, column covers—aluminum plate
Alclad 3004	Alclad A93004	Sheet metal flashings, siding, column covers—aluminum plate
3105	A93105	Sheet metal flashings, flat panel work—must be painted
5005	A95005	Sheet metal work; superior surface finish, aluminum plate
5050	A95050	Sheet metal work, aluminum plate, hardware
5052	A95052	Sheet metal work, column covers, aluminum plate
5083	A95083	Extruded strutural shapes, fabrications requiring welding
5086	A95086	Extruded structural shapes, fabrications requiring welding
5154	A95154	Extruded structural shapes, fabrications requiring welding
6005	A96005	Extruded structural shapes; good strength characteristics
6061	A96061	Structural alloy, good surface quality; Aluminum plate work and extrusions
6063	A96063	Architectural extrusions; excellent finish and surfacing
6463	A96463	Architectural extrusions; excellent finishing characteristics

System (UNS) format. The Aluminum Association (AA) alloy number is indicated just below the UNS number. Table 1-3 provides a list of the common architectural alloys available in wrought form.

UNS Alloy A91100

AA Alloy 1100
Non-heat-treatable alloy

Elemental Constituents

Aluminum   99.00%
Silicon   1.00%

Other elements, such as iron and copper, may be present in trace amounts. The aluminum content will never be below 99.00%.

This alloy of aluminum is often clad over other, less corrosion-resistant alloys. Similar to A91100 is the alloy A91145, which is also used to clad stronger, but less corrosion-resistant, alloys of aluminum. Alloy A91145 is listed with an even higher purity than A91100, at a minimum aluminum content of 99.45%.

Advantages
This is a highly pure aluminum alloy. It has excellent corrosion-resistant characteristics, is very ductile, and can undergo severe forming operations. Deep drawing and spinning operations can be performed with little difficulty. Alloy A91100 can be readily polished and anodized. When polished, alloy A91100 will hold its reflectivity longer in relatively corrosive environments because of its purity level. Polishing brings out a very white reflective surface, which can be anodized. When anodized, this alloy develops a very clear oxide film.

Common Tempers
> A91100-H14
> A91100-H18

Disadvantages
As indicated in Table 1-4, this metal has very low strength. It is soft and can be easily scratched or dented. Its hardness criterion is lower than that of copper.

Available Forms
> Sheet
> Strip
> Plate
> Extrusion
> Bar

TABLE 1-4. Temper Designations for Alloy A91100

TEMPER	YIELD STRENGTH		BRINNEL
	psi	MPa	Hardness
O	5,000	34.5	23
H12	15,000	103.3	28
H14	17,000	117.1	32
H16	20,000	137.8	38
H18	22,000	151.6	44
H112	3,000	20.7	23

UNS Alloy A93003

AA Alloy 3003
Non-heat-treatable alloy

Approximate Elemental Constituents

Aluminum	96.8%
Silicon	0.6%
Iron	0.7%
Copper	0.4%
Manganese	1.5%

Advantages
This alloy is also very high in purity. Alloy A93003 has excellent corrosion-resistant characteristics and is very ductile. Severe forming operations, such as deep-drawing and spinning operations, can be performed with little difficulty.

This alloy can be polished and can receive anodizing well. Alloy A93003 is used for aluminum panel work, aluminum plate work, and general sheet metal. This alloy is often painted with a decorative KYNAR® or other paint finish.

This alloy can be given a cladding of a more corrosion-resistant alloy such as A91100. Known as Alclad A93003, the clad sheet is often used to fabricate siding and roofing panels intended to remain unpainted.

Common Tempers

    A93003-H14
    A93003-H18
    Alclad A93003-H14
    Alclad A93003-H18

Disadvantages
This metal is relatively soft and, as indicated in Table 1-5, has low strength. There are other tempers available for this alloy that provide greater strength characteristics, but at a sacrifice of workability.

TABLE 1-5. Temper Designations for Alloy 3003

| TEMPER | YIELD STRENGTH | | BRINNEL |
	psi	MPa	Hardness
O	5,000	34.5	28
H12	12,000	82.7	25
H14	17,000	117.1	40
H16	21,000	144.5	47
H18	24,000	165.4	55
H112	5,000	34.5	25

Available Forms

Sheet
Strip
Plate
Extrusion
Bar

UNS Alloy A93004

AA Alloy 3004
Non-heat-treatable alloy

Approximate Elemental Constituents

Aluminum	96.2%
Silicon	0.3%
Iron	0.7%
Copper	0.3%
Manganese	1.5%
Magnesium	1.0%

Advantages

This alloy is also very high in purity. Alloy A93004 has excellent corrosion-resistant characteristics and is very ductile. Severe forming operations, such as deep-drawing and spinning operations, can be performed. This alloy is stronger than the A93003 alloy aluminum. Although it can be polished and will receive anodizing, this alloy is not typically anodized except to enhance its corrosion-resistant characteristics.

Alloy A93004 is used for aluminum panels, industrial siding, architectural aluminum plate work, and general sheet metal. This alloy is considered when the material is to be painted with a decorative KYNAR® or other paint finish.

Common Tempers

A93004-H32
Alclad A93004-H32

TABLE 1-6. Temper Designations for Alloy 3004			
TEMPER	YIELD STRENGTH		BRINNEL
	psi	MPa	Hardness
O	10,000	68.9	45
H32	25,000	172.3	32
H34	29,000	199.8	63
H36	33,000	227.4	70
H38	36,000	248.0	77

Disadvantages

This alloy will not anodize consistently. It should be protected with paint or Alclad if it is left exposed to weather. Table 1-6 gives the temper designations for this alloy.

Available Forms

Sheet
Strip
Plate
Bar

UNS Alloy A93105

AA Alloy 3105
Non-heat-treatable alloy

Approximate Elemental Constituents

Aluminum	96.2% to 97.3%
Silicon	0.6%
Iron	0.7%
Copper	0.3%
Manganese	0.3% to 0.8%
Magnesium	0.2% to 0.8%
Chromium	0.2%
Zinc	0.4%

Advantages

This alloy is high in purity; however, amounts of manganese and magnesium can vary. Alloy A93105 has good corrosion-resistant characteristics. The alloy is ductile, but because of the variations in alloying constituents, some forming operations may not perform as well as others. This alloy is more economical than the other alloys of aluminum because it is made specifically from recycled material. The impurities are accumulated from the recycling process.

This alloy will show streaking when anodized because of the variations in impurities. Painting is recommended for all exterior surfaces.

TABLE 1-7. Temper Designations for Alloy 3105			
TEMPER	YIELD STRENGTH		BRINNEL
	psi	MPa	Hardness
O	5,000	34.5	—
H12	15,000	103.4	—
H14	18,000	124.1	—
H16	21,000	144.8	—
H18	24,000	165.5	—
H25	19,000	131.0	—

Alloy A93105 is used for aluminum panels, architectural aluminum plate work, and general sheet metal. This alloy is considered when the material is to be painted with a decorative KYNAR® or other paint finish. Residential siding is made from this alloy of aluminum.

Common Temper

A93105-H25

Disadvantages

This alloy will not anodize consistently. It should be protected with paint. Its surface quality is not up to the levels of the other aluminum forms. Table 1-7 gives temper designations for Alloy 3105.

Available Forms

Sheet
Strip

UNS Alloy A95005

AA Alloy 5005

Non-heat-treatable alloy

Approximate Elemental Constituents

Aluminum	97.15% to 97.75%
Silicon	0.3%
Iron	0.7%
Copper	0.2%
Manganese	0.2%
Magnesium	0.5% to 1.1%
Chromium	0.1%
Zinc	0.25%

Advantages

This alloy is high in purity. The surface finish of sheet forms is of very high quality. A95005 has excellent corrosion resistance and good workability. The

TABLE 1-8. Temper Designations for Alloy A95005

| TEMPER | YIELD STRENGTH | | BRINNEL |
	psi	MPa	Hardness
O	6,000	41.3	28
H12	19,000	130.9	—
H14	22,000	151.6	—
H16	25,000	172.3	—
H18	28,000	192.9	—
H32	17,000	117.1	36
H34	20,000	137.8	41
H36	24,000	192.9	46
H38	27,000	179.1	55

surface can be anodized with superior results. This alloy is used for architectural metal work requiring a good, smooth, consistent surface.

Common Tempers

A95005-H32
A95005-H34

Disadvantages

This alloy is slightly more expensive than the A93000 series alloys. Cold working the harder tempers can be slightly more difficult. See Table 1-8 for temper designations for Alloy 95005.

Available Forms

Sheet
Strip

UNS Alloy A95050

AA Alloy 5050

Non-heat-treatable alloy

Approximate Elemental Constituents

Aluminum	96.45% to 97.15%
Silicon	0.4%
Iron	0.7%
Copper	0.2%
Manganese	0.1%
Magnesium	1.1% to 1.8%
Chromium	0.1%
Zinc	0.25%

TABLE 1-9. Temper Designations for Alloy A95050			
TEMPER	YIELD STRENGTH		BRINNEL
	psi	MPa	Hardness
O	6,000	41.3	36
H32	17,000	117.1	46
H34	20,000	137.8	53
H36	24,000	192.9	58
H38	27,000	179.1	65

Advantages

This alloy is high in purity. The surface finish is of fine quality. A95050 has excellent corrosion resistance and good workability. The surface can be anodized with superior results. This alloy is used for architectural metal work requiring a good, smooth, consistent surface.

Common Temper

A95050-H32

Disadvantages

This alloy is slightly more expensive than the A93000 series alloys. Cold working the harder tempers can be slightly more difficult. Table 1-9 gives temper designations for Alloy 95050.

Available Forms

Sheet
Strip

UNS Alloy A95052

AA Alloy 5052

Non-heat-treatable alloy

Approximate Elemental Constituents

Aluminum	95.9% to 96.7%
Silicon	0.25%
Iron	0.4%
Copper	0.1%
Manganese	0.1%
Magnesium	2.2% to 2.8%
Chromium	0.15% to .3%
Zinc	0.1%

TABLE 1-10. Temper Designations for Alloy A95052

TEMPER	YIELD STRENGTH		BRINNEL
	psi	MPa	Hardness
O	13,000	89.6	47
H32	28,000	192.9	60
H34	31,000	213.6	68
H36	35,000	241.2	73
H38	37,000	254.9	77

Advantages

This is one of the stronger aluminum alloys in common architectural use. The alloy can be formed and shaped with standard sheet metal fabrication equipment. Alloy A95052 has good corrosion-resistance characteristics and will take polishing well.

Common Temper

A95052-H32

Disadvantages

This alloy will not anodize well; a slightly yellowish tone develops. Welding can be performed, but not with standard techniques. See Table 1-10 for temper designations.

Available Forms

Plate
Sheet
Strip

UNS Alloy A95083

AA Alloy 5083
Non-heat-treatable alloy

Similar Alloy:

UNS Alloy A95086

AA Alloy 5086
Non-heat-treatable alloy

Approximate Elemental Constituents

Aluminum   92.55% to 94.25%
Silicon   0.4%

Iron	0.4%
Copper	0.1%
Manganese	0.4% to 1.0%
Magnesium	4.0% to 4.9%
Chromium	0.05% to 0.25%
Zinc	0.25%
Titanium	0.15%

Advantages

This alloy should be considered for welding structures needing to develop good fatigue strength around the weld zones. Alloys A95083 and A95086 have good corrosion-resistant characteristics.

Common Tempers

A95083-O
A95083-H321
A95083-H343
A95083-H111
A95083-H112

Disadvantages

This alloy is more expensive than the A95005 or A95050 alloys. It has lower corrosion resistance than the A95005 or A93000 group alloys. It should not be anodized for appearance sake. Table 1-11 lists temper designations for Alloys A95083 and A95086.

Available Forms

Plate
Sheet
Strip
Extrusion

TABLE 1-11. Temper Designations for Alloys A95083 and A95086

TEMPER	YIELD STRENGTH		BRINNEL
	psi	MPa	Hardness
A95083-O	21,000	144.8	—
A95083-H116	33,000	227.5	—
A95086-O	17,000	117.2	—
A95086-H34	37,000	255.1	—
A95086-H112	19,000	131.0	—
A95086-H116	30,000	206.8	—

UNS Alloy A95154

AA Alloy 5154
Non-heat-treatable alloy

Approximate Elemental Constituents

Aluminum   94.5% to 95.5%
Silicon   0.25%
Iron   0.4%
Copper   0.1%
Manganese   0.1%
Magnesium   3.1% to 3.9%
Chromium   0.15% to 0.35%
Zinc   0.20%
Titanium   0.20%

Advantages
This alloy has good corrosion resistance, particularly in and around coastal environments. A95154 has good welding characteristics and good workability in the softer tempers.

Common Tempers

A95154-O
A95154-H32
A95154-H34
A95154-H36
A95154-H112

Disadvantages
The appearance characteristics of this alloy are not as good as those of A95005 or A95050. This alloy should not be considered for anodizing if finish appearance is the important factor. Temper designations are given in Table 1-12.

TABLE 1-12. Temper Designations for Alloy A95154

| TEMPER | YIELD STRENGTH | | BRINNEL |
	psi	MPa	Hardness
O	18,000	124	58
H32	30,000	207	67
H34	33,000	227	73
H36	36,000	248	78
H38	39,000	269	80
H112	17,000	117	63

Available Forms

Plate
Sheet
Strip
Extrusion

UNS Alloy A96061

AA Alloy 6061
Heat-treatable alloy

Similar alloy:

UNS Alloy A96005

AA Alloy 6005
Heat-treatable alloy

Approximate Elemental Constituents

Aluminum	96.0% to 97.5%
Silicon	0.4% to 0.80%
Iron	0.7%
Copper	0.15% to 0.40%
Manganese	0.15%
Magnesium	0.8% to 1.2%
Chromium	0.04% to 0.35%
Zinc	0.25%
Titanium	0.15%

Advantages

A96061 is a very popular architectural alloy with good corrosion-resistance characteristics. Formability is best in the O and T4 tempers.

When aluminum is called on for its structural characteristics, this alloy is usually considered because of its yield strength in the T6 temper. This alloy can be heat-strengthened. Welding can be performed without difficulty.

Common Tempers

A96061-T4
A96061-T6

Disadvantages

Alloy A96061 is very difficult to cold work in the T6 temper. It will crack under simple forming operations.

This structural alloy of aluminum can be joined by welding; however, there is a loss of strength resulting from the welding process in the metal around the weld. The stress-reduction factor is as high as 30%. This is not a

TABLE 1-13. Temper Designations for Alloy A96061

| TEMPER | YIELD STRENGTH | | BRINNEL |
	psi	MPa	Hardness
O	12,000	82.7	30
T4	16,000	110.3	65
T6	35,000	241.3	95

specific characteristic of this alloy alone, but is a characteristic of aluminum in general.

See Table 1-13 for temper designations.

Available Forms

Plate
Sheet
Strip
Extrusion

UNS Alloy A96063

AA Alloy 6063

Heat-treatable alloy

Approximate Elemental Constituents

Aluminum	97.65% to 98.5%
Silicon	0.2% to 0.6%
Iron	0.35%
Copper	0.1%
Manganese	0.1%
Magnesium	0.45% to 0.9%
Chromium	0.1%
Zinc	0.1%
Titanium	0.1%

Advantages

This is a high-purity alloy used frequently in architectural metal applications. It is available in the extrusion form only. Alloy A96063 can receive anodizing finishes of various colors. The surface of this material can be polished. Heat treatment is used to increase the strength characteristics of this alloy.

Common Tempers

A96063-T1
A96063-T4
A96063-T5
A96063-T6

TABLE 1-14. Temper Designations for Alloy A96063

| TEMPER | YIELD STRENGTH | | BRINNEL |
	psi	MPa	Hardness
O	7,000	48.3	25
T1	9,000	62.1	42
T4	10,000	69.0	—
T5	16,000	110.3	60
T6	25,000	172.4	73

Disadvantages

This alloy is not available in sheet. Forming of extruded shapes can crack the material in the cold-worked state. Table 1-14 gives temper designations for Alloy 96063.

Available Form

Extrusion

UNS Alloy A96463

AA Alloy 6463

Heat-treatable alloy

Approximate Elemental Constituents

Aluminum	98.0% to 98.9%
Silicon	0.2% to 0.6%
Iron	0.15%
Copper	0.2%
Manganese	0.05%
Magnesium	0.45% to 0.9%
Zinc	0.05%

Advantages

This extruded alloy possesses excellent corrosion-resistance characteristics. It will accept anodizing and can be brightened to enhance the surface.

Common Tempers

A96463-T1
A96463-T5
A96463-T6

Disadvantages

This alloy is available in extrusion form only. Cold forming of T6 temper alloy is difficult and the material may crack. Temper designations are listed in Table 1-15.

TABLE 1-15. Temper Designations for Alloy 6463

TEMPER	YIELD STRENGTH		BRINNEL Hardness
	psi	MPa	
T1	9,000	62.1	25
T5	16,000	110.3	42
T6	25,000	172.4	74

Available Form

Extrusion

Aluminum Castings

Aluminum is an excellent material for creating intricate ornamental and functional castings. As with the wrought forms of aluminum, there are numerous alloys specifically designed for casting. In architecture, there are a number of common alloys considered for casting applications, and there are proprietary alloy mixes certain foundries use more frequently in their particular operations. For architectural castings, seven are listed, variations of which are common in the industry. The choice of alloy should be based on the ability of the foundry and the end use of the casting. Some alloys can be cast into more intricate detail than others, and some are more corrosion resistant. Cast aluminum forms should be able to take a polish and, in some cases, should be able to receive an anodizing coating. The user should review the choice of alloy with a foundry and with the end finish in mind.

For the cast alloys of aluminum the nomenclature uses similar categories to designate the main element constituents. The UNS system of alloy designations is described here in terms of the Aluminum Association system. The AA uses a different method to describe the cast alloys. Instead of a four-digit alloy definition, cast alloys are designated by a three-digit number followed by a decimal point and then a single digit to define the product form.

XXX.X

The first three digits are separated into categories as follows:

1xx.x	Alloys with 99% aluminum purity
2xx.x	Alloys with copper as major alloying element
3xx.x	Alloys with silicon and copper or magnesium
4xx.x	Alloys with silicon as major alloying element
5xx.x	Alloys with magnesium as major alloying element
6xx.x	(Unused category)
7xx.x	Alloys with zinc as major alloying element
8xx.x	Alloys with tin as major alloying element
9xx.x	Other elements alloyed with aluminum

The single digit to the right of the decimal point is used to define the product type. The numeral "1" indicates ingot, and the numeral "0" indicates casting. An ingot is a simple cast form developed for further hot working operations such as extrusions or hot rolling. The digits following the first defining number represent the specific alloy type or the level of purity. An alphanumeric classification preceding the cast alloy designation indicates a modification to an existing alloy or a new alloy.

The UNS attempts to put the designation systems of all alloys into perspective. For the cast alloys, the UNS system adds the first two alphanumeric characters, "A0" to the classification and eliminates the decimal point. For example, in the AA system we would have the following: **Alloy 513.0.**

The digit 5 stands for magnesium as the major alloying constituent. This alloy contains 4% magnesium. The next two digits represent the alloy type. The last digit, to the right of the decimal point, represents this as a casting and not an ingot alloy. The UNS designation for this alloy is: **Alloy A05130.**

For aluminum castings there are a number of design considerations to be taken into account. For example, the choice of casting process is determined by a number of additional factors. For architectural uses, finish quality and detail are two of the main factors to consider. These, of course, need to be weighed against cost of the finish product. Aluminum can be cast using sand, permanent mold, and die casting techniques. Each of these techniques is operated within certain parameters and has advantages over the others. (Table 1-16 summarizes the casting techniques for various architectural alloys.)

Sand casting is considered when the quantity of parts to be cast is not great. Sand casting has the advantage of quick and inexpensive die development. The dies can be considerable in size, but the casting tolerance must be great. The finish surface obtained from sand casting is coarse and typically requires postcasting cleanup, which adds to the overall cost. The

TABLE 1-16. Casting Techniques for Various Architectural Alloys

ALUMINUM ASSOCIATION ALLOY NO.	UNS ALLOY DESIGNATION	CAST TECHNIQUE USED	EXAMPLES OF PRODUCT USES
308.0	A03080	Sand, Permanent	General purpose, coarse surface
360.0	A03600	Die	General purpose, fine surface
380.0	A03800	Die	General purpose, fine surface
443.0	A04430	Sand, Permanent	Ornamental, hardware, shapes
513.0	A05130	Sand	Ornamental, grilles
518.0	A05180	Die	Coastal uses, hardware, grilles
713.0	A07130	Sand, Permanent	General purpose, high strength

detail achievable is dependent on the quality and practice of the foundry. In general, the detail can be very good on large sections without major extensions. Extensions can be added by bolting or welding sections together.

Permanent mold casting is another common aluminum casting technique. Permanent mold casting is more expensive than sand casting, but can achieve better detail and repeatability. Permanent mold casting uses the lost wax method. A ceramic shell or metal shell is created to allow repeated casting of identical parts. The finish quality received is very good, requiring minor clean-up of sprue connections (feeders) and seams in the mold. The cost of producing the mold is much higher than that of the sand casting technique. Repetition of parts is necessary to spread out this cost.

Die casting is another technique used to cast aluminum. This is the most expensive technique and is used when the quantity of casting is high and the size of the casting is relatively small. Die casting requires the manufacture of a metal mold. The molten aluminum is forced into the mold under pressure. The cooling rate can be controlled by heating the metal die; thus, better structural characteristics can be achieved. The surface quality of a die cast part is smooth and clean. The die itself can be polished to give a polished finish surface to the aluminum casting.

UNS Alloy A03080

AA Alloy 308.0

Alloying Constituents (%)

Aluminum	90.0
Silicon	5.5
Copper	4.5
Magnesium	0
Zinc	0
Manganese	0

This is a general-purpose sand or permanent mold casting alloy. This alloy has good fluidity, which benefits the development of good detail. The surface finish is grainy and will not take to polishing well.

Alloy A03080 is a poor choice for quality anodizing. It can be considered for painted or concealed applications. The corrosion resistance of this alloy is low, therefore it is not recommended for external uses unless it is coated.

UNS Alloy A03600

AA Alloy 360.0

Alloying Constituents (%)

Aluminum	88.0
Silicon	10.0
Iron	2.0

Additional small amounts of copper, manganese, and magnesium may be present.

This is a general-purpose die casting alloy. It has good fluidity, which benefits the development of good detail. The surface finish, because of the casting technique used, can be very smooth. The alloy, however, does not polish or take visual-quality anodizing well. This alloy has good corrosion resistance and is used in the manufacture of frying skillets, among other things.

UNS Alloy A03800

AA Alloy 380.0

Alloying Constituents (%)

Aluminum	82.0
Silicon	8.5
Iron	2.0
Copper	3.5
Zinc	3.0
Manganese	0.5
Nickel	0.5

This is a general-purpose die casting alloy. It has good fluidity, which benefits the development of good detail. The surface finish, because of the casting technique used, can be very smooth. The alloy, however, does not polish or take to visual-quality anodizing well. This alloy is used for the manufacture of various die cast parts used in everyday equipment.

UNS Alloy A04430

AA Alloy 443.0

Alloying Constituents (%)

Aluminum	94.1
Silicon	5.3
Magnesium	0.6 maximum

This is a sand or permanent-mold cast alloy used to form custom ornamental hardware. It has good fluidity and can achieve decent levels of detail. The surface does not polish well, and the alloy makeup does not take to good-quality anodizing, when appearance is a consideration. The corrosion resistance of this alloy is very good. Often this alloy is considered for the fabrication of cast parts used in coastal marine atmospheres. Consider painting this alloy to improve performance.

UNS Alloy A05130

AA Alloy 513.0

Alloying Constituents (%)

Aluminum   94.2
Silicon   0
Magnesium   4.0
Zinc   1.8

This alloy is used to cast ornamental panels and grilles requiring moderate detail. Its fluidity is very poor because of the lack of silicon, so fine detail is difficult. This alloy can be polished and anodized with ease. Its corrosion resistance is superior to most other alloys. Welding together sections of this aluminum alloy is very difficult.

UNS Alloy A05180

AA Alloy 518.0

Alloying Constituents (%)

Aluminum   89.45
Silicon   0.35
Copper   0.25
Iron   1.80
Magnesium   8.00
Zinc   0.15

This alloy is a die cast alloy with some improvement to fluidity over the A05130 alloy. It is used for the manufacture of hardware parts and has decent corrosion resistance. This surface can be polished and will take anodizing well.

UNS Alloy A07130

AA Alloy 713.0

Alloying Constituents (%)

Aluminum   94.2
Silicon   0
Copper   0.7
Magnesium   0.4
Zinc   7.5

This cast aluminum alloy has poor fluidity but good strength. It can be cast by either the sand or permanent-mold method. This alloy has good corrosion resistance and can be polished and anodized to good levels of visual quality.

Care in Fabrication—Castings

Aluminum castings must overcome a series of conditions which, if not totally understood, can lead to the detriment of successful products. For all metals, when they move from a liquid state to solidification as a solid cast part, shrinkage occurs. For aluminum, this change in overall volume can be as much as 8%. Depending on the shape of the casting, shrinkage can have major effects on quality or can render the casting useless.

To compensate for shrinkage, molds are given risers and feeders to certain regions to allow for a replenishment of molten metal as the shape begins to solidify. Without these, cavities can develop, separations in the mold can occur, and cracking may happen where stresses develop. There should be adequate feeders and risers to supply metal at an even, nonturbulent rate in order to allow the entire casting to solidify at close to the same rate. For sand castings, the fabrication should be removed from the mold as soon as practical to allow for further shrinkage without binding around the sand form. In addition, for sand castings, the molds should be vented adequately to allow air to escape. Sand is usually moist and, as the molten metal is poured, steam develops and must escape quickly.

Porosity will occur in cast molds, particularly when the sections are heavy and thick. The use of chill bars in the mold will assist the thicker regions to solidify at the same rate as the thinner sections. Chills are made of metal or graphite and work to pull the heat away, producing solidification at a quicker rate. If the mold solidifies at a uniform rate, cavitation in the casting is reduced.

Alclad Alloys

Pure aluminum is highly corrosion resistant; however, it is relatively weak. Alloying is necessary to develop strength, but at the cost of corrosion resistance. To help counteract the decrease, the Alclad process can be performed on the alloyed sheet. This process consists of coating a less corrosion-resistant sheet with a more corrosion-resistant surface. The surface can be a high purity aluminum or another alloy form with better corrosion resistance.

Some of the common architectural alloys are predesignated as Alclad. The aluminum core is bonded metallurgically to a pure aluminum or other alloy aluminum on both surfaces. Once bonded, the Alclad aluminum alloy acts structurally as a composite. Alclad is applied to sheet and plate forms of aluminum by spraying the surface with molten aluminum, which is applied at 500°F to 800°F (260°C to 427°C). The coated sheet is passed through pressure rolls that further bond the higher purity aluminum to the base metal. The coating increases the overall thickness of the sheet from 0.0002 mm to 0.0010 mm.

The pure aluminum coating acts as cathodic protection to the core alloy while maintaining the strength and forming characteristics of the base material. For the heat-treatable alloys with copper and zinc as the main alloying elements, Alclad coatings provide improved corrosion resistance with a slight reduction in overall strength. For the non-heat-treatable alloys, such as the A92000 and the A93000 groups, Alclad coatings can develop improved corrosion characteristics for aluminum sheet or plate intended to be left exposed without a paint coating.

Improved corrosion resistance is achieved by restricting the corrosion process to the thin Alclad layer. As corrosion attacks the surface of the Alclad material, it moves laterally along the clad layer, rather than perforating the sheet. The life of the clad coating is dependent on the severity of exposure, the environment, and the thickness of the Alclad surface material.

Alclad coatings are inexpensive, yet valuable, treatments for exposed aluminum surfaces. Siding panels, used without paint or anodized coatings, can be provided with Alclad and still have a long service life. An Alclad coating holds a reflective luster for years. Because corrosive attack to the surface operates laterally, weathering is more even and uniform than with uncoated aluminum sheet.

Precautions—Alclad Coatings

The use of Alclad coatings is not widespread in architecture. The metal suppliers do not keep a stock of the material, and some order requirements are excessive. Large production runs are necessary to keep the cost of processing low. Therefore, there is a need to verify availability and quantity limitations.

Alclad coatings are soft. If the intention is to use a coating in a natural mill finish, the metal must be protected from scratching. The mill processor must be required to cover the metal with a protective PVC coating. Without the coating, excessive scratches from stacking and handling the sheets will render the material's surface worthless. Brake forming of Alclad sheets requires protection of the dies with a thick plastic sheet.[1] Without such protection, small creases develop along either side of the brake.

Alclad coatings are very reflective. When Alclad coatings are applied to the aluminum sheet, they are further bonded to the surface by passing the sheets through polished rolls. The polished rolls impart a smoothing effect on the aluminum surface. Depending on the polished quality of the rolls, the Alclad surface will be given a level of reflectivity. The enhanced reflectivity will show minor surface dents and imperfections as well as capacitor discharge stud welding on the reverse side of the sheet.

AVAILABLE FORMS

Sheet and Strip

Aluminum is available in sheets and strip forms, as well as coils. Thicknesses of sheets and strips are as follows:

---

[1] A popular protective sheeting material used for press brake dies is known as "rhino hide." This is a thick, malleable plastic wrap that is held between the die and the metal to be formed. For the upper die, the rhino hide is held by loops attached to the bolts running the length of the die holder.

Minimum Thickness*	0.010 inches	0.254 mm
Maximum Thickness	0.190 inches	4.826 mm
Maximum Width**	120 inches	3048 mm
Standard Widths	72 inches	1829 mm
	60 inches	1524 mm
	48 inches	1219 mm

*Aluminum is available in thickness less than those indicated. However, as thicknesses decreases, it begins to reach foil thicknesses and properties.
**This maximum width is available on special order and for certain alloys. The standard widths are more readily available in the various alloys.

Plate

Aluminum is available in plate form with the following dimensions:

Minimum Thickness   0.250 inches (6.35 mm)
Maximum Thickness   up to 3 inches (76.2 mm)

Width and length limitations vary. Check with mill representatives on these limitations.

Plate surface quality is superior to that of other hot rolled metals. The surface can be ground and polished to produce architectural quality surfaces.

Extrusions

The preferred method of fabricating long, thin shapes of identical cross section is the extrusion process. Many metals can be extruded into simple cross-sectional shapes that are solid throughout. Such shapes as round bars, rectangular bars, angles, hexagons, octagons and other polygons are easily achieved. More intricate and complex cross sections are also possible and far more economical to produce using the extrusion press. Aluminum, of all the metals, is the most flexible for this form of fabrication.

Long architectural shapes and sectioned assemblies are common and economical uses of this form of fabrication. Clean, smooth surfaces with sharp corners and identical cross sections from one part to the next make extrusions the optimum choice of many architectural members, most notably the curtain wall members used structurally to hold glass or metal panels onto office structures. Curtain wall extrusions are developed into structural shapes that are both functional and beautiful. Extrusions can be assembled into larger shapes that act together as supporting members and perform an ornamental function. Another important benefit of extrusions is the reduction of waste, since excess material can be melted down and reused.

Extrusions are made by pressing a heated billet of metal—a solid cylinder of metal—through a hardened steel die with a cross section of the shape cut into the die (Figure 1-1). The resulting metal is extruded out of this cross section in lengths typically of 12 to 24 feet (4 meters to 8 meters). Longer lengths are possible, but depend on the amount of metal pushed and the logistics of the extruder. Once pushed through the die and still hot from the process, the extruded aluminum is stretched slightly and cut to size to

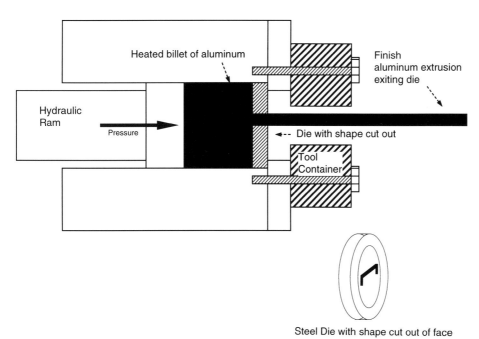

Figure 1-1. Cross section of a typical extrusion press.

remove the ends which are usually deformed. Deformations at the ends occur because of a differential in stresses holding the shape. The waste metal is recycled back into a new billet.

All alloys of aluminum can be extruded. Approximately 75% of all architectural extrusions use the alloys A96061 and A96063. Alloy A96061 is considered to be a structural alloy with yield stress level of 35,000 to 40,000 psi when used in the T6 temper. Both of these alloys find wide use as curtain wall framing members on architectural structures. They have good strength, excellent corrosion resistance, and can be anodized.

Certain characteristics of extrusions must be understood before they are developed into a particular design. Due to the extrusion process, the cross section of the shape must be consistent throughout the piece. All screw guides, slots, and legs must run along the length of the fabrication. The extrusion can take the form of a tube with an open space or spaces running down the center. The cross section of the tube can be round, rectangular, oval, diamond shaped, or any other geometric form as long as certain parameters are followed.

The metal must flow through the cross section cutout on the die. Off-balance cross sections pose difficulties in processing. The metal will flow best through the area of least resistance, which is the area of greatest cross section. The pressure of the push must be equal across the cross section from the center of the piece, or bending will result.

Holes are difficult to accomplish in certain shapes because of the way the metal must flow around the shape. Holes are produced in aluminum extrusions by using a dual die setup. The die is composed of two parts: the cap, which shapes the outside of the aluminum extrusion, and the mandrel, which

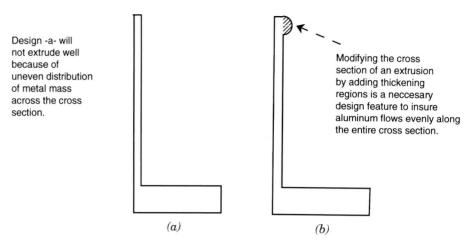

Design -a- will not extrude well because of uneven distribution of metal mass across the cross section.

Modifying the cross section of an extrusion by adding thickening regions is a neccesary design feature to insure aluminum flows evenly along the entire cross section.

(a)   *(b)*

Figure 1-2. General guidelines for extrusion design considerations.

forces aluminum to flow around it and forms a void in the cross section. Hollow dies are more expensive because of the added complexity and setup they entail. There are numerous standard hollow shapes available at a considerable savings over the cost of custom fabricating a die.

Large flat shapes, after forming, may require flattening by rolling through flattening rolls. To address this situation, consider adding grooves or ridges in the center of the flat section (Figure 1-4). These can occur on the reverse, non-exposed side, if one exists, or they can be developed into the shape's design. Ridges will aid the metal flow as extrusion occurs. However, these ridges or grooves can create the problem of structural streaks appearing on the opposite face. The streaks develop as the extrusion cools and are merely cosmetic.

Another limitation of extrusions is the size of the cross section. Extrusion presses vary from one extruder to the next. However, all cross sections must

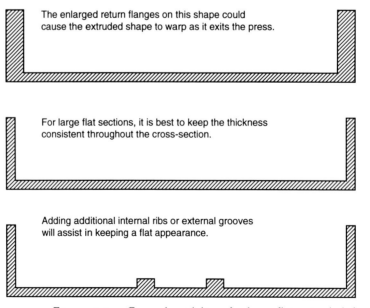

The enlarged return flanges on this shape could cause the extruded shape to warp as it exits the press.

For large flat sections, it is best to keep the thickness consistent throughout the cross-section.

Adding additional internal ribs or external grooves will assist in keeping a flat appearance.

Figure 1-3. General guidelines for large flat extruded shapes.

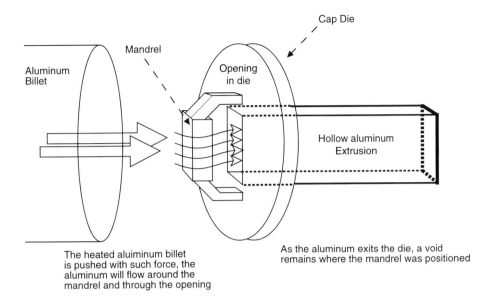

Figure 1-4a. Extrusion process for hollow shapes.

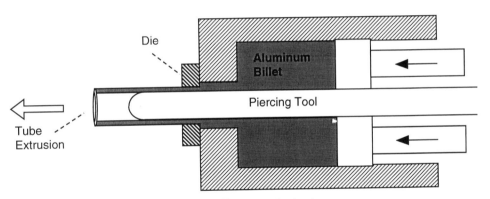

Figure 1-4b. Extrusion process for hollow round tube forms.

fit within the circle size of the extrusion press. Some extruders are limited to 6-inch (153-mm) diameter dies, while others can extrude up to 15-inch (381-mm) shapes. When the dimensions of the finish part exceeds the cross section of the die, the part must be built up from multiple extrusions that may snap or bolt together.

Extruded aluminum is a very economical means of fabricating complex shapes. Cost limitations are based on quantity requirements. The dies for certain shapes are expensive. However, for large quantities the one-time die cost can be spread over the entire order. For small quantities, unless the extruder already has a die, the die cost may make extruding the part commercially impractical.

Bars and Tubing

Because of the ease of extruding, an endless variety of shapes and dimensions are readily available in all the aluminum alloys. For tubing and piping,

Figure 1-5. Various extrusion shapes.

wall thicknesses can be created for the necessary strength characteristics or stiffness criteria.

FINISHING OF ALUMINUM

All alloys of aluminum are available in a variety of finish surfaces, both mechanically and chemically produced. Certain finishing techniques, such as anodizing and die impregnation, are unique to the alloys of aluminum. These techniques use the singular porosity of the aluminum oxide film to develop colors in the aluminum surface. Other finishes use processes similar to those used for other metals. These are the mechanically applied finishes. On aluminum sheet, plate, and extrusions, mechanical surface treatments have had limited usage, except as decorative trim on appliances or car interiors. The reason is partly the softness of the metal surface.

Mill Finishes

The initial surface of aluminum fresh from the mill is known as the "as-fabricated" finish or the "mill finish." The mill finish is a common architectural surface. For aluminum extrusions and castings, the as-fabricated finish is the surface available. There are three varieties of as-fabricated finishes. When ordering mill finish or as-fabricated finishes, request "Architectural Quality Surface." This qualifier gives the supplier an idea of the final intention for the metal.

As-Fabricated Finishes

Unspecified
Specular
Nonspecular

Unspecified

An unspecified surface can be in any form the mill wishes to provide. In the case of hot rolled and heat-treated alloys, surface oxide and mill oil may be present. For cold formed sheets or plates, this finish is determined by the quality of the rolls used to impart the specific cold working temper. The unspecified as-fabricated finish is the natural finish provided on cast and extruded products. For extrusions, the finish has tight, fine lines running the length of the product. The surface should be an even, consistent color and is often considered ready for further anodizing or paint pretreatment. Extrusion dies, carefully machined and maintained, will reduce the occurrence of striations and deep grooves in the finished extrusion. For extrusions, this finish is available in two forms. You must specify one of the forms in order to signify the expected quality.

Aluminum Extrusion As-Fabricated Finishes

Structural Quality
Architectural Quality

Extrusions of Structural Quality will show minor surface imperfections such as streaks and grooves. The Architectural Quality is more uniform. The surface has very fine machined lines created by a finely milled die. For cast products, this finish is the surface imparted by the type of mold used. Sand cast surfaces are rough textured, often blotched with oxides. Smooth surfaces can be obtained by the use of ceramic or metal molds.

When specifying or ordering a smooth finish, qualify the surface for Architectural Quality. This request will minimize the occurrence of pits, streaks, cold structural streaks, stop marks, excessive oxides, and other defects which characterize poor quality surfaces.

Specular

A specular surface finish is available only on sheet or plate wrought products. This is an Architectural Quality surface imparted by highly polished rolls as the sheet or plate is cold rolled to the required temper. The surface is reflective and contains a consistent luster, free of oxides. The resultant sheet or plate is ready for further fabrication, embossing, or finishing.

Upon receipt from the mill this finish requires protection during storage. Water will damage the surface, creating grayish-white to black stains. Surface quality can also be damaged by scratches or gouges and surface coining as sheets are decoiled and stacked. The specular finish is available on Architectural Quality sheets in three basic varieties.

1. *Mill One Side Bright.* This finish is a smooth, bright, surface on one side. The other side is mill finished or unspecified.

2. *Standard One Side Bright.* This surface is a very reflective, almost mirror surface finish, produced by passing the sheet through highly polished rolls. The other side is unspecified.
3. *Standard Bright.* This specification provides a mirrorlike, reflective finish surface on both sides of the sheet.

Nonspecular

A nonspecular finish falls between the two previously mentioned finishes. It provides an even, consistent surface without the discoloration of surface oxides, but also without the reflective luster of the specular finish. Additional surface finishing, such as etching or chromate pretreatment for paint applications, is typically performed on this mill surface. Soft tempers and annealed and heat-treated alloys are often provided in this finish from the mill.

Satin and Mirror Finish Surfaces

Architectural uses of the mechanically applied satin and mirror finishes are limited. With aluminum, these finishes tend to oxidize over time, loosing the luster provided during the polishing process. Coatings, such as clear lacquers, decrease the luster generated by this finishing and thus reduce their decorative effect. Lacquers often show yellowing or dark zones when applied clear over the aluminum surface. Anodizing for protection also reduces the brightening effect because anodizing etches the surface as it develops the deep oxide pores. Brightening posttreatments do not last when fabricated parts are exposed to the environment or to excessive handling. Maintaining polished surfaces is difficult and rarely cost-effective when compared with stainless steel.

Figure 1-6. Reflectivity of mill finish, "as-fabricated" specular aluminum.

Figure 1-7. Reflectivity of "as-fabricated" specular Alclad aluminum.

Mechanically polished finishes are used in some decorative interior applications requiring low cost and light-weight construction. Light fixture reflectors and decorative trims and accents are commonly constructed from the polished aluminum finishes. The common enhanced finishes applied to aluminum are discussed in the following paragraphs.

Directional Satin Finishes

Directional finishes on aluminum are similar to those provided on the stainless steels. These finishes are characterized by grit lines running the length of the part. Grit lines are induced onto the surface by wheel or belt polishing with different levels of grits. There are three basic degrees of directional finishes:

> *Fine Satin*—aluminum oxide grit of 320 to 400 size
> *Medium Satin*—aluminum oxide grit of 180 to 220 size
> *Coarse Satin*—aluminum oxide grit of 80 to 100 size

Other directional finishes applied to aluminum surfaces are the hand-rubbed texture and the brushed texture. A hand-rubbed texture, because of the more labor-intensive nature of the process, is relatively expensive. The finish can be described as having fine, nearly parallel, lines produced by rubbing with stainless steel wool. The brushed finish is applied by using rotary stainless steel wire brushes. It is characterized by short, deep parallel scratches. The hand-rubbed finish and the brushed finishes are available on small fabrications or on highlighted areas only. These processes are slow and tedious, requiring a certain amount of skill and artistry.

The Nondirectional Satin Finishes—Abrasive Blast and Shot Blast

A nondirectional surface finish is achieved by abrasive blasting the surface with glass beads, sand, steel shot, aluminum oxide, or a wide variety of other blasting media. The degree of coarseness is determined by the type of medium used and the nozzle pressure of the blasting equipment.

For a sheet aluminum product, nondirectional finishes induced by abrasive blasting is not recommended. For highlighting surfaces, this finishing technique may work, but the distortion this process produces may not be very appealing. Abrasive blasting tends to stretch the aluminum in ways difficult to predict. For castings, the abraded surface needs to be coated or anodized. Uncoated nondirectional surfaces will retain fingerprints.

The shot blast finishes are achieved by firing stainless steel and steel shot from an air blast compressor system or a centrifugal system. The steel shot are solid spheres of various sizes; size is determined by passing them through a sieve of a particular size. This type of finishing produces a reflective surface by creating a minute dishing effect on the aluminum. There are various shot sizes, as determined by the Society of Automotive Engineers (SAE). The diameter of standard shots and the corresponding SAE shot numbers are listed in Table 1-17.

Angel Hair and Distressed Surfaces

Angel hair is a custom finish, induced onto an aluminum surface by using 180 to 320 grit abrasive wheels or Scotch-Brite[2] pad abrasives. This finish is characterized by small scratches induced onto the surface at various angles. Although these are radial scratches, they overlap to produce a nondirectional appearance. When performed on large panels and viewed from a distance, the different intensities and the reflective nature of the scratches produced a lightly mottled effect. This finish is performed on the natural surface, uncoated and without anodizing or brightening. Anodizing will greatly reduce the effect of the etching process of the finish.

Custom Finishes

Special order finishes, such as machine turned and grinding swirl finishes, can also be applied to aluminum. These are finishes provided by custom metal manufacturers for specific artistic or ornamental treatments. Certain finishes produced by using grinding wheels and stainless wire brushes can be created on the aluminum surface. Using a grinder in a random fashion can, in the hands of a craftsperson, produce elegant surfaces of alternating swirls. The swirls can be repeating circles, sometimes known as "engine turn," or random brushes of various sizes and biases.

These custom finishes are usually dependent on the abilities of their applicator. A light directional polish is applied to the surface by passing the sheet through a polishing roll. A grinder is applied by hand or by using a

---

[2]Scotch-Brite is a nonmetallic abrasive pad manufactured by 3M Corporation of St. Paul, Minnesota.

TABLE 1-17. Shot Blast Sizing and Corresponding Aluminum
Association Finish Designation

AA CODE	SAE SHOT NO.	SCREEN SIZE OPENING (INCHES)	SCREEN SIZE OPENING (MM)
M45	70	0.017	0.43
M45	110	0.023	0.60
M45	170	0.033	0.85
M46	230	0.039	1.00
M46	280	0.047	1.18
M46	330	0.055	1.40
M46	390	0.066	1.70
M46	460	0.079	2.00
M46	550	0.079	2.00
M47	660	0.094	2.36
M47	780	0.111	2.80
M47	930	0.132	3.35
M47	1100	0.157	4.00
M47	1320	0.187	4.75

CNC (computer numeric control)-assisted frame to control the movement
and bias of the grinding head. As the grinder passes across the surface, small
areas receive the texture in a slightly curved fashion. Alternating the grinder
disc edge will apply the curved brushing in similar fashion to a painter using
a small brush. Larger swirls can be produced, as well as multiple grades of
disc grits.

Figure 1-8. Ornamental aluminum light fixtures with satin finish texture. Artist: R.H.
Fischer Artworks

Small areas and prewelded assemblies can also receive the custom swirl ground patterns. It is important to have control samples to work from if any level of consistency is required.

Mirror Finishes

The mirror finishes are produced by polishing the smooth surface of aluminum with buffing wheels. Buffing the surface creates a highly reflective, smooth surface. The reflective surface is initially very mirrorlike. However, surface tarnish will quickly develop on exterior exposures. Interior exposures will last longer if they are not in a place where they can be handled. There are two basic quality levels of the mirror finishes: smooth specular finish and specular finish.

A smooth specular finish is very bright and reflective. This finish is produced by buffing and polishing the aluminum surface until all scratches and grit lines are removed. Electropolishing assists in achieving a highly reflective finish of this type. A specular finish is also very reflective, but some minor grit lines are apparent upon close inspection. These finishes are often used for light reflectors and other interior decorative treatments. The specular finishes can be lightly anodized. Anodizing etches the surface, creating a more matte texture. Chemical brightening can be used to further enhance these finishes and seal the surface. Such finishes are not recommended for exterior uses.

Embossing

Aluminum can be readily embossed with matching rolls. Embossing imparts a texture by passing smooth, thin-gauge metal through engraving rolls, which press a pattern into the metal surface. Embossing stiffens the metal surface, reducing the "oil canning" tendencies of the thin sheet. The embossing rolls

Figure 1-9. Ornamental spun aluminum chandeliers with satin finish texture. Artist: R.H. Fischer Artworks

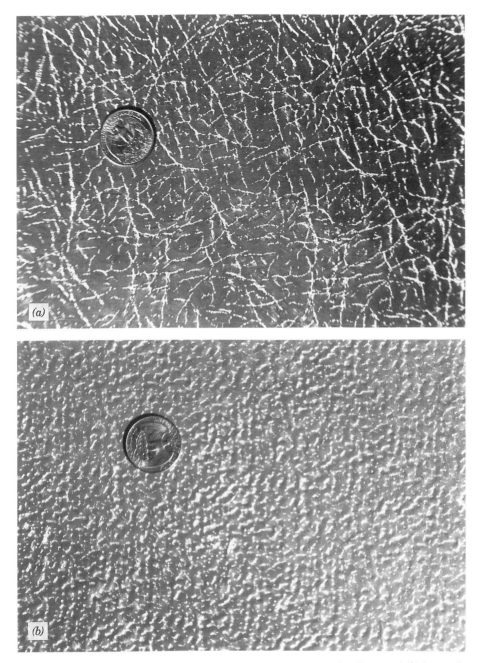

Figure 1-10. Examples of embossed aluminum textures. (*a*) 'leathergrain' (*b*) 'stucco'.

flatten the metal for further roll forming into corrugated panel shapes or for forming flat sheets into sheet metal configurations. Embossing can occur in unfinished or painted aluminum sheets. It will not work well on anodized sheets because of the brittle nature of the anodized surface. When embossing is used on unfinished aluminum, the surface enhancements conceal many small imperfections and are often used in exterior applications where no other finish is provided. Embossing is performed while the thin metal is still in coil or sheet form. The surface enhancement is very economical.

Embossed finishes are standard for thin-gauge siding panels and flat sheets that are to be made into exterior trim or fascia. Common embossed

---

TABLE 1-18. Aluminum Association Standard Mechanical Finish Designations

AA CODE	TYPE	DESCRIPTION
	As Fabricated (Mill Finish)	
M10	Unspecified	Finish provied on extrusions and castings. Quality left to the mill supplier of the metal.
M11	Specular	Reflective smooth finish. Initial finish selected for anodizing or paint finishing.
M12	Nonspecular	Nonreflective, smooth surface finish. Commonly used for architectural wrought products.
M1X	Other	Not currently specified.
	Buffed Finishes (Reflective)	
M20	Unspecified	Any degree of buffing or mirror polishing can be specified.
M21	Smooth Specular	Very reflective surface. Initial surface for anodizing. Used for reflective ornamental trim.
M22	Specular	This reflective finish is similar to M21, but with minor grit lines apparent.
M2X	Other	Not yet specified.
	Directional Textured Finishes	
M30	Unspecified	Satin texture determined by the supplier.
M31	Fine Satin	Fine directional satin finish produced by abrasives in the 320 to 400 grit range.
M32	Medium Satin	Directional satin finish produced by abrasives in the 180 to 220 grit range.
M33	Coarse Satin	Coarse directional finish produced by abrasives in the 80 to 100 grit range.
M34	Hand Rubbed	Hand-rubbed directional satin produced by rubbing the surface with steel wool.
M35	Brushed	Directional finish applied by rotating stainless steel wire brush.
M3X	Other	Not yet specified.
	Nondirectional Textured Finishes	
M40	Unspecified	Nondirectional polished determined by fabricator or metal supplier.
M41	Extra Fine Matte	Very fine abrasive blast with glass beads or 200 mesh aluminum oxide.
M42	Fine Matte	Abrasive blast with 100 to 200 mesh aluminum oxide.
M43	Medium Matte	Abrasive blast with 40 to 50 mesh aluminum oxide.

TABLE 1-18. (Continued)

AA CODE	TYPE	DESCRIPTION
M44	Coarse Matte	Abrasive blast with coarse, 16 to 20 mesh abrasive.
M45	Fine Shot Blast	Cast steel shot of SAE size 70 to 170.
M46	Medium Shot Blast	Cast steel shot of SAE size 230 to 550.
M47	Coarse Shot Blast	Cast steel shot of SAE size 660 to 1320.
M4X	Other	Not yet specified.

textures are stucco, leather grain, and a small diamond or bubble pattern. Other textures and patterns are also available.

Table 1-18 describes the various standard finishes produced by mechanical means. These standard finishes correspond to an Aluminum Association code for specification purposes.

WEATHERING

When correctly used, the natural mill finish aluminum coating will provide a beautiful, long-lasting, corrosion-resistant barrier to the environment. Natural mill finish aluminum will weather to various shades of gray; the metallic luster of the installed surface will gradually fade. Over time, and usually first at sheltered areas of exposure, the surface will obtain a slightly roughened appearance. Some shallow pitting may develop, appearing as small, fuzzy, gray deposits of aluminum oxide mixed with pollutants.

In rural and arid environments weathering of natural mill finishes will proceed in even gray tones. In coastal environments and urban industrial exposures natural mill surfaces will often exhibit a mottled gray appearance. The more vertical exposed regions will turn a uniform gray color, while the sheltered regions will develop the mottled tones.

The low-strength alloys—those with a higher purity level—are best for weathering appearance. The higher-strength alloys, such as the A92000 series, contain copper as an alloying constituent. Copper adds strength but reduces corrosion resistance. Low-strength alloys, such as A93003, were used on the Cincinnati Union Terminal Building, constructed in 1932. The weathered surface appears as a uniform gray color, showing no signs of excessive corrosion. The higher-strength alloys, those containing copper and magnesium, tend to weather more in the mottled tones. If these alloys are used in the natural mill tones, consider pure aluminum Alclad coatings. The Alclad coating will weather more uniformly and provide superior corrosion resistance.

With natural mill finishes, the color initially provided is a bright silver-white metallic tone. The bright tone will remain for several years before graying out, depending on the pollution level and the amount of rainfall. Caution should definitely be observed while storing and handling the material prior to final installation on a wall or roof. Moisture, even small amounts, when allowed to rest on the surface will develop into dark, almost black, streaks and spots. Removal of such dark spots is impractical over large surfaces. The materials should be stored in clean, dry areas. Venting the storage area is necessary to dry out condensation that may form on the metal surface.

CHEMICAL FINISHES

The Matte and Bright Chemical Treatments

Chemical treatments produce matte and reflective mirror surfaces. The matte treatment is often a precursory treatment for other subsequent processes such as anodizing or painting. The *matte finish* for aluminum comes in three common variations, each of which is produced by etching the surface with a caustic solution.

Fine   Produced by immersion in mild alkali solution
Medium   Caustic etch of surface by sodium hydroxide solution
Coarse   Produced by immersion in sodium fluoride and sodium hydroxide solution

These finishes are prone to fingerprinting. The surface produced acts like a sponge, soaking up oils or moisture from machinery or handling. If this finish is applied prior to painting or anodizing, consider reducing the process time to prevent contamination to the surface. If these finishes are used without additional chemical treatments, such as anodizing or chromate pretreatments for painting, the surface may acquire a chalky or smutty appearance. Etching the aluminum surface removes a small portion of the oxide coating, leaving a matte tone. The oxide will naturally develop on the etched surface but initially the surface is porous.

A clear anodized treatment over a medium or fine etch is hardly discernible from the nonanodized etched surface. Therefore, it is recommended to clear anodize after etching if the resulting appearance is a matte, low-gloss tone.

Bright finishes created by chemical means are uncommon for exterior architectural uses. Generally, bright finishes are used on small fixtures or decorative trim. Lighting reflectors are a common use for this highly reflective kind of finish. Often, chemically brightened sheet material is embossed. Embossing works to hide scratches and surface marks that would be readily apparent on smooth surfaces.

Bright finishes are available in two standard classifications: highly specular finish and diffuse bright finish.

A highly specular finish is an extension of the smooth specular finish discussed previously. A smooth specular finish is achieved by dipping the aluminum sheet or part into a heated bath of phosphoric acid, nitric acid, and water.

A diffuse bright finish is produced by first producing a matte finish, then chemically brightening the surface by electropolishing. The resulting finish has a bright frosty appearance, still reflective but without the mirror effect.

These finishes are sometimes provided to aluminum to enhance further treatments such as anodizing. They are not commonly used in architecture, but rather as ornamental enhancements for consumer products.

One formulation for brightening the surface of aluminum uses various acid concentrations in a heated bath. The level of dilution is critical. As the bath is heated, the concentration will change as the water evaporates. If the water content changes too much, the surface may etch, or brightening will be inhibited.

Concentrated phosphoric acid	80% by volume
Nitric acid	5% by volume
Acetic acid	5% by volume
Water	10%
Heat the solution to 220°F	

Note: Exercise extreme caution when using acids. Follow all instructions for handling and discarding acid solutions. Wear eye protection and protective clothing. Acid fumes can be toxic.

Table 1-19 lists the various chemical treatments recognized by the Aluminum Association. These treatments are usually considered pretreatments with the expectation that additional surface enhancements will follow.

Anodic Coatings

Anodic coatings on aluminum are widely accepted as a premier architectural metal treatment. Anodizing is a controlled extension of the natural oxide layer on aluminum. Unique to aluminum, anodic coatings offer the designer a variety of color options in conjunction with an enhanced surface barrier. This barrier is both more corrosion- and abrasion-resistant than the much

TABLE 1-19. Aluminum Association Chemical Treatment Designations

AA CODE	TYPE	DESCRIPTION
	Nonetched and Cleaned	
C10	Unspecified	Any cleaned aluminum surface is acceptable.
C11	Degreased	Treatment with organic solvents.
C12	Chemically cleaned	Cleaned with nonetching chemical cleaners.
C1x	Other	Yet to be specified.
	Etched Chemical Treatment	
C20	Unspecified	Any degree of etching is acceptable.
C21	Fine Matte	Trisodium phosphate etching.
C22	Medium Matte	Caustic soda etching (sodium hydroxide).
C23	Coarse Matte	Sodium fluoride and sodium hydroxide.
C2X	Other	Yet to be specified.

thinner oxide layer that develops naturally. The film is also very porous. During the anodizing process, the top 0.001 inches (25 μm) of aluminum is converted to Al_2O_3. The aluminum continues to thicken as the anodizing coating develops. For architectural uses, this thicker oxide coating is from 0.000225 inches to 0.0007 inches (0.0057 mm to 0.0178 mm) in total depth, whereas the naturally developing oxide film is but a millionth of an inch in thickness. Coatings beyond 0.003 inches (0.076 mm) will fracture and chip easily. The color, if translucent, can be milky.

The initial development of the anodizing process was patented in Britain in 1927. The process consisted of a sulfuric acid bath accompanied by an electrical charge to develop the oxide coating. A similar process is still in use today to produce a clear anodized finish or as a base for further dye impregnation.

In architectural uses, this process of enhancing the surface of aluminum did not come into common use until the early 1950s when aluminum became an architectural metal of choice for many external metal structures. Today nearly every exterior use of aluminum in the unpainted form is anodized.

All aluminum alloys can be anodized. However, some alloys, those that contain relatively high amounts of manganese, silicon, and copper, cause a level of opaqueness in the film. The electrochemical processes used to generate the anodic film require several steps of controlled surface treatment of the aluminum.

First, the aluminum surface is thoroughly cleaned to remove all foreign matter, such as soil and oils, in a nonetching solution. Following the cleaning the aluminum part or sheet is etched in sodium hydroxide (caustic soda) to remove some minor surface irregularities and provide a medium matte texture. The actual anodizing of the aluminum surface occurs next, when the aluminum sheet or part is immersed in an acid solution, which acts as an electrolyte. An electric current is passed through the suspended aluminum sheet and the electrolyte, creating a positive charge in the suspended aluminum part or sheet. Oxygen migrates out of the solution and combines with the aluminum surface to develop the aluminum oxide coating. The oxide film is monitored to arrive at the desired thickness. Hydrogen is also released from the solution and migrates toward the cathode side of the tank. Figure 1-11 shows the basic anodizing process.

Today, use of computerized monitoring equipment has vastly improved thickness, and thus color control, achieved in the anodizing process. Transparency depends in part on the film thickness and alloy makeup. The longer the alloy is submersed, the thicker the oxide film develops and the darker the aluminum surface becomes. Actually, current and time together control film thickness. Voltage controls the anodic film structure, in particular the pore size and cell diameter. The transparency of the coating can be varied by adding additional treatments to the oxide film. Additives can also reduce the dissolution effect of the acid used to develop the electrolyte, thus allowing faster oxide film development. (The pore structure of anodized aluminum is shown in Figure 1-12.)

The final step in the anodizing process is the hardening and sealing of the surface by use of deionized boiling water or metal salt sealers. Sealing is required to close the pores of the oxide film and provide uniformity to the

Bank of China Tower. Hong Kong Central.
Large, flat clear anodized aluminum panels outline the monolithic glass curtainwall.
Architect: I.M. Pei and Partners

Bartle Hall Expansion. Kansas City, Missouri.
25-mm thick extruded aluminum tube shapes welded into a massive sculpture.
Finish on the aluminum is 'as fabricated'.
Designer: R.M. Fischer

RCA Building.
New York, New York.
Cast aluminum panels
decorate the upper
floor spandrels and parapet.
The panels are approximately
12-mm in thickness. Finish is
a weathered anodized coating.
Architect: Eliot and
John Walter Cross

Bartle Hall Expansion. Kansas City, Missouri. Sculpture installed.

Grady Hospital.
Atlanta, Georgia.
Decorative aluminum
grilles created by
various sections
of extruded tube shapes
curved and welded then
coated with a Kynar 500
spray-applied coating,

Grady Hospital. Atlanta, Georgia.
Decorative aluminum panels created by a series
of high pressure forming operations.
Panel thickness is 3.17-mm and the finish is
a KYNAR 500 spray-applied coating.
Architect: Kaplan McLaughlin Diaz

AMA Headquarters.
Chicago, Illinois.
Decorative cast aluminum
ceiling panels with a white
KYNAR 500 finish.
Architect: Kenzo Tange

Detail.

Museum of Science and Industry. Tampa, Florida.
Alclad coated aluminum panels, 4.75-mm thick.
The bright reflective finish is produced by passing the aluminum sheet through polish rolls
producing a specular appearance.
Architect: Antoine Predock

Cyprus Club. San Francisco, California.
Decorative polished copper surfaces.
Designer: Jordan Mozer

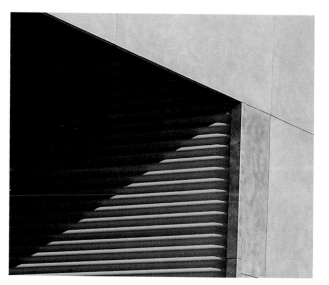

Missouri Public Utilities. Kansas City, Missouri.
One-sided specular finish aluminum with an
'angel hair' texture.
Architect: Rafael Architects

Missouri Public Utilities. Kansas City, Missouri.

Advanced Laser
Technologies.
Iowa City, Iowa.
Commercially pure copper
flat seam panels.
Architect:
Frank O. Gehry Associates

American Heritage Museum. Laramie, Wyoming.
Blackened copper flat seam panels.
Architect: Antoine Predock

United Missouri Bank. Kansas City, Missouri.
Polished copper bar stock shaped to create a custom handrail.
Architect: Linscott, Heylett, Wimmer and Wheat

United Missouri Bank. Kansas City, Missouri.
Spun copper light sconce, polished and lacquered.
Architect: Linscott, Heylett, Wimmer and Wheat

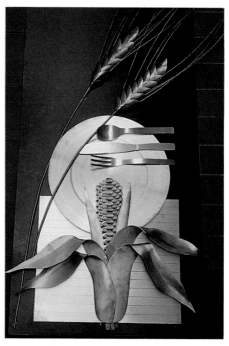

Farmland Industries. Kansas City, Missouri.
Copper sculpture.

Bank. Tucson, Arizona.
Curved copper vaulted ceiling
with a specular mirror finish.
Copper panels are approx-
imately 2.3-mm thick.

St. Lukes Hospital. Kansas City, Missouri.
Curved standing seam copper roofing.
Note the golden interference colors occurring
on some panels after a few weeks of
atmospheric exposure.
Architect: Horner and Blessing Architects

1220 Washington Building. Kansas City, Missouri.
Copper cladding applied over an internal steel
frame to create a metal and glass storefront.
The steel is isolated from contact with the copper.
Architect: BNIM Architects

Private residence. Kansas City, Missouri.
Walls clad in flat seam copper panels. Natural weathering of the copper
brings out various hues of purple, yellow, and brown.

Detail.

Midland Theater Kiosk.
Kansas City, Missouri.
Various copper alloys used to
produce subtle color tones in
this ornamental kiosk
and entryway. The canopy
and door cladding are
muntz metal, the column
capitals are cast silicon bronze,
and the bases are spun
commercial bronze.
Architect: BNIM Architects

Close up detail.

The dome canopy under
fabrication. The features are
cast and stamped copper alloys.

San Francisco Mart. San Francisco, California.
Panels constructed from perforated 2.5 mm polished
muntz metal attached over mirror polished stainless steel.
Architect: Pfister Partnership

San Francisco Mart.
San Francisco, California.

Custom Planter for Mackey Mitchell Architects.
St. Louis, Missouri.
Commercial bronze and muntz metal sconce.
Finish produced by glass bead blasting

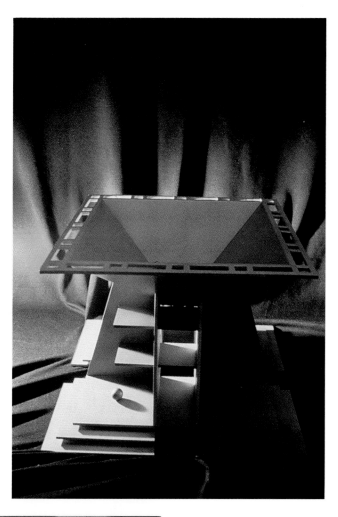

120 North LaSalle Building.
Chicago, Illinois.
Muntz metal sign band
and tubular uplight for
the building entryway.
Architect: Murphy-Jahn
Architects

Barnett Bank Building. Jacksonville, Florida.
Detail of perforated muntz metal over mirror stainless steel.
Architect: Spillis Candela and Partners

KCPL Building. Kansas City, Missouri.
Etched Nickel Silver cladding on elevator doors.
Architect: Holt, Price and Barnes

KCPL Building. Detail.

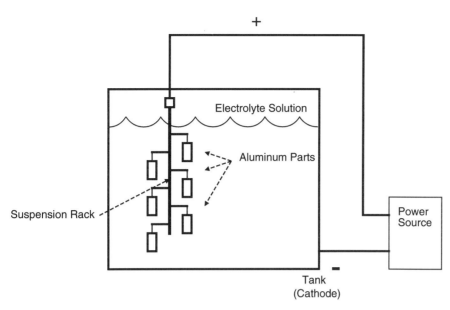

Figure 1-11. Basic anodizing process.

final surface. With the integral anodizing process, pure deionized water or, for better results, potassium dichromate is used. Potassium dichromate improves the corrosion resistance of the oxide film. With the die-impregnated anodizing processes and the two step processes, nickel acetate or nickel-cobalt acetate is used to seal the pores and keep the dye in place.

For good color uniformity and quality control in anodizing, the concentration of acid must be maintained to ± 5 grams per liter and aluminum concentrations to within 8 to 12 grams per liter. In addition, good quality anodizing requires careful temperature control. Fluctuations in temperature affect uniformity more than variations in any other component, with the

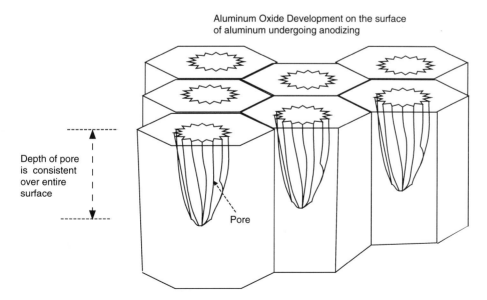

Figure 1-12. Depiction of microscopic pore development during anodizing process.

exception of the alloying constituents. Temperatures must be maintained to ± 2°F from the original set point.

The best color and clarity of the oxide film can be obtained when anodizing the more pure aluminum alloys. Alloy A91100 develops a beautiful, clear anodized coating.

Temper of the aluminum part and thickness of the anodizing film also affect the finish produced. In general, the harder the sheet, the greater the variations within the oxide film; the thicker the oxide film, the less clarity. It is difficult to achieve a good, consistent anodized film on castings. The grain development in the cast and the nonhomogeneous nature of the alloy mix can produce variations in color tone.

Welding, even with extensive grinding and polishing, will appear as a different color and tone.

There are two main forms of the anodizing process, coil anodizing and batch anodizing. Batch anodizing requires the suspension of each part or rack of parts into the acid bath. Aluminum extrusions and formed parts can only be batch anodized. The Duranodic® and Kalcolor® processes are batch anodizing processes. Electrolytic deposit and overdyeing processes can also be batch anodized. Coil anodizing is available for sheet and thin plate aluminum. Coil anodizing can be performed only by the electrolytic deposit process and the overdyeing processes. Coil anodizing is the more economical process for anodizing sheet and thin plate forms of aluminum.

For aluminum extrusions, alloy 6063 is the material of choice. This alloy, used on more than 90% of aluminum extrusions, receives color anodizing more effectively than other alloy types. The amount of iron in the alloy will determine the degree to which a matte or bright anodized surface tone is obtainable.

Health and Environmental Concerns Related to Anodizing

An anodized finish on aluminum is chemically stable and nontoxic. The process is nonhazardous and involves no volatile compounds; it involves water-based chemistry which is easily treated. Acids and other corrosives are used in the anodizing process, but most anodizing facilities adjust the pH of the wastewater and remove much of the aluminum solids left in solution. The aluminum solids are considered nonhazardous and are often resold to municipal waste treatment facilities to use in phosphate removal. Other by-products of the anodizing process are isolated and resold for use in a variety of industries such as cosmetics, fertilizer, and newsprint.

The process itself produces no listed hazardous products. The waste is considered a low-metal sludge. No air pollutants, or VOCs, are created during the anodizing process.

Clear Anodizing

One of the more common and most economical anodizing treatments is clear anodizing. Clear anodizing produces a transparent, clear oxide film. It has a soft satin texture in a very uniform metallic aluminum silver color. Clear anodizing can be applied to sheet, plate, extrusions, castings, and most fabricated parts.

Clear anodized coatings are produced by the sulfuric acid process. The initial degreasing and cleaning of the aluminum in an etching tank is followed by the caustic etching process or the bright dipping process to develop a matte or mirrorlike finish, respectively. The aluminum sheet or part is immersed into a sulfuric acid solution. A charge is applied to transform the aluminum part into a positive anode (thus the term *anodizing*). The tank containing the acid solution acts as the cathode. Oxygen migration occurs from the electrolyte solution to the aluminum surface, creating an oxide layer growth. The rate of film development is controlled and will continue until the charge is removed.

As the oxide film develops, minute pores are created perpendicularly through the film. When the desired film thickness is achieved, the aluminum is submerged in a hot deionized water bath or metal salt solution to seal the pores of the oxide film. (Deionized water is distilled water with a high level of purity.)

There are three common techniques for sealing the surface of an anodized film. The most common is a hydrothermic sealing process that immerses the anodized part in a bath of hot deionized water at 210°F (99°C). The pH is maintained at between 6.0 and 6.4. This technique creates an hydrated aluminum oxide molecule at the pore. The hydrated molecule is larger than, and sufficiently seals, the pore.

Another common technique uses nickel acetate at a lower temperature of between 175°F and 185°F (79°C and 85°C). Nickel is precipitated into the pores of the oxide along with the hydrated molecule, which basically clogs the pore. This technique is used when dies are applied to the porous sheet of aluminum. The pores suck up the dye, and the nickel closes the route of escape from the pore.

A third technique uses a cold seal, at 90°F to 94°F (32°C to 35°C), of nickel ions to accelerate the sealing process. The nickel joins with the

Figure 1-13. Reflectivity of clear anodized aluminum.

aluminum and fills the pore. This process is also used in the color sealing of dyes into the pores of the aluminum oxide.

Color Anodizing Hardcoat

There are three basic techniques of inducing color into the anodizing film. One of the oldest architectural techniques is the development of *integral* color anodic coatings. The integral color anodizing coatings are limited to the grays, champagnes, and bronze to black color range. These colors offer excellent durability and fade resistance when exposed to ultraviolet radiation. The main reason for their durability is the thickness of the oxide film and the lack of coloring agents. They achieve color by chemical reactions involving alloying constituents within the aluminum (Figure 1-14).

The integral colors are obtained in a similar acid bath technique as used for clear anodized coatings. The main differences, other than certain current and temperature variations, are alloy-specific controls. Certain alloys will develop only colors in the black or dark bronze ranges. The alloying elements within the metal chemically combine with the electrolyte to produce a color in the film. As the anodic film develops, particles of the alloying material are diffused into the film. When the sulfuric acid method is used in conjunction with a specific alloy control, the process is called integral color hardcoating. The Duranodic®, Kalcolor®, and Permadonic® processes are proprietary integral color processes. Table 1-20 lists various alloys and the colors they can produce from the integral process.

Other techniques, using proprietary processes, also produce these integral color anodic coatings. Many of the proprietary processes use mixtures of acids in solution with modified temperature and current densities. In all cases, specific alloy controls are necessary to obtain the correct colors. No dyes are added to obtain the colors. Integral color anodizing is a cold process. That is, coloring is performed at relatively low temperatures. The integral coloring process is a power hog, requiring extensive amounts of electricity; thus it is more expensive. Difficulties arise when the acids in the bath dissolve away the aluminum contacts used to provide the electrical charge. Corroded contacts can create differential current across the sheet and, thus, problems in color uniformity. Frequent maintenance is necessary to ensure that the contacts are in proper working condition.

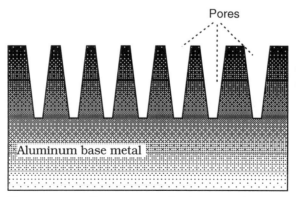

Figure 1-14. Cross section of color created in pores of the integral anodizing process.

ALLOY	MAXIMUM FILM THICKNESS		COLORS
	inches	mm	
A91100	0.0017	0.043	bronze tint
A91100	0.003	0.076	clear, green-trinted gray
A93003	0.002	0.051	clear
A95005	0.0035	0.089	clear, bronze
A95052	0.0035	0.089	clear, bronze
A95052	0.0053	0.135	black
A95083	0.0035	0.089	clear, bronze
A96061	0.003	0.076	bronze
A96061	0.0035	0.089	black
A96063	0.004	0.102	dark gray, green-gray, bronze
A96063	0.0050	0.127	black
A96105	0.0035	0.089	clear, green-gray, bronze
A03600	0.0005	0.013	clear
A03800	0.0005	0.013	clear

TABLE 1-20. Architectural Alloys and Integral Anodizing Colors

Electrolytically Deposited Color—Two-Step Method

Another technique used to induce color into aluminum is electrolytic deposition. This proprietary process is an acceptable and common coloring technique for exterior architectural projects. This anodizing technique introduces a coloring agent into the pores of the metal (Figure 1-15). The coloring agent is a metalized salt such as tin or nickel. Known also as the two-step oxide coating, this process produces colors ranging from champagne and light bronze to the black. These colors are obtained by using nickel or cobalt salts. A range of red colors can also be obtained by introducing copper salts, and blues by using molybdenum salts. Golds and yellows can be obtained by using the more expensive gold or silver salts.

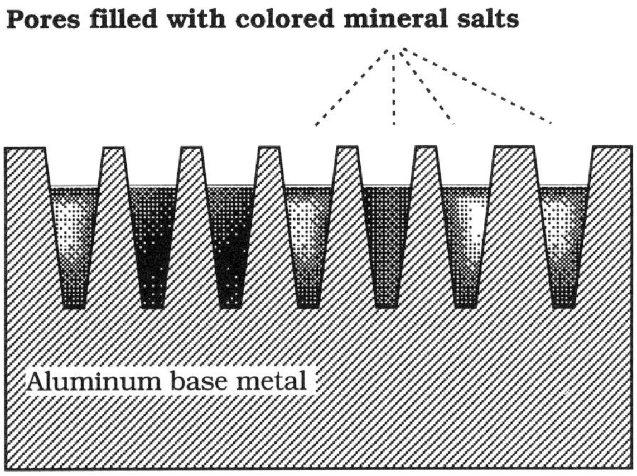

Pores filled with colored mineral salts

Aluminum base metal

Figure 1-15. Cross section of color induced using the "two-step" anodizing process.

After the sulfuric acid process (step 1), before the pores in the coating are sealed, the aluminum part is immersed in a bath of inorganic mineral salts (step 2). The mineral salts are electrolytically deposited to the base of the pores. The mineral salts provide the color to the coating. The pores are sealed by immersion in a hot deionized water bath. The mineral salts are not subject to ultraviolet radiation and when sealed will provide a long-lasting, durable film. It is necessary to produce thick films particularly when the surface will be exposed to corrosive environments. Films as thick as 3.0 mils are often used for saltwater atmospheres or areas subject to abrasion. Good pore structure is also critical in achieving quality architectural surfaces. The two-step process is less expensive than the integral process because less electricity is needed to achieve the coloring of the aluminum. Alternating current is also used in the salt deposition process in lieu of the direct current requirements of the integral processes. The two-step technique is not as alloy-specific as the integral process. Control parameters are not as narrow, thus reducing the occurrences of scrap parts resulting from color inconsistencies.

Organic Dye Impregnation

Another technique of imparting color to an aluminum coating is impregnation of the unsealed pores with organic dyes or pigments (Figure 1-16). Organic dyes will change when exposed to ultraviolet radiation. Therefore, these dyes are not typically used for exterior applications. Organic dye impregnation is not in common architectural use because of the problems with fading and durability. Gold is the only color produced by dye impregnation sufficiently resistant to the effects of ultraviolet radiation.

Very decorative treatments can be obtained using a technique of overdyeing the porous, presealed anodized surface. Best suited for interior uses, striking colors can be obtained by using single or multiple dyes before sealing the surface. The final seal uses solutions of water-repellent compounds to set the dye colors.

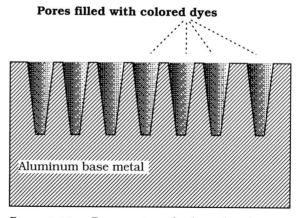

Figure 1-16. Cross section of color induced using the dye impregnation process.

Combined Organic Dyeing and Electrolytic Deposition

A relatively new proprietary technique uses a combination of the electrolytic deposition process and the organic dye process to produce a wide range of colors. Once the electrolytic deposition of the mineral salts is performed, organic dyes are introduced prior to sealing the pores. The process, which has been successfully used in Europe for exterior architectural projects, can produce very vivid colors in tremendous array of tones. A thick anodic coating is produced and then saturated with the organic die before hardening and sealing.

Other Anodizing Processes

Chromic anodizing is a process used to develop thin oxide films. It is not typically considered for architectural purposes except as a pretreatment for paint applications. Chromic anodizing produces a very thin corrosion-resistant film. The process retains some chromium, which in turn assists in further corrosion protection.

Further variations on this finish are the chromate conversion coatings. These coatings are formulated from chromic acid or chromium mineral salts. When correctly applied, they provide additional protection to the aluminum surface and act as an excellent base for the application of finish paints.

When a chromate conversion coating is applied to a clean aluminum surface, the colors produced are very iridescent. These are not typically considered finish coatings, but thin chromium oxide films. The surface feel is almost tacky but very adherent. A molecular bond joins the coating to the aluminum surface. The tacky surface assists the conversion layer's adhesion to the final coat of paint. Some of the common trade names for various chromate conversion coatings are Alodine, Chromicoat, and Bonderite.

The Alodine conversion coating etches the surface slightly and rebuilds the aluminum oxide layer with a chromic phosphate film. Bonderite develops a chromium aluminum chromate film.

Troubleshooting—Integral Hardcoat Anodizing

Color control in the integral anodizing process is determined by alloy variations and temper, concentration of acids, and time and temperature of the bath. Some variation in color is unavoidable because of the number of process variables. Aluminum producers will not provide guarantees on the anodizing quality of their materials. Computerized monitoring used by most modern anodizing facilities, however, will aid in reducing the variations in color.

It matters little what kind of form or shape is being anodized. Plate and extrusions all demonstrate color variations. Clear anodized, bronze anodized, and, to a lesser degree, black anodized colors will vary within a particular batch.

For every project using anodized aluminum it is recommended a color range be submitted for control purposes. The anodizing facility will use the range as its control for color variation. Colors falling outside this range are rejected. Often, this range can be quite significant, and when the colors of two adjoining sections happen to be the opposing outer limits of the color range, the results are less than desirable. It is important to have the range samples prepared from the aluminum batch or coil material to be used. Providing stock range samples will not give very accurate range parameters. Some projects have resorted to having the anodized panels and extrusions sorted and color matched to reduce, or at least control, the occurrence of color variations. This technique is an expensive proposition.

Different alloys within a batch will not color the same. Welds within a fabricated part will not color the same as the base material, regardless of the type. Mig welds, using matching alloy filler metals, have variations in grain and metal constituents at and around the weld zone. There is no guarantee that the weld will color in the same way as the base metal. Scratches, surface streaks, and other imperfections will not be concealed or removed by the anodizing process. Any surface imperfections will be accentuated by the anodizing process. If the sheet or fabrication is scratched, then grinding, followed by fine sanding, will prepare the surface for anodizing. The surface must be smooth and consistent.

Light and dark streaks, known as structural streaks, are caused by chemical and physical differences within the metal alloy. Streaks caused by alloying imperfections in the sheet will show up in the anodizing process. There is little that can be done if streaking occurs. Specifying anodizing-quality alloy sheet will minimize this problem. Streaking of another kind will also occur if sections of the aluminum fabrication trap the acid anodizing solutions and do not allow thorough rinsing. The methods of hanging the part for immersion and of draining trapped solution are critical to ensure good results. Assemblies to be anodized must have drainage holes to eliminate the trapping of anodizing solutions. No non-aluminum parts should be included in the assembled part being anodized. Non-aluminum parts will contaminate the anodizing tank.

White powder, adhering to the anodized surface, is a cosmetic problem caused by the mineral content of the hot water seal or nickel salts left from the color seal. The powder can be removed by soft scrubbing with a light abrasive. If the powder has a yellow tint, the electrolytic color tank is contaminated. If the powder has a green tint, the nickel seal tank is dirty or the pH in the tank is too high. An iridescent powder on the surface is caused by a desmutting tank contamination.

Temperature changes, if extreme enough, can cause crazing resulting from the differential expansion of the anodized film and the base metal. Crazing will also occur when postforming of the anodized aluminum is performed. The anodized surface will not stretch as the base aluminum bends. Small cracks develop through the anodized film along the length of the bend. In clear anodizing, the exposed aluminum at the break line does not contrast adversely with the undamaged surface. The exposed aluminum develops the protective oxide film and will offer some defense against corrosion. In dark anodized finishes, the exposed aluminum contrasts negatively.

Quality problems for the anodizing company may begin with the quality of metal obtained. The surface of the aluminum sheet or extrusion received from the mill can have defects that will adversely affect anodizing quality. Die lines resulting from the extrusion process may be too deep. When anodized, the die lines may appear intensified and show as color lines running the length of the extrusion.

Excessive buffing of the surface of an aluminum sheet can create small pits if the buffing compound is too thick. The surface may anodize with a "burned" appearance if the polishing operation is not performed evenly across the sheet. The alloy and temper, provided by the mill to the anodizing facility, will affect the resulting surface appearance. If the alloy contains an excess amount of zinc, more than 0.05%, the anodized sheet or extrusion will have a spangled appearance. This defect often is not apparent until anodizing occurs.

Iron, silicon, and magnesium surface concentrations will affect the brightness of the part. Some extrusions or sheets may appear bright; others will have a rougher, duller surface. Brighter tones or matte surfaces are dependent on the amount of iron. Magnesium-silicide precipitation can occur during the extrusion process, and when it does, the surface will appear rough.

Tempers at the T52 level will cause the surface to smut when anodized. Smutting appears as dark "wash" streaks across the surface. Dead-soft tempers will not etch well and will show smutting or a grainy appearance when anodized.

Other anodizing problems develop at the anodizing facility. Sheets, extrusions, and fabricated parts require racking capable of transferring the electrical current to the part as it is immersed in the tank. Care and planning should be undertaken when the part is attached to these conductive racks. Wires and clips are often used to secure the part. A Faraday effect occurs at the proximity of the attachment, which causes the aluminum directly adjacent to the hangers to reject the color. The attachment of the part should be located at a place that will not be exposed to view or that will be later removed.

Troubleshooting—Two-Step Method

Variable color, not a desirable result, can occur at the ends of a sheet undergoing anodizing. The variations are due to the incorrect voltage setup, which causes differential pore development. Adequate voltage regulation will ensure good pore structure. Color shade can be incorrect if the temperature of the anodizing process is not maintained. Nonuniformity across the surface of the anodized part is generally a result of incorrect immersion bath chemistry.

Other Chemical Coloring Techniques

Aluminum can be artificially weathered. The finish created has a darker, pewterlike appearance, slightly mottled. The treatment uses selenious acid solutions similar to that used to age copper. Along with the selenious acid, copper sulfate, phosphoric acid, flouboric acid, and nickel sulfate are used in solution to treat the surface of the aluminum. Some of these treatments are preformulated and proprietary. The artificially weathered aluminum surface will not weather well out of doors. Clear acylic lacquer coatings can seal the surface against further aging for approximately three years.

This treatment can also be used to turn aluminum black. Wiping or spraying on applications of the solution, in minutes begins to blacken the aluminum surface. The procedure is to allow the surface to blacken beyond the tones needed and then rinse thoroughly with water. Finally, the surface is rubbed with steel wool to even out the tone and lighten it to the degree desired.

Organic and Inorganic Coatings

Aluminum, when painted, couples a long-term corrosion-resistant material with endless color variations. One of the prime benefits of painted aluminum is evident when the paint finish is scratched. The base aluminum quickly generates the protective oxide film without producing a rust stain.

Aluminum requires pretreatment before paint will adhere. Pretreatment can be any of a number of forms, depending on the type and quality of the paint finish to be applied. Chemical etching is an excellent pretreatment to finishing with most paints. Chromate conversion coating is another common pretreatment for most aluminum surfaces. Alkaline chromate pretreatment processes offer additional corrosion resistance and a good base for most paints. Chromium phosphate pretreatment and chromium chromate pretreatment provide a good base for many powder coatings and solvent-type paints on aluminum.

Most common paint finishes on aluminum can be formed without damage to the coating. Roll-forming operations and brake-forming operations, which do not severely shape the metal, can easily be performed without damage to the paint surface. Painted aluminum, in thicknesses greater than 0.040 inches (1 mm) will not accommodate simple forming without minor damage to the paint film. Often, what occurs are small micro-cracks at the edge of a bend. If the primer has been correctly applied, the paint will not peel or flake at the cracks. The tiny cracks pose minor concern, because the exposed base aluminum will quickly oxidize and inhibit corrosion. Cracks through dark-painted aluminum surfaces, however, will appear as fine lighter-colored lines. These require touching up to conceal the contrasting color. With the thicker gauges, consider postpainting the formed parts to conceal the large cracks and to prevent peeling along the bends when the primer is too brittle.

The American Architectural Manufacturers Association (AAMA) has established guideline specifications to assist in the selection of organic coatings for aluminum extrusions, panels, and fabrications. The specifications require that the applied paint surface meet certain minimum criteria of performance, depending on the intended use.

When an architectural aluminum fabrication is to be exposed in the interior of a building, or used for light commercial exterior work, AAMA 603.8 would be referenced. AAMA 603.8 is called "Voluntary Performance Requirements and Tests Procedures for Pigmented Organic Coatings on Extrusions." This specification allows for less stringent requirements because of the more protective environment of the interior. It also allows for using more economical paint coatings for light commercial work.

AAMA 605.2, "Voluntary Specification for High Performance Organic Coatings on Architectural Extrusions and Panels," contains more stringent requirements for exterior aluminum products. Exterior aluminum is exposed

TABLE 1-21. Comparisons of the AAMA 603.8 and 605.2
Specifications

Uses	AAMA 603.6 Interior	AAMA 605.2 Exterior
Coating Thickness	0.8 mil	1.2 mil
Pre-treatment	None	Conversion coatings
Abrasion Resistance	None	Falling sand
Chemical Resistance	Muriatic acid, mortar	Muriatic acid, nitric acid
Color Retention	1-year Florida exposure	5-year Florida exposure
Chalk Resistance	None	Maximum #8
Film Adhesion	Dry and wet adhesion	Dry, wet, boiling water
Erosion Resistance	None	Less than 20% after the 5-year Florida exposure

to a much harsher environment. This specification also sets minimum levels of performance and testing for architectural aluminum coatings. (See Table 1-21.)

CORROSION CHARACTERISTICS

Aluminum alloys are well known for their excellent corrosion resistance. Painted aluminum surfaces when scratched, exposing the base metal, will not show the characteristic red rust of steel. The exposed base metal quickly develop a tight, clear protective film of aluminum oxide, which resists further corrosion by most atmospheric pollutants.

With natural mill finish aluminum and the Alclad material, sheltered surfaces will exhibit more corrosive activity than more openly exposed surfaces. The openly exposed surfaces benefit from the natural cleaning effects of rain. The sheltered areas collect corrosion particles which, when not diluted or cleaned away, continue to develop.

Aluminum is subject to certain chemical corrodents and galvanic attack by contact with other metals. The most common corrosion of aluminum is surface corrosion caused by a uniform attack on the entire exposed surface of the metal. This type of corrosion is also known as tarnish. When unattended, the surface will appear rough and possibly develop pits over time. It will look frosted and blotchy, often with dark gray spots intermixed with a grayish-white color. This type of corrosion is superficial and objectionable only from an aesthetic standpoint. Structurally, the surface tarnish, a tight adherent mixture of surface oxides and atmospheric pollution, will act to protect the base aluminum from further damage. A buildup of deposits is frequently seen on horizontal surfaces of unpainted aluminum alloys.

The white or gray powdery deposits, sometimes found within the surface frost, are signs of a pit developing, usually just underneath the deposit. Pitting is due in part to the discontinuities of the alloying process. During the fabrication of an aluminum alloy, either extrusions or wrought shapes, the metal develops microscopic grains with clearly defined boundaries. The boundary elements are of different composition than the grain centers. Polarity can develop between the grain boundary and the grain center, generating an intergranular attack. This is a very weak and slow corrosion effect, one that requires X-ray or ultrasonic methods to detect.

Exfoliation is a form of intergranular corrosion, which first appears on the surface as small blisters. The blisters eventually break away, and the metal flakes off in layers. Exfoliation occurs when the corrosion process travels laterally along the grain boundaries. Intergranular corrosion also can occur around welds. Welding will cause the aluminum to undergo heat treatment in the proximity of the weld. The heat-treatment process can separate the metal grains into anodic and cathodic poles. Some of the metal constituents can come out of solution. When this occurs, corrosion cells can develop.

Galvanic corrosion is a common form of corrosion on all metals, including aluminum. When aluminum is in close proximity to metals of different electrochemical potentials, galvanic corrosion can occur. The joining of aluminum with steel fasteners can develop galvanic coupling between the two different metals. The initial stages of this type of corrosion is a ring of white or red deposits around the fastener. Galvanized steel, when concealed from direct exposure, will not adversely affect uncoated aluminum. The zinc in the galvanized coating sacrifices itself while protecting the aluminum and steel. Zinc is close to aluminum on the electropotential scale (Table I-2) and will provide protection to the aluminum base metal by sacrificial action. Similarly, cadmium-plated steel parts are also considered satisfactory for coupling with aluminum. The cadmium coating is near the aluminum in electromotive potential. Copper will stain aluminum and create pits, which will lead to the development of intergranular corrosion. Separate aluminum and copper with bituminous paint. Do not allow moisture coming from copper to drain over the aluminum surface—this will pit and stain the material.

Stainless steel of the chromium-nickel alloys (S30000 series), under most circumstances, will not adversely affect aluminum. Stainless steel clips and fasteners are the materials of choice and are commonly used with aluminum curtain wall assemblies exposed to urban environments.

To protect against galvanic corrosion, keep the aluminum from contact with other metals by separating them with neoprene or other nonhygroscopic material. Asphalt-impregnated materials, such as felts or wax film paper, work effectively to separate the metals. Most paints will also work well. Do not use lead-base paints; these will corrode aluminum. Scratches through the paint can concentrate the effect of galvanic corrosion by isolating the attack to the small area of exposed metal.

Filiform corrosion is another type of nonstructural surface corrosion that painted aluminum may experience. This type of corrosion, also called "underpaint corrosion," appears as a small wormlike track, similar to the worm track on a tree's surface when the bark is removed. Corroding particles enter through the paint surface and travel along the aluminum/paint interface in a chaotic line. Once the corrosion cell develops, aluminum hydroxide corrosion products weave an indiscriminate line under the paint. Damage, however, is superficial only.

More severe forms of corrosion are stress corrosion cracking, fretting corrosion, and corrosion fatigue. Stress corrosion cracking occurs when the aluminum part undergoes constant stressing while exposed to corrosive environments. This type of corrosion is rare in most architectural uses since severe stress is not a condition. The aluminum material is weakened irreversibly by corrosive materials combining with the base metal.

Fretting corrosion, or erosion corrosion, occurs when metal parts are rubbed together. Constant rubbing removes the protective oxide film that develops, exposing the metal to further corrosion. Other materials constantly abrading the aluminum surface will also cause this type of corrosion. In exterior architectural exposures aluminum will expand and contract. Constant rubbing of an abrasive surface during this thermal movement can remove the protective oxide layer, allowing corrosive materials to attack the aluminum surface.

Corrosion fatigue is a type of corrosion resulting from the cyclical stresses generated by expansion and contraction during exposure to corrosive environments. This is a slow and nonreversible process, developing brittle metal parts. Architectural metal fabrications such as metal roofs can develop this type of corrosion if detailing and assembly do not allow free movement of the metal during thermal cycles.

Aluminum will also corrode when it is in direct contact with other nonmetallic materials. Materials containing lime, concrete, or other masonry materials, when wet, will corrode aluminum. Separate these materials from the aluminum with bituminous paints or layers of felts, and protect aluminum surfaces during construction from splatters of concrete or masonry particles. Aluminum in contact with moist or wet wood products should be protected with a coating of bituminous paint. The wood can also be painted with two coats of aluminum paint or lined with a layer of asphalt-saturated felts.

Anodized Aluminum and Corrosion

Anodized aluminum is more corrosion resistant than natural, unfinished aluminum. There are, however, certain contaminants that will affect anodized aluminum, causing discoloration and fading. Cooling towers located near or under anodized aluminum can emit salts in the condensate. As the condensate collects on the aluminum skin and evaporates, salt compounds or alkaline compounds used to clean the cooling tower piping will discolor the anodized aluminum.

During construction, prevent contact with wet concrete or mortar on an anodized surface. If contact occurs, immediately rinse the wet materials from the surface. If the concrete has dried, care should be taken to prevent scratching. Try heavy rinsing with water while gently brushing the surface with a soft-bristle nylon brush.

All aluminum parts will stain when water is allowed to wet the surface and slowly dry. Major problems arise when moisture has entered between faces of aluminum sheets or panels. On natural aluminum materials such as Alclad siding or coping sheets, dark gray to black stains will quickly develop. On anodized material, watermarks and grayish blotches can develop. The removing and cleaning of surfaces blotches is very difficult and usually more expensive than replacement.

Under magnification, anodic coatings exhibit microscopic cracks. These cracks, also known as surface crazing, result from differential thermal movement of the anodized surface film and the base metal. This crazing is not considered detrimental to the overall protection the anodized film provides.

In the formation of preanodized sheet material, the surface will also crack. The anodized surface is very hard and lacks the ductility of the base metal. When undergoing simple forming operations, this surface film will craze audibly. If the forming is severe enough, the formed edge will crack and tear.

Minor cracking should not be considered a problem. The underlying aluminum will rapidly develop its natural oxide coating. On clear anodized aluminum, an edge crack is hardly discernible. On dark anodized surfaces, however, an edge crack is apparent. Filling the crack with a dark-colored paint usually makes it disappear, and the paint will offer some minor protection. Dark-colored anodized surfaces should be preformed if at all possible, before anodizing.

MAINTENANCE AND CLEANING

Anodized finishes offer long-term, relatively maintenance-free coatings for aluminum. As with most exterior surfaces, however, periodic maintenance is required to sustain the appearance in near original condition. Harsh chemicals and general neglect will affect the overall appearance of an anodized surface. Urban environments may dictate more frequent cleaning, as do coastal environments where condensation on an aluminum surface builds up salt residues along with the dust and dirt particles from the air.

Caulk joints within curtain wall panels and horizontal ledges tend to concentrate dirt particles and stain adjacent aluminum surfaces, even in regions of moderate rainfall. Air-conditioning condensate units can create excessive accumulations of salts and dirt as the evaporating water collects on the sides of metal walls and accumulates at projections. These conditions, if left unattended, will cause premature aging of the anodized finish.

Periodic maintenance of exterior aluminum surfaces should occur with the frequency of glass window wall maintenance. Starting at the top of the building and working down, spray water and a mild detergent to remove most surface dirt and film. Use a sponge or soft brush to remove stubborn particles and rinse with clean water. Squeegee excess water and allow to air dry. If oil or adhesive is present, remove it with MEK or an equivalent solvent. Care should be taken when using industrial solvents, both for the personnel using the solvent and for the surrounding environment. Some solvents attack gaskets and sealants.

Lacquers should never be used on exterior aluminum surfaces. They will yellow and crack in a short period of time and require more expensive cleaning techniques to remove.

Basic Cleaning Procedures

1. *Mild Conditions (Standard)*
   Use commercial soaps and detergents with a damp cloth and hot water.
   Rinse thoroughly and squeegee dry.
   Use a nylon-bristle brush to remove any adherent particles, with particular emphasis on sills and overhangs.
2. *Stubborn Conditions*
   Use commercial solvents and emulsion-type cleaners with a detergent to remove oil and grease.
   Follow up with a hot and detergent-soaked cloth.
   Rinse thoroughly and squeegee dry.
3. *Heavy, Difficult Conditions*
   Use a nonetching solvent cleaner. Add a mild abrasive cleaner.
   Use a damp cloth saturated in hot detergent and add the mild abrasive cleaner, passing the cloth in the apparent direction of the grain in the aluminum.
   Use power water blasting equipment if necessary for very adherent materials.
   Rinse thoroughly and squeegee dry.

Water Stains

Water should not be allowed to sit on the surface of unfinished aluminum products. It will cause the aluminum to oxidize, creating dark stains on the surface. These stains cannot be easily removed. Store aluminum in dry places until it is ready to use. Stacks of aluminum sheets, if wetted, must be quickly separated and dried. Water allowed to remain between the sheets will stain severely. A water stain is superficial and poses no structural performance loss; however, for exposed architectural use, such stained material is worthless. Water allowed to remain between stacks of aluminum creates interesting black and gray fractal patterns of aluminum oxide. These are extremely difficult to clean. Mechanical and chemical means are available to remove or reduce such stains, but these are usually impractical because of their high cost and difficulty in achieving good results. Remove light water stains with stainless steel wool and oil. Use an aqueous solution, 10% by volume of sulfuric acid and 3% by weight of chromic acid, at 180°F (82°C) to remove mild stains. This technique will slightly etch the aluminum surface and requires experience with handling corrosives.

Aluminum oxide is colorless and does not stain adjacent materials. Because it is on the positive pole of the electropotential chart, aluminum has little effect on other metals "downstream" or directly adjacent.

Maintenance of Mechanical Finishes

All the mechanical finishes on aluminum require some level of maintenance to keep surface oxides from developing if no further processing is performed. Left exposed, particularly in exterior environments, polished finishes tarnish after a period of time. Texturing of the metal using embossing dies can conceal some of the discoloration or tarnish. Further chemical treatments can also stabilize the surface of the aluminum to prevent the development of surface oxides and tarnish.

FABRICATION CONCERNS

The ability to cold form aluminum using conventional sheet metal equipment can be affected by the alloy makeup and the temper of the sheet. Alloy A96061-T6 sheet forms or plate forms will crack under relatively simple forming operations. Anodized aluminum in any of the alloys will crack or craze at simple bends.

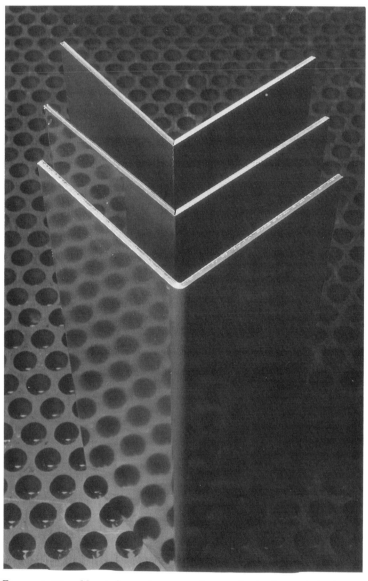

Figure 1-17.   V-cut shapes in comparison to brake formed aluminum sheet.

The A93000 series alloys will cold work with little difficulty. Spinning and deep-drawing operations can be undertaken with little resistance. Bend radii are determined by the temper and thickness of the aluminum sheet. V-cutting of the aluminum can reduce the bend radius requirements, but with a loss of strength at the corner (Figure 1-17). V-cut edges should be reinforced.

WELDING AND JOINING

Aluminum alloys can be welded using various techniques, primarily MIG and TIG. The area around the weld will take a reduction in stress of approximately 30%, which, compared with welded steel connections, is a major stress reduction factor. Aluminum structural welding must incorporate into the design of the weld a fatigue stress factor. Welds should be contoured to allow for the stressing and unstressing of the joint without sharp transitions or abutments.

The American Welding Society has a series of code requirements that dictate the weld and structural characteristics that must be followed for good structural welding.

Welded parts can be anodized; however, the heat generated in and around the weld changes the surface characteristics of the metal. When anodized, these surfaces appear lighter in color.

Aluminum can also be soldered. Special fluxes are required, and the temperature of soldering is higher for aluminum than for other sheet metal materials. Alloys of aluminum containing more than 1% magnesium or 5% silicon are difficult to solder. It is also difficult to achieve a clean, strong solder joint in alloys with high copper or zinc content. Alclad aluminum has improved soldering characteristics owing to the purity of the coating.

Low-temperature solders, those containing tin or lead as a base along with zinc, melt at temperatures in the 300°F to 500°F (149°C to 260°C) range. These are the easiest to apply, but provide reduced corrosion resistance. The mechanical strength they provide is also low. These solders do not wet the joint well. Intermediate temperature solders, 500°F to 700°F (260°C to 371°C), contain zinc and tin. These have better corrosion-resistance characteristics and will wet the joint better. The temperature required to solder can warp the sheet if it is not restrained.

The soldering operation for aluminum requires cleaning the surface of all heavy oxides and foreign particles. Oxides can be removed by wire brushing or ultrasonic dispersion during the soldering process, using a vibrating soldering tip. Fluxes can also be used, but they must be neutralized. Zinc chloride fluxes will work to remove the oxide film. However, these fluxes, if not neutralized, will corrode the aluminum around the joint.

2

Copper

INTRODUCTION

Copper, in both the pure and alloy forms, has served humankind longer and in more varied ways than any other metal. It was not until the second half of the twentieth century that copper was replaced as the primary metal used in construction. Today, iron has the widest use, in its alloy forms of steel, and aluminum ranks second. Copper, in third place, is still a significant architectural metal.

Used frequently as roofing and wall cladding material for major public and private buildings, copper adds natural, alluring beauty to enhance design. Its distinctive color speaks of quality that no other metal, short of gold, possesses. Desired for their durability and corrosion resistance, copper and the various copper alloys have no equal.

Only the element silver has a higher electrical conductivity among metals. Copper has a specific gravity of 8.96. This means that it is a relatively heavy metal, almost nine times the weight of an equivalent volume of water. In relation, iron has a specific gravity of 7.87, and gold 19.32.

Copper has a very low strength-to-weight ratio, making it a good, dense cladding material by nature of its specific gravity, but a poor structural material because of its low yield strength. The yield strength can be improved by cold working the metal, but not significantly.

Copper is one of the third transition elements, which are characterized by being heavy, hard, and inert. Other third transition elements are silver, gold, mercury, cadmium, and zinc. These metals are relatively rare in comparison with other metals, and because of the inert nature of their atomic composition, these elements sometimes occur naturally in pure form.

92

The worldwide supply of copper is said to be in the neighborhood of 5.8 trillion pounds. To date only 0.7 trillion pounds have been mined. With the high scrap value of the metal, 95% the value of primary ore, people go to extremes in salvaging copper. In 1994, recycling of copper totaled 2,977 million pounds (1,350 million kg) in the United States alone. This reflects nearly 44% of the total U.S. consumption for that year.

Using copper as a predominant design material over a building's surface, the architect, wielding the tool of the sheet metal craftsman, develops geometry and form into a design. Copper is valued by architects because of the clean, distinct lines that can be achieved from thin metal skin coverings. As a naturally aging material, copper provides a beautiful, durable, and protective surface. When installed correctly the metal will last for decades, even centuries.

Thin sheet metal fabrications that form tight fitting skins over a building enable the designer to define clearly and enhance the geometry of the structure. Copper's malleability allows folded and seamed edges, and the ability to solder copper enables clean watertight joints, free of exposed fasteners. Intricate shapes can be achieved by hammering the metal over forms or stamping into design molds. Severe forming of pure forms of copper are achievable because of the slow rate of work hardening.

Copper in its alloy forms, the brasses and bronzes, provides elegance in various tones, textures, and colors that no other metal can come close to offering the designer. From the soft, light red color of pure copper to the yellow-gold of Muntz metal, copper alloys provide variations that can be further enhanced by applications of statuary and chemical conversion finishes. No other metal provides the designer with so many options of natural color tones and the ability to enhance and develop these colors through chemical treatments. The brasses and bronzes used in architecture have characteristics specific to their alloy compositions. These characteristics are treated in detail in Chapter 3, "The Major Copper Alloys."

HISTORY

Copper was discovered in Europe and central Asia during the New Stone Age about 8,000 years ago. It was probably first located in the form of pure metal nuggets or colorful copper oxide ores. Large lumps of nearly pure copper have been found near the shores of the Great Lakes of the United States and Lake Tanya in Ethiopia. One such very large lump, weighing in at 2,977 pounds (1350 kg), was found near Lake Superior, on the shores of the Ontonagon River. The local indians worshiped this massive, nearly pure copper specimen as a gift from the gods and thus left it generally intact. It can be seen in the Smithsonian Museum of Natural History today.

Mineral forms of copper, green malachite and blue azurite ores, contain copper in a very colorful oxide state. Early man probably found the colorful stones of interest and collected them. The dark green stones with patterns of darker and lighter lines could be cut into beautiful jewelry. Decorative jewelry made from copper has been uncovered in archaeological sites in the Middle East dating back to 4,000 B.C.

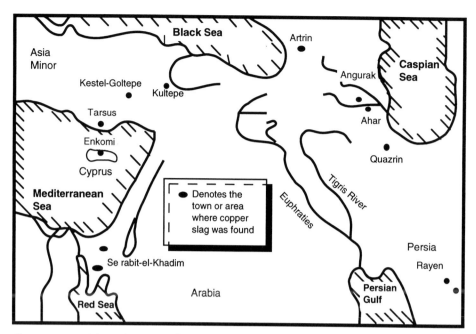

Figure 2-1. Map of region where copper was mined between 5,000 and 1,000 B.C.

Early humankind also sought the metal for its ability to be shaped. Unlike any stone or piece of wood, copper could be made into tools with definitive purposes. Edges could be repaired and resharpened. The frozen remains of a Stone Age man were found in the Alps in the late 1980s.[1] His culture dated back to between 6,500 and 5,700 B.C. The copper axes found alongside him were tools for survival and were probably his most important possessions.

The origin of copper technology is believed to have been in the lands bordering the Eastern Mediterranean Sea, the area known as the Fertile Crescent (Figure 2-1). This region, now present-day Turkey, yielded copper ores that could be reduced in fire to more pure forms. The ancient Sumerians were well known for their early knowledge of extracting and forging copper into finished implements. Copper slag from smelting operations was found in and around a settlement on the Anotolian Plateau of Turkey. This settlement dated back to 6,500 B.C.

Cooper was used extensively by the Egyptians in the fifth millennium B.C. The Egyptians mined the copper ore and smelted the ore by roasting it in wood-fired furnaces. The landscape around these early mines is still scared by the denuding of the nearby forests for fuel for the smelting operation and the deposition of the slag waste from the process.[2] Ancient practices of smelting copper have been recorded in illustrations in Egyptian tombs dating back 4,000 years ago. In early smelting, the copper ore was heated in furnaces sometimes stoked by using foot bellows or by winds that

[1]"The Iceman, Lone Voyager from the Copper Age." *National Geographic.* Vol. 183, No. 6 (June 1993), pp. 36–67.
[2]John Delmonte. *Origins of Materials and Processes.* Lancaster, PA: Technomic Publishing Company, 1985, p. 219.

could be concentrated to feed a very hot flame. The slag would crack and pop and the copper would coalesce into ingots at the bottom of the kiln, or into tiny droplets suspended in the thick slag waste material. The ancient Egyptians were proficient in working with copper. Cooking utensils, ornamentation, and tools were fabricated from the refined metal. The great Pyramid of Khufu was assembled from 2,300,000 blocks of stone, cut and finished by copper tools—the only metal tools available at the time. Cyprus, an eastern Mediterranean island, was widely known for its abundance of copper mines. The word "copper" itself comes from the Latin *cuprum*, which is derived from the ancient name for Cyprus.

The ancients would seek out basic ores such as cuprite, malachite, melaconite, and azurite, which could be reduced to a more pure metal state by roasting with charcoal. These minerals were identifiable to early peoples because of their striking colors and their availability at the surface of the earth. Cuprite is red copper oxide, melaconite is black copper oxide, malachite is green copper carbonate, and azurite is blue copper carbonate. Soon the readily available and easily identifiable copper compounds became scarce. Simple smelting operations, which initially proliferated, shut down because of the lack of reducible ore.

Mining became necessary to obtain the copper sulfide ores. These ores required a more elaborate smelting operation—expelling the sulfur by heat reduction and further charcoal reduction to yield the copper. This operation created two devastating problems: one was the need for major quantities of wood to burn, thus causing deforestation of the areas around a smelting operation, and the other was due to the sulfur reduction and the hazards associated with the gas generated. Sulfur is removed from the ore as sulfur dioxide gas, which is very hazardous to both humans and plant life in the surrounding area. Smelting sites required for this operation were always located in areas of steady and dependable crosswinds, such as on ocean shores.

From these smelting operations, "blister" copper was produced. Blister copper, approximately 90% pure, was used as a valuable trade item. Blister copper was used by coppersmiths, further melting down and forming it into nails, pins, decorative objects, and objects of war. Blister copper in the form of small copper nuggets were bagged in 48.5 pounds (22 kg) animal skin sacks called "oxhides" and used as a measure of currency for trade. Many of these oxhide sacks have been found in sunken ships along the Mediterranean shores.

The Romans were the first to use metal as a major architectural material. The Pantheon, originally constructed in 27 B.C. by Marcus Agrippa, then rebuilt in 120 A.D. during the reign of Emperor Hadrian, had a copper roof covering the outer skin of the main dome and cornice, and bronze ornamentation on the inner surface of the dome. Copper cladding used at the cornice of the central opening of the dome was still in service up to the beginning of World War II.

Sheets of copper were sand cast and hammered to small, relatively thin copper panels of various sizes. The sheets were probably lapped and formed to the configuration of the dome. The Romans were instrumental in developing soldering and brazing techniques. They formulated fluxes to clean the copper oxide from the surface, and they used soldering irons to apply solder to seal joints.

The continued use of copper as a roofing material was enhanced during the Renaissance when the popularity of Gothic architecture spread throughout Europe. Major architectural buildings of the time were the majestic churches constructed of stone and timber. The steep pitched roofs were often covered in metal, either copper or lead. The ability to form the soft metal to the required shape of the roofs, as well as its natural beauty made copper the first choice. The cathedral at Hildeshiem, Germany, was roofed with a thin copper skin in 1230. The roof was installed in short plates and seamed together by craftsmen of the time, by folding and malleting the joints together. The original roof remained in good working condition until the cathedral was destroyed by bombs of World War II—more than 700 years later.

Copper was first "rolled" in England back in the seventeenth century. Rolling reduced the thick copper slab to thinner sheets. These sheets were relatively thick in comparison with the thin sheet we use today for metal roofing, and they were much smaller in size. This first rolling would come to revolutionize the roofing industry by making available a long-lasting material that could be formed into a thin building skin. Christ Church in Philadelphia, Pennsylvania, was constructed in 1737. The roof was clad with imported English copper sheets. The builder and the architect wanted a roofing material of great durability and longevity. The metal roofing was of standing-seam geometry with loose-locked, hand-formed standing seams. The transverse seams were interlocked and unsoldered. The thickness of the copper used was equivalent to 20-ounce copper of today. The copper roof lasted until a fire destroyed the church in 1967. In 1795 a roofing skin of English copper sheets was installed over the First Bank of the United States, also in Philadelphia. The sheets were approximately 24 inches by 48 inches and were formed into interlocking standing seams.

The Statue of Liberty, fully restored in 1986, was fabricated from 0.098-inch (2.5-mm) thick copper sheets attached over an interior iron truss. Moisture and pollutants allowed to enter the copper skin over the years has caused some of the iron truss supports to undergo severe galvanic corrosion. The copper, in general, was found to be in excellent shape. It had developed a rich, adherent patina layer, which provided sufficient protection from the corrosive atmospheres of industrial New York and New Jersey.

PRODUCTION AND PROCESSING

Today the majority of the domestic copper is obtained from the sulfide ores—chalcocite, covellite, chalcopyrite, bornite, and enagite—mined domestically in Arizona, Utah, Montana, the northern peninsula of Michigan, and Canada. Copper is also produced in massive quantities in Chile, Germany, South Africa, Namibia, and New Guinea, as well as in the Ural mountain region of Russia.

Production of copper and its subsequent reduction to a more pure form are accomplished by means similar to the ancient methods, with the addition of purifying techniques as well as great technological advances in efficiency. First, the mineral ore is concentrated by floatation and other physical separation techniques. Then the ore is roasted in special furnaces to remove volatile materials and organic contamination. The ore is not melted at this point. Smelting, which is the melting of the ore in a controlled atmosphere to reduce the contamination, is the next step. In this operation the roasted ore is placed in a reverbatory furnace to remove the major contaminant materials, such as iron. Smelting produces a concentration of approximately 30% copper in a block form known as a matte. Converting the molted copper matte to a purer copper concentration is achieved by passing air over and through the matte. The oxygen in the air combines with molten sulfide contaminants, which are reduced to the form of sulfates, a form that is less difficult to remove. What remains of the matte is 98% to 99% pure copper. This relatively pure matte is still referred to as "blister" copper. Impurities at this point are oxygen and some minor sulfur contaminants.

Further reduction of the copper is achieved by a process known as electrolytic reduction. Electrolytic reduction makes the blister copper an anode, or electrically positive. The copper is then plated to a cathode of commercially pure 99.9% copper. The electrodeposited copper is then melted and cast to arrive at the material used extensively today in construction. This material is known by the colloquial term "electrolytic tough pitch" copper.

In the 1960s a process known as "flash smelting" gained in recognition as a means of producing high-grade copper matte. This process has a marked increase in efficiency and improved environmental impact. The process involves leaching copper from the low-grade ore pile by sprinkling an acidic solution over the waste. The acid dissolves the copper particles, even minute particles, in the waste. The solution is collected, and the copper is extracted with other organic particles. The extracted particles are dissolved in sulfuric acid, creating a powerful electrolyte solution. The copper is removed from the electrolyte by deposition onto a cathode of copper. The copper cathode increases in size from approximately 17 pounds (7.7 kg) to 200 pounds (90.8 kg) as the copper comes out of solution and deposits on the surface of the cathode. This process of ore recovery is practiced by many of the modern mining facilities around the world.

ENVIRONMENTAL CONCERNS

Copper has the highest recycling rate of any engineering material. The rate of recycling equals the rate of copper that is currently mined. Wiring uses mostly newly refined copper; however, in the architectural market, 75% of the metal comes from recycled materials. Premium-grade copper scrap retains 95% of the value of refined ore. No other common architectural metal has such a high scrap value.

Copper mining consumes large amounts of energy, considering the fuel needed to power the mining equipment. In processing copper ore, approximately one third of the total energy consumed goes into mining, while the remaining two thirds is used in the ore refinement and concentration operation. By-products produced in the mining operation are large quantities of waste or slag materials. In the United States, copper mining operations result in greater quantities of overburden waste than produced by the mining of all other architectural metals combined. For every 1,800 pounds (815 kg) of copper produced, approximately 200,000 pounds (90,498 kg) of overburden is removed. Efforts are being make to contain and restore the damage done to the environment by the mining operation by redeposition of the waste slag. Waste slag is basically rock and earth removed during the mining operation. Unfortunately, much of the mining of copper is taking place in Third World regions where the destruction of forests and streams is irreparable. The International Copper Association is addressing the situation by investing millions of dollars in research to establish means of reducing the environmental impact of all production operations.

In the copper smelting and refining operations, the by-product is sulfuric acid. In most cases efforts are being made to recover the acid for other industrial uses. The hope is to make this aspect of production waste-free.

Copper and copper salts are eco-toxic. Copper is used to cover the hulls of boats in order to prevent the growth of algae, barnacles, and other aquatic organisms. Copper nails, driven into trunks of trees or branches will kill the trees. Runoff from exterior copper roofs or cladding will prevent the growth of plants in the local area of high concentration, but has little if any effect on animal life.

Mammals, including humans, need a minute level of copper in their systems to survive. Copper is essential for the development of certain enzymes the body needs to function. Much of this needed mineral is obtained from certain food sources. In rare instances, copper has shown some toxicity to humans. Alcohol products used for consumption that are produced by homemade stills fabricated from copper coils have caused toxic reactions resulting from a high intake of copper. In India, a rare condition of copper toxicity in children has been recorded. Wilson's disease is another extremely rare disorder associated with copper metabolism in the liver. Currently, there is no data available on the carcinogenicity of copper or on toxic exposure levels of copper salts.

Copper has been and still is used for the transfer of potable water to homes and businesses throughout the world. Copper cookware has been used for decades. There is little data supporting any detrimental effects attributed to copper from the intake of copper or copper salts acquired from these sources.

Copper salts have an EC (European Community) classification of R22, but copper itself is unclassified. In the United States, copper is not registered as a toxic or hazardous material.

THE COMMERCIALLY PURE COPPER ALLOYS

Copper is extensively used as an exterior architectural metal because of its natural color and aging quality, as well as its excellent durability and corrosion resistance. The copper alloys used in most building construction are the commercially pure forms classified as cold rolled temper and specified under ASTM B370. These commercially pure alloys are "electrolytic tough pitch" and "fire refined tough pitch" coppers. The alloys are 99.9% pure copper and have good soldering and brazing characteristics.

The copper industry does not currently have an alloy classification number. The Copper Development Association previously classified the architectural copper alloys as C11000 and C12500. Alloy C11000 was the specified copper alloy known as electrolytic tough pitch copper. Resistance from mills to the alloy number classification has brought confusion into the market. Some of the mill producers did not wish to have a tight alloy definition used, requiring a purity of two significant figures. Thus the specification that is suggested relates to the American Society of Testing Methods (ASTM) classification B370. This specification defines a purity level of 99.9%. Using the alloy classification C11000 is still accepted and understood, but in regard to accuracy, it is not correct.

Alloy C12500 is called the "fire refined tough pitch" copper alloy. This alloy contains small amounts of silver and other trace elements. Copper, with a little silver, is considered to make up 99.88% of this alloy. Aside from its reduced purity, this alloy performs similarly to C11000 and falls within the ASTM B370 specification.

The commercially pure alloys featured in ASTM B370 are available in three basic types of sheet copper: cold rolled temper, soft temper, and lead coated.[3] Soft temper copper should not be used except where extensive forming is required. Soft copper is cold rolled copper that has been fully annealed to lower the temper. For metal roofing and other relatively flat planes, soft copper will show minor impacts, such as hail marks. Soft copper is very malleable and can be formed easily. This type of copper was used more frequently in past years when thicker sheets were more common and roof seams had to be hammered to lock them into position. Soft copper was also used on ornamental cornice work to decorate many urban structures in the early twentieth century. The soft copper could be easily stamped into elaborate shapes over wood molds. Hand forming is possible with this temper copper sheet. Very little spring back will occur when the sheet is bent or folded, but thin sheets will not hold their own weight when lifted along an edge or corner. The surface appearance of soft copper is dull, lacking the luster of cold rolled copper.

Cold rolled copper is most often used for architectural roofing and fascia construction because of its increase in yield strength and hardness, more than twice that of soft copper. Repeated passes through rolls while in the nonheated state will work to harden the copper sheet. *Cold rolled temper* is the term given the copper alloy sheet that has received the lightest possible pass through rolls without lubrication. The term is a brass mill colloquialism referring to the passing of cold copper sheet through unlubricated rolls to give it a burnished effect. The appearance is bright and shiny. This passing between rolls reduces the copper thickness and work hardens the copper sheet. The degree of rolling is based on the desired thickness of the sheet copper. Copper is reduced by rolling to a nominal weight per square foot of material. The comparable B&S (Brown and Sharpe) gauge equivalent for copper and the corresponding thickness is presented in Table 2-1.

---

[3]Lead-coated copper, as of this writing, is finding resistance in most cities and communities due to the lead coating. Other alternative coatings exist, such as tin-coated copper and pewter-coated copper, which have a similar, but brighter, gray appearance. See Chapter 7, "Metallic Coatings."

TABLE 2-1. Various Standard Thicknesses of Copper Sheet			
COPPER WEIGHT OZ/SF	NEAREST EQUIVALENT B & S	NOMINAL THICKNESS (INCHES)	NOMINAL THICKNESS MM
90	8	0.129	3.28
88	9	0.119	3.02
80	10	0.109	2.74
72	12	0.097	2.46
40	14	0.054	1.37
32	17	0.043	1.09
24	20	0.032	0.82
20	21	0.027	0.69
16	23	0.022	0.55
12	26	0.016	0.41
10	27	0.014	0.34
8	29	0.010	0.27

TEMPER

When a metal is repeatedly passed through rolls at room temperature, the metal work hardens, which is to say that it becomes harder, offering resistance to each subsequent pass through the rolls. The processor of copper sheet will continually pass the copper through the rolls to decrease the thickness, thus increasing the copper surface, and to increase the hardness of the copper sheet. If further reduction is required, the coil is placed in a furnace and annealed to reduce the hardness.

Annealing is the controlled heating of the copper alloy until a critical temperature is reached. This critical temperature occurs when the strained metal grains, which were stretched and hardened during the rolling and reducing process, disintegrate and form into many small new grains. As the annealing temperature is increased, these tiny grains grow and absorb one another. The metal becomes more ductile as the grain size increases, allowing the copper coil to be further reduced. The metal can also remain at a particular annealed temper instead of undergoing additional cold rolling. *Temper,* designating a metal's hardness, is a relative term specific to a particular metal.

The hardness, or temper, can be of great importance to the fabricator and finisher of the final copper piece. When forming operations are to be performed, it is critical that the correct temper is used. Hardness can lead to brittleness or require a greater bending radius on the material. For most copper flat sheet fabrications, such as roofing and flashing work, the cold rolled temper is adequate. The cold rolled temper equates to the Eighth Hard Temper classification. If major finish polishing work is required, then Quarter Hard Temper, which has the characteristic fine grains, should be considered. In Europe the mills supply Quarter and Half Hard Tempers for architectural metal applications. The harder surface copper with the tight

TABLE 2-2. Various Tempers Available for Copper

TEMPER	ASTM CODE	YIELD (ksi)	YIELD (MPa)	VARIOUS USES
Soft Temper	O60	10.0 to 11.0	69 to 76	Ornamental, severe forming
Eighth Hard	H00	28.0	193	Sheet metal roofing, flashing
Quarter Hard	H01	30.0	207	Sheet metal roofing, flashing
Half Hard	H02	36.0	248	Flat fascia, deep-drawn forms
Hard	H04	45.0	310	Perforated panels, little forming
Spring	H08	50.0	345	Not used typically, difficult to form
Extra Hard	H10	53.0	365	Not used typically, difficult to form
Hot Rolled	O25	10.0	69	Not used typically, poor surface quality

grain structure apparently is the metal of choice for the fabricator of metal roofing and flashings.

The various tempers are produced at the mill by reducing thicker sheets to varying degrees and annealing the sheets to the degree of temper desired by subsequent rolling through the metal rolls. For the Eighth Hard Temper, the approximate Rockwell Hardness B-Scale reading is 50. A list of the common tempers available can be found in Table 2-2.

AVAILABLE FORMS

Sheet

Copper coils and flat sheets are available in standard widths as follows:

24 inches	(610 mm)
30 inches	(762 mm)
36 inches	(914 mm)
39.37 inches	(1,000 mm)
49.21 inches	(1,250 mm)

Other widths, less than those indicated, are available as slit coils. Copper is available in lengths decoiled to specification or in coil stock. Sheet copper specifications are covered in:

ASTM B370-92 Standard Specification for Copper Sheet and Strip in Building Construction
DIN 1751 (German)
DIN 1791 (German)

These specifications discuss cold rolled copper sheet in flat lengths and coils used for building construction.

Plate

Plate copper specifications are covered in:

> ASTM B24 Standard Specification for General Requirements for Wrought Copper and Copper Alloy Plate, Sheet, Strip, and Rolled Bar QQ-C-576

Copper is available in plates of thicknesses greater than 0.188 inches (4.77 mm). The surface of the plate is of hot rolled quality and will have numerous imperfections. Further polishing and grinding may not eliminate the imperfections and can reveal further pits and dark particles concealed just below the surface. There is no domestic manufacture of wide plate material. Maximum domestic width is 36 inches (914 mm). European and Japanese sources can supply plate in widths of 49.21 inches (1250 mm).

Bars

Bars are available in round, square, and rectangular shapes and rectangular shapes with rounded edges. The standard length is 144 inches (3658 mm). Longer lengths are available on special order.

For round bars, available diameters range from 0.063 inches to 10.00 inches (1.60 mm to 254 mm).

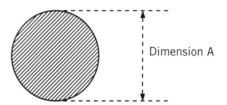

For square bars, dimension A can be from 0.25 inches to 4.00 inches (6.35 mm to 102 mm).

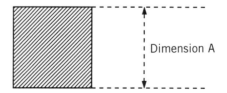

Rectangular shapes are available in thickness from 0.063 inches to 4.00 inches (1.60 mm to 120 mm) by widths of 0.5 inches to 6.00 inches (12.7 mm to 152 mm).

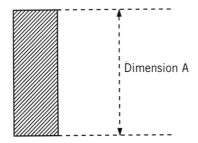

Casting

Copper can be cast using sand casting and ceramic mold techniques. Continuous casting and other production method castings are also used. Copper is rarely cast into architectural fabrications. High-purity copper is a requirement of the electrical industry, which needs the high conductivity of the pure metal.

High-purity copper alloys are difficult to cast. They have good fluidity, but they shrink. The surface of the high-purity alloys is rough and coarse. These alloys have a higher melting temperature than the other copper alloys. Alloy series C80100 to C81200 contain 99.3% copper and are considered high-purity alloys. Alloy series C81400 to C82800 have high copper content and additive elements for specific characteristics.

FINISHES AND WEATHERING CHARACTERISTICS

The commercially pure copper alloys are a light red color in the new, unoxidized mill state, which on exposure to the atmosphere quickly develops into interference colors and, eventually, to brown tones. Interference colors can appear as varying hues as the copper initially weathers. These colors can be tones of purple, yellow, and dark blue. As the copper continues to weather, the interference colors change to darker, more even, brown tones. These colors can remain on a vertical or low-slope copper surface for a year or more before even darkening of the surface will occur. If combustion by-products are present in the atmosphere, or if the exposure is near the sea, the characteristic gray-green patina will develop. In normal atmospheric conditions, the beginnings of the gray-green color will take from 5 to 10 years, depending on sulfur and moisture levels reaching the surface of the metal.

Mill Finishes

Like other architectural metals, the copper alloys produced at the mill in sheet, plate, bar, or rod are provided in an "as-fabricated" or "mill" state. This as-fabricated state is the surface and texture induced on the metal by the rolling and reduction of the sheet. The as-fabricated condition also denotes the surface of the metal when it is extruded or drawn, prior to any secondary polishing. This is the most economical finish, since it leaves the metal in an initial, unprocessed condition.

In copper sheet reduced and cold rolled at the mill to a specific temper, the as-fabricated surface is bright and very metallic in appearance. The bright finish is induced in the metal by cold rolling the sheet in highly polished and clean steel rolls. The smooth rolls burnish the copper surface as it rapidly moves along the coil line. The metal surface is free of oxides but often acquires rolling lines by passing through these unlubricated rolls. Generally, there are no excessive oxides in cold rolled products. Excessive oxides would be a sign of improper storage of the materials. A cold rolled sheet should have a bright luster, with slight rolling lines or mars along its length.

The soft temper copper is slightly duller than the cold rolled temper. The surface appears grainy in comparison with cold rolled temper sheets. Such a sheet does not receive polishing well and should be used only in the as-fabricated finish.

Hot rolled products, such as heavy plate and rods, have relatively dull surfaces in the as-fabricated form. They often contain some oxides and appear darker in color. The surface can be rough and contain slight pits or inclusions. Further grinding and polishing will help to conceal the inclusions, but the softness of the copper makes finishing to a mirror surface difficult with hot rolled products.

Extruded copper is usually clean of dark stains and oxides, but it can have die marks running down the parts. Die marks are induced by the fine imperfections in the die used to extrude the shape.

In sheet and plate form the as-fabricated finishes are delivered unprotected to the finishing fabricator or secondary processor. They are furnished flat and straight and well skidded against damage. All as-fabricated parts should be free of major scratches, wear marks, and water stains. The softness of the copper surface allows copper parts packed together to dent and mar the surfaces, even on thick extrusions or bar forms. It may pay to order individually wrapped copper parts to prevent marring from adjacent materials.

There are three variations of the as-fabricated surface that can be requested of the mill producer.

1. *Unspecified.* This finish is left up to the particular mill. The metal is provided with the surface appearance and characteristics induced into the metal during its refinement into a wrought product. In cold rolled products this can be a shiny, reflective surface or somewhat dull if the steel rolls are dull. On hot rolled products mill oil will be present on a dull, discolored sheet. Cast products will be rough and coarse in texture. This is the most economical form. Bars and tubing are supplied in this form.

2. *Specular.* The mill will provide a reflective finish produced by passing the sheet or plate material through highly polished steel rolls. This finish may be on one side or on both sides. Most thin cold rolled copper sheet arrives with this finish. Copper sheets receive a final temper pass on cold unlubricated rolls. This tends to brighten and smooth the surface. The specular finish is the cold rolled finish on copper sheet.

3. *Matte.* This finish is a dull texture on the surface of copper alloy sheet, plate, extrusion, or casting. It can be produced by hot or cold rolling,

followed by annealing of the metal. This is the finish provided on soft temper copper and on most copper extrusions. The sandblast texture provided on copper castings is also considered a matte finish.

Secondary Finishes

There are two basic types of standard finishes that are commonly applied to copper alloys after the mill finishes. The first type, the more economical of the two, is the directional texturing finish. This finish is characterized by tiny scratches induced in the surface of the metal. It is similar to the #3 or #4 finishes in stainless steel. Often known as "satin" finish, this surface treatment produces a matte tone that reflects light in a diffuse manner. This tends to conceal the surface deflections, or "oil canning" of the metal sheet, by spreading the reflection. Satin finish is known as "directional," because of the tiny parallel lines or scratches in the surface of the metal. These scratches are applied to the surface by passing a polishing roll over it or passing the sheet under a fixed polishing roll. The edges of the sheet are clamped, and a coarse-grained abrasive wheel is passed over the surface to produce a parallel grit line. There are six versions of this finish. These are not usually specified for thin sheet metal fabrications intended for exterior use, such as a copper roof or copper flashing, but are used on ornamental features in the interior of the buildings. Reference Table 2-3 for a list of the common finishes applied to copper wrought products.

These finishes are produced on metal surfaces clean and free of oxides. The finishes can be selectively applied to the surface over specular finishes, thus creating interesting decorative treatments.

Specular (Mirror) Finishes

The specular or "mirror" finish is characterized by a bright metallic luster of high reflectivity. This finish is induced in the surface by first polishing with successively finer abrasive wheels or belts, sometimes of leather saturated with abrasive materials, followed by final polishing with 320 grit abrasive. The reflective surface is then buffed with rouge and large cloth wheels, bringing out the high reflective luster of the metal.

Custom Finishes

Nondirectional satin or matte finishes can be induced in copper sheet and fabrications through a number of different techniques. One technique, known as *abrasive blast*, is characterized by a matte luster without scratch lines. This finish is produced by high-pressure spraying of fine particles against the metal surface. Varying degrees of texture with different matte lusters are obtainable by altering the media used. The finish produced by this method is usually slightly mottled or irregular. Distortion of thin metal sheets can occur because of the impacting of the abrasive media and the transfer of energy into the sheet. The finish is best applied after the copper part is fabricated or firmly laminated to a substrate. Thick plates or sheets should be the base metal for these finishes in order to resist the shaping. Start with low-pressure spraying to reduce work hardening and shaping of the copper sheet. Interesting effects can be obtained by masking certain sections of a part and blasting the

TABLE 2-3. Various Finishes Available on Copper

FINISH DESCRIPTION	DISTINGUISHING CHARACTERISTIC	APPLIED BY	RELATIVE COST
Satin Finish— Fine	Fine parallel grit lines	Scotch-Brite pad or wheel or 150 to 180-grit belts	Moderate
Satin Finish— Medium	Parallel grit lines Deeper than fine	100 to 150-grit abrasive belt or disk	Moderate
Satin Finish— Coarse	Parallel grit lines Coarse, rough feel	80-grit abrasive belt or disk	Moderate to low
Satin Finish— Uniform	Long parallel grit lines For small planes of view	80-grit abrasive belt run on bars, tubes, extrusions	Low
Satin Finish— Hand Rubbed	Tiny surface scratches Nonparallel	Pumice and solvent rub or brass wire brush	Moderate
Satin Finish— "Angel Hair"	Variable scratches run in different directions	Abrasive disks or wheels applied at angles	High
Satin Finish— "Engine Turn"	Various circles or circular grit pattern—fine texture	Small abrasive wheels or blocks on a CNC	High
Satin Finish— Swirl	Multidirectional fine texture swirls	Abrasive disk applied manually at radials	High
Satin Finish— Distressed	Multidirectional fine and coarse scratches	Abrasive disk applied manually sharp angles	High
Matte Finish— Abrasive	Matte texture, low reflectivity	Abrasive blasting of the metal surface	High
Mirror— Specular	Mirror finish, not as bright Highly reflective	Buffing with large cloth wheels to a lower degree	High
Mirror— Smooth Specular	Bright mirror finish. Highly reflective	Buffing the surface with large cloth wheels	High

exposed areas with abrasives. Removal of the masking leaves a contrasting surface of matte and specular finishes or matte and directional satin.

Using different abrasives on various sections of a part can produce subtle differences of tone. Experimentation and the use of control samples should be the norm when working with these custom finishes. In addition, caution is needed when touching these finishes after the abrasive application; they fingerprint easily. Clear coatings of lacquer, wax, or silicone are necessary secondary steps.

Radial blending, using a circular grinder, can create a satin finish characterized by small, fine arcs in different biases to one another. The result is a nondirectional finish exhibiting subtle light and dark regions when viewed at a distance, sometimes known as *"angel hair."*

This finish can be modified into the "swirl" and "distressed" finishes by deepening the coarseness and directional application of the grinding disk. Such finishes are almost an art form. Each has its little variations, depending on the craftsperson applying it.

Electropolishing

Copper can be electropolished to brighten dull and oxidized surfaces. The higher the purity of copper, the better the finish quality will be. Electropolishing uses phosphoric acid in an agitated heated bath. Nitric and sulfuric acid are also used to brighten the surface of copper. An electrical current is passed through the part or sheet. Electrochemical action removes microscopic surface irregularities to the order of 10,000 Å. This level is of the same magnitude as visible light. So as the surface irregularities are selectively dissolved into the electrolyte, the roughness of the metal decreases and the surface reflectivity increases. To brighten a copper surface, only a few minutes of treatment are necessary before a reflective tone develops.

Embossed and Coined Finishes

Patterned or embossed finishes are available on flat, thin sheets of different copper alloys. These finishes are induced in the metal surface by passing the sheet, usually still in the coil form, between patterning rolls. These rolls have a design imprinted into them, one being the negative of the other. As the metal passes through the rolls, the design is imprinted into the metal. Designs vary, but the pattern always repeats or is continuous along the length of the coil. Sometimes the design occurs on only one roll, the other roll being smooth. This technique is known as "coining" of the metal surface. Coining produces a shallow design pattern, which is pronounced on one side of the sheet.

In embossing the sheet the surface of the metal is stressed, and thus additional stiffness is imparted to the sheet. Therefore, thinner sheets can be used. By breaking up the surface into patterns, distortions and surface imperfections can be concealed. This finish is available only on flat sheet metal.

Precautions—Copper Finishes

Copper is so soft that most of the mechanical finishes can be easily applied. Copper can be mirror polished with relative ease when using the correct buffing equipment and abrasive rouge. The surface of copper, however, develops oxides quickly upon exposure to the air. Copper surfaces, even the as-fabricated finishes, fingerprint immediately when touched. Once any of the mechanical finishes are applied, coat the sheet with a clear plastic film.

Be certain that all trapped air bubbles are out of the plastic. This will allow handling while further fabrication is proceeding. When the fabricated part is installed, remove the plastic and clean all oxidation that may have developed along the edges or in places where the air or moisture has reached the copper surface. Coat the metal with wax, or use a lacquer to seal the clean surface.

Patina—Naturally Developing

Patinas are the blue-green or gray-green copper sulfide conversion coatings that develop when copper oxides and certain salts combine. The natural green patina that forms on copper is essentially copper sulfate. All metals change from energy-rich to energy-poor conditions as they seek a natural equilibrium. Most metals are found naturally in chemical combinations of oxidized, sulfated, or carbonated forms of mineral ores. An exposed copper surface continues to oxidize until chemical limits are reached. These limits correspond to the natural mineral forms of the metal. The patina that forms naturally on copper is chemically the same as the natural mineral brochantite. This patina, after 70 years, is basically mineralized and, like brochantite, is extremely resistant to further environmental changes.

Natural patina is produced by sulfur compounds, which are found in all urban environments owing to combustion products released into the atmosphere. Coal-fired power plants are a major source of sulfur compounds today. If there is a lack of atmospheric sulfur, the patina will not form. In urban industrial regions the basic copper sulfate forms, whereas in country-side regions copper carbonate makes up the patina formation and coastal regions develop a copper chloride patina.

It should be noted that the term *verdigris* refers not to a natural patina, but to copper acetate. This pigment develops by chemical reaction of copper alloys with acetic acid. Verdigris is soluble and is not of the same composition as copper patina. Verdigris is a man-made coating and will hold up only if used on copper fabrications exposed to limited environments, such as the interior of buildings.

Moisture on a copper surface promotes the initial copper oxide formation. The copper sulfate formation must develop from the initial copper oxide. As the copper sulfate begins to develop, it is soluble and will be partially washed off by rain. The runoff, rich in copper salts, can stain adjoining surfaces such as limestone or concrete. Once the sulfate is converted to the basic mineral form, it becomes less soluble and develops the final protective copper sulfate layer. Further staining of adjoining surfaces is reduced.

The more insoluble form of copper sulfate may take from 5 to 30 years to develop. This process is dependent on the sulfur content in the surrounding air (pollution) and the moisture levels to which the copper is exposed.

In severely polluted regions, particularly those exposed to acid rain, copper will develop dark to black specks and streaks. Sometimes the surface will darken over large areas of the copper surface, bypassing the development of the brown copper oxide condition. Eventually, within a few years, the hues of the entire surface will blend together and form the characteristic patina.

As a copper surface oxidizes naturally, it passes through a series of color changes. These changes can develop interference colors of yellow, purple, blue, and red. The oxide layer grows, turning a richer brown. This layer is absorbing sulfur from the air, and eventually a green tint develops. Over years the green tint becomes deeper and richer.

Time Required to Develop Natural Patina

Industrial Seacoast	5–8 years
City—Industrial	8–12 years
Typical City Region	15–20 years
Bucolic, Mountain	> 30 years
Dry Desert Region	indefinite

Current successes in reducing the sulfur content in the air, produced by industry, has, on the one hand, benefited humankind and the natural environment, but has slowed the growth of patina on copper. Today, in many regions of the United States and Europe the development of the patina is taking two to three times as long to develop—if it develops at all.

Chemical Reaction—Copper Patination

1. $2Cu + \frac{1}{2}O_2 + H_2SO_3 » Cu_2O + H_2SO_3$
2. $2Cu + 2H_2SO_4 + 0_2 » 2CuSO_4 + 2H_2O$
3. $2Cu_2O + SO_2 + 1\frac{1}{2}O_2 + 3H_2O » CuSO_4.3Cu(OH)_2$

There are a number of factors that determine the rate of patination growth:

- Duration of initial surface wetness
- Temperature of copper surface while wet
- Sulfur level in the atmosphere
- Slope of copper surface

After the first few months, the moisture and temperature of the copper surface will play a smaller role and the corrosive environment at the metal surface will govern. The slope of the roof will affect the duration of moisture on the surface and the tendency of corrosive particles to remain on the surface. The greater the slope, the slower the patination development.

A copper surface should be initially clean and free of all dirt and foreign matter. Use linseed oil to wipe over the surface and create an even brown tone. The linseed oil will quickly wash away, and the even copper surface will age uniformly.

Artificially Induced Patinas

For many years, the acceleration of copper patinas by chemical means has been attempted with varying degrees of success. The reason for the difficulty is the range of variables that must be confronted. Temperature, moisture, surface area, method of application, surface of the metal, and time of drying all greatly influence the results. Lack of adhesion of the patina film and color instability are the typical modes of failure. The patina conversion films that are generated by the acceleration processes are water soluble and photosensitive in their initial stages, making them highly susceptible to changing weather conditions. Large surface areas, such as metal roofs, pose the most difficulty in consistently achieving adequate results.

Patinas Applied in Situ

If a patina is desired on a roof after it is installed, many natural barriers have to be overcome. A large copper roof is exposed to many surface contaminants, such as oil from forming machines, perspiration from the workers installing the roof sheets, and general atmospheric dirt and moisture. In addition, large roof areas make it difficult, if not impossible, to achieve a consistent treatment over the entire area. What often occurs is a mottled patina without a stable attachment to the metal surface. Subsequent rains and ultraviolet radiation break down the simple chemical bond and wash the patina away. It is necessary to have a clean roof surface prior to the application and stable weather conditions during and after the treatment. Changes in humidity and temperature can ruin the chemical conversion process.

Frank Lloyd Wright used chemical acceleration to produce patinas on many of his structures. He favored an ammonium chloride formula to achieve the blue-green patina. The formulated mixture was applied by spray or brush. The mixture of the components of the solution could occur no more than 24 hours prior to spraying on the metal surface. Two applications were required, 48 hours apart, with a clear water spray 24 hours after the second application. Dry weather was required throughout the entire process.[4]

Failure typically occurs through oxides washing off the surface and staining adjoining materials. The roof, or surface, is left with a speckled, inconsistent green and brown coating. There have been many more failures, or at least unsatisfactory results, than successful patina applications on copper roofs.

Different methods have been promoted over the years, that use paints containing copper sulfates or chlorides. The paints were applied, giving a green "patina" appearance. The idea was for the paint to slowly wear away, exposing portions of the copper to weather naturally. This did not work in most cases, and the roof or other copper fabrication showed black streaks intermixed with green paint.

Other methods used acids, such as muriatic acid. (Muriatic acid is diluted hydrochloric acid.) This acid is used to clean the oxide layer around a joint intended to be soldered. Many a sheet metal worker knows that if the acid is not neutralized, a green patina coating quickly develops around the solder joint. The use of acids to generate a patina is not recommended because of damage the acid will cause to surrounding materials and the potential hazards to the environment, and to the worker applying the acid.

Prepatination Processes

Prepatination of copper sheets in a controlled environment has proven to be more successful than application on-site. When the patinated sheets are installed, some of the fragile layer of copper sulfate is removed, particularly at

---

[4]Copper Development Association Inc. *Copper/Brass/Bronze Design Handbook, Sheet Copper Applications.* Greenwich, CT: 401/OR, pp. 55–56.

the folded edges. These "copper"-colored edges soon oxidize and disappear from view. There are a number of methods for prepatination of copper sheets; one can use the same methods as those for in situ application, or for even better results, seek out the manufacturers of copper sheets.

Forced patina techniques were developed by corrosion expert John P. Franey of ATT Bell Labs, who worked on the instant patina system to match new sections of the Statue of Liberty with older sections. Two techniques were developed, vapor deposition and chemical seeding. The result of either technique is thin, stable copper sulfate layers that are indistinguishable from the natural-forming patina.

In the vapor deposition method, the object is cleaned and then enclosed in an airtight gas chamber filled with hydrogen sulfide, distilled water, and warm air. The hydrogen sulfide combines with the copper to form a sulfur copper compound on the surface of the metal. After a few days of exposure in this environment the metal surface has developed a "natural" patina, stable and insoluble in water.

The other system is a process being considered for patent by Bell Labs. This system induces a patina by chemically seeding the metal surface with corrosion cells and growing the patina around these cells. The chemical seeding is done in a closed chamber like that used in the vapor deposition system and produces a stable copper sulfide patina that is chemically indistinguishable from a naturally developed patina.

There are European and domestic firms now offering prepatinated copper roofing sheets. The copper sheets are provided with the rich green patina already applied and ready to be fabricated into copper standing seam panels or other metal panel systems. The green coloring is very consistent from one sheet to the next and has shown good resistance to further change resulting from environmental exposures. The system was developed by Dr. Hans Jürgen Laubi with Oxidwerk of Switzerland and is licensed to major copper sheet processors around the world.

Additional Coloring Options

Different colors can be generated on copper surfaces through chemical treatments similar to those used on the copper alloys. Most of these, when used on exterior surfaces, will eventually develop the natural green patina. In preparation for any of the coloration processes, the copper surface should be cleaned and degreased. This can be performed by dipping quickly in caustic soda or by mechanically brushing the surface. A word of caution: In all the processes, hazardous chemicals are produced, and vapors from chemical reactions require appropriate protective clothing and eye protection, as well as a ventilated space.

The color can be black, brown, red, orange, or variations on the green patina mixed with browns. All the coloration processes require time and experimentation to achieve a metallic, lustrous tone. Some basic examples follow.

Black Color

A rich, matte black color can be generated on a copper surface by immersing a clean copper sheet or fabrication in a hot bath of caustic soda. Other black tones can be produced by immersing the sheet in a boiling bath of copper nitrate, followed by a cold bath of potassium sulfide.

The black surfaces will remain for a long period of time, eventually developing a brown claylike appearance. A black surface is created by a thick interference film of copper oxide.

Brown Color

A rich brown color with orange or reddish tints can be produced on copper by immersing a sheet or fabrication in boiling solutions of copper sulfate and copper or sodium acetate. Other chemicals can be added to modify the resulting appearance.

Clear Coatings—Interior

Quite often a design using copper and copper alloys requires the color, either the natural polished finish or the statuary or patina coloring, to be maintained at some desired level: the rich dark brown statuary finish on a cast statue, for instance, or the bright reflective brass color of a Muntz metal column cover. The designer may wish this color to remain and not develop into the gray-green patina that nature ultimately will achieve.

Certain clear coatings were developed to impede further oxidation and thus sustain the desired color tones. These coatings either dry at room temperature or require baking at temperatures sufficient to catalyze the hard finish coat.

Many factors affect the performance and appearance of these clear coatings, some of which are handling, dust, abrasion, humidity, ultraviolet radiation, and pollution. It may be preferred under certain circumstances to use a coating that can easily be removed for refinishing at a later date. Other more durable coatings will, as deterioration occurs, create a more unsightly condition with spotty decay intermixed with glossy areas. Removal and repair of this surface is difficult and costly.

There are many additives common to the clear coatings that provide different properties beneficial to the performance and results of the application. Ultraviolet absorbers are beneficial when the coating is used in an exterior environment. These additives combat the effects of ultraviolet radiation on surface films. Chelating agents, such as benzotriazole, are useful additives that develop a barrier at the metal surface, absorbing oxygen and resisting tarnish. All additives increase the cost of the coatings, but are often the determining factor for a successful application. It is usually much more expensive to clean and recoat.

As with the statuary and patina finish coatings, the single most important requirement is a clean surface. In patina or statuary surface finishes imperfections in the clear coatings are not as obvious or as critical. Minor surface tarnish will not be visible in most instances. It is important for these surfaces to be passivated by the coatings so that further oxidation does not occur.

Many architectural applications require the shiny, untarnished brass or copper surfaces to remain visible. These lustrous metallic surfaces reflect all imperfections in the coatings. Imperfections such as "orange peel," surface tarnishing, and fingerprints are usually deemed inappropriate and require cleaning of the clear coating and reapplication. Coating highly polished surfaces of a metal with any of the clear coatings can be a nightmare to the unfamiliar operator. Urethane lacquers can go on smoothly and evenly, only to congeal and orange peel as they dry. Brown tarnish spots that were not present prior to the application of a clear coating can pop out under the coating when degradation of the copper surface is caused by reactions of solvents remaining in the paint.

Dust particles in the air can settle on the still-tacky clear coating during the time necessary to dry. The particles become imbedded in the surface as the film dries. Clean-room filtration systems are necessary to minimize dust contamination. This can be a costly requirement in the application of clear films.

Tarnish inhibitors should be applied as soon as possible after the polished surface is brought to the required luster. Dust should be kept to a minimum both during and after application of the clear coating. When a satin polish is desired, use only silicone carbide pads (Scotch-Brite[5]) or stainless steel pads. Steel pads, steel wool, and other wool pads may contain chemicals that will affect the surface or leave steel in the surface to react with the copper. Pumice stone in a slurry of 5% oxalic acid solution, rubbed on with a clean cotton cloth, is a good means of achieving a finish surface polish that will not generate a corrosive reaction and will remove surface tarnish prior to the application of a protective coating.

All polish residue must be removed prior to application of tarnish inhibitors and final clear coatings. Buffing compounds can be removed by degreasing in solvents such as butyl cellosolve or tri-chloro ethylene. There are other solvents, less harmful to the environment, which will also work to remove the buffing compounds without leaving streaks or films. Caustic soda is one example; however, this tends to color the copper.

Moderate temperature and low humidity must be sustained during the coating process. High humidity will adversely affect the coating application. Discoloration of copper alloys resulting from high curing temperatures for thermoset coatings is often a problem. It is recommended that low curing temperatures and short cure times be used whenever possible.

Clear Coatings—Exterior

Like other surface coloration processes used on copper alloys, the clear coatings pose problems when used on large exterior surfaces. Numerous attempts at controlling oxidation and keeping the initial lustrous coating of newly installed, untarnished copper have had qualified success. One of the largest applications of clear coat protection was the copper shell, hyperbolic dome over the Mexico City Sports Palace built for the Summer Olympics in 1968. The lacquer used for this roof was INCRALAC®, developed by the International Copper Research Association.

INCRALAC® is an air-dry acrylic base lacquer that contains benzotriazole to inhibit underfilm corrosion. It is a good exterior clear coating that has shown some success over the years.

---

[5]"Scotch-Brite" pads are nonmetallic abrasive pads manufactured by the 3M Company, St. Paul, Minnesota.

The procedure for applying INCRALAC® is as follows:

1. Prepare the metal surface.
   a. Degrease with alkali base solvent or vapor.
   b. Abrade the surface with Scotch-Brite pads.
2. Apply the INCRALAC® by misting over the entire surface.
3. Allow 30 minutes drying time.
4. Apply a second wet coating of INCRALAC®.

CORROSION CHARACTERISTICS

Staining—Copper Runoff to Other Materials

One of the concerns that designers have when copper is used as an architectural material is the possibility of staining adjacent materials with green corrosion products. Copper in its natural form, allowed to weather, will stain porous materials when the runoff of copper-bearing moisture is allowed to pass over or deposit onto the porous material. Some of the copper oxides that develop as copper is exposed to the atmosphere are soluble and will run off the copper fabrication as moisture from rain or dew travels across the surface.

Dew, condensing on a cool metal surface, will absorb copper salts into solution. As the dew collects and then slowly runs off the metal skin, it deposits the copper salts onto the warmer surrounding surfaces as the moisture evaporates.

Limestone, stucco, concrete, and other light-colored porous materials can pick up and absorb the moisture. After a short period of time, the surface of these porous materials will show the stain associated with the green copper sulfate.

Initially, copper oxide is water soluble, but clear. The green coloration develops from the formation of the copper sulfate or copper chloride compound. As copper sulfate forms on the surface of the exposed copper, a small portion will wash off and stain adjacent surfaces.

The staining of adjoining surfaces by copper runoff is accompanied by staining from soot. Soot deposits on metal wash off in rainstorms or are carried by dew forming on the cool metal surface. Soot can be deposited in this manner by any metal roof, regardless of type or material makeup.

Extended drip edges and guttering of the runoff will greatly reduce and impede staining by soot and copper electrolyte onto the materials directly adjacent. Elimination of ledges that collect staining materials or building a positive slope into the ledges will help. As the staining materials collect, they are often deposited at joints in the ledges. The joints or seams in the metal act as local diverters of moisture, particularly dew. The joints also act as local depressions, which can collect soot. Sealants, which are often face applied at these joints, can collect soot either by means of a slight tackiness or by actually attracting soot because of a slight electromotive charge difference.

The sealing of stone and brick surfaces is another solution, but this requires periodic maintenance. It is preferable to handle the staining problem with careful consideration of details in the design. Painted wood and most exterior woods do not show such staining. Nor will the staining be apparent in many dark-colored brick or stone materials.

Staining by copper usually occurs on materials directly adjacent to the copper or where downspouts expel the collected moisture. Copper will stain concrete sidewalks where drips and sills allow the moisture to drip onto the lighter-colored surfaces. When a surface is stained by green copper sulfate, the stain is permanent. There is no recommended way of removing it short of destructive sandblasting or painting.

Staining—Other Materials onto Copper

Rust Stains

Another type of staining occurs on a copper surface. Iron rails or galvanized metal caps located above or near copper surfaces can cause a dark brown stain to develop on the green patina surface. This stain results from iron oxide (rust) washing onto the copper surface, where it then becomes a part of the copper sulfate coating. The iron compounds form iron sulfate, which is yellowish to dark brown in color. The resulting stain is unattractive and impossible to remove without damage to the copper sheet. The removal method that does the least amount of damage to the copper sheet is abrasive blasting with walnut husks. This method removes the outer corrosion layer but leaves a bright, slightly hammered appearance on thin copper sheet. Abrasive blasting with materials such as walnut husks performs adequately on thicker objects such as bronze statues or brass panels.

The damage produced by these rust stains is only superficial to the copper as far as performance is concerned. Aesthetically, the damage is major. To restore the original patina, roof sheets should be removed and replaced with either prepatinated sheets or in situ patination of new sheet copper.

Oak Trees

Copper can be stained by certain organic substances allowed to rest on the surface. The effect is lasting if the organic material is allowed to rest on the surface as the copper oxide initially develops. Oak leaves, which contain acids that leach out of the dying leaves, is one of these substances. Even trees near walls of copper will cause some modification to the oxide development. Actually, this is a natural effect and can be considered quite beautiful. The copper will go through the reddish and purple hues resulting from interference films created by the mildly acidic effect of the leaves. Eventually, the colors will blend, unless the leaves are allowed to remain on the surface.

Other woods that produce acidic solutions are sweet chestnut, red cedar, and Douglas fir.

Cedar Shingles

Wood shingle roofs draining into copper gutters will impede the development of the patina. After a while the shingles lose their acidity and pose no threat to the durability of the copper. Acids leaching from red cedar will impede the growth of an oxide film for a period of time. Initially, the copper stays bright, particularly around zones where moisture from the cedar roof is collected and concentrated. Line corrosion will occur along the junction of cedar roofing and copper valleys or copper edge strips. The effect is basically a thinning of the copper as the oxides are removed, replaced, and removed again by the acidic action of the cedar. The copper will, nonetheless, outlast the effects created by the cedar shingles. Copper gutters and valleys have been used on numerous cedar shake shingle roofs without ever perforating or showing any signs of premature deterioration.[6]

Gypsum and Cement

Gypsum, cement, and mortar, in immediate contact with the metal surface, will discolor copper. The discoloration caused by salts leaching from these materials is only superficial and does not produce long-term detrimental effects on copper.

Copper is used as a superior throughwall flashing material for masonry walls. Applications using copper as throughwall flashing have shown excellent results over time, in effect proving copper's resistance to the corrosive effects of masonry materials.

Bitumen

Bitumen, used in built-up roofing and roofing repair, is corrosive to copper. The oxidation of the bitumen creates corrosive acids that will dissolve the underlying metal. Mopped-in molten bitumen applications should not be used under copper roofing. There are some forms of bitumen laminates with smooth plastic facings and adhesive backs that adhere to the roof insulation or underlying support material. When applied behind a copper skin they provide an excellent, impervious barrier. Building paper should be used to separate the back of the copper sheet from the smooth plastic facing. The building paper keeps the hot copper sheet from adhering to the bitumen.

Fire-Treated Wood

Fire-treated woods, if they contain ammonia salts, will corrode copper. Care should be taken to prevent the contact of copper with ammonia-bearing substances. Other treated woods, such as marine plywood, can be used as underlayments, but should be separated from the metal by felts and building paper.

Asphaltic Repair Materials

Asphaltic and aluminum-bearing materials used to repair damaged copper surfaces will create localized corrosion cells and are ineffective repair materials. As with bitumen, when asphaltic materials decay, acids are generated that can damage the copper.

Use copper patches soldered down to the copper surface. This is a relatively simple process. The patches eventually will match the surface of the existing copper. If the damage is to a seam or joint, rework the joint rather than apply mastic materials.

---

[6]The author has lived in two homes with cedar shake shingles. Both homes had copper guttering. On one of the houses, the cedar shake shingles had been replaced twice in twenty years while the copper gutters and copper valleys remained in excellent condition beyond this time period.

Water

Fresh waters promote the development and formation of the protective oxide films on copper. Rates of corrosion are very low. For this reason, copper is used in water lines and tanks. In seawater copper holds up better than most metals. It has the ability to be antifouling and is used to coat the underside of boats to restrict the growth of barnacles.

Sealants

Acrylic-based, neoprene-based, and nitrile-based sealants will corrode copper. Most silicones and urethane sealants are preferred. Used with copper, a sealant must have sufficient elasticity to expand and contract with the thermal movement of copper and copper alloys. Silicones with good elasticity have shown the best results. Butyl sealants do not have sufficient elasticity. Have any sealant to be used with copper tested for compatibility, particularly with the copper salts that will develop.

Some discoloration will occur immediately around the copper/sealant interface. The discoloration is superficial. Sealants tend to pick up soot from the air; dirt and dust collect around sealants, making the joints appear larger and darker.

Once a joint has been sealed, further application of sealant becomes the only practical means of maintenance. Once the sealant fails, the joint cannot be soldered unless all the sealant is removed. This can be a monumental, if not impossible, task.

Dissimilar Metals—Galvanic Corrosion

In architectural applications, galvanic corrosion of other metals in close proximity to copper is an important consideration. Copper salts will corrode and discolor other metals in relatively short order, and all designs using copper should take galvanic corrosion seriously.

Copper is considered a noble metal; that is, a metal that will not displace the hydrogen ion in an acid solution. Copper will not dissolve in an acid except by oxidation processes. These processes occur naturally in all atmospheric exposures. Metals will corrode through an electrochemical reaction which develops when certain conditions are met. The effects of this corrosion vary from one metal to the next, but, basically, metal dissolves or goes into solution as metal ions when an acidic electrolyte is present. An acidic electrolyte is created with seawater or with rain containing atmospheric pollutants.

The hydrogen ion in the acidic electrolyte is displaced by the metal ion and is deposited at the cathode region of the reaction. When oxygen is present, the copper displaces the hydrogen as a hydroxide ion by reducing the oxygen. Copper goes into solution and combines with the hydroxide to form a metal hydroxide, oxide, or salt, such as copper oxide, which causes the brown color that initially forms on the copper surface. If sulfur is present, for instance, as atmospheric sulfur, the salt that forms is copper sulfate, the characteristic green patina. These corrosion products, copper oxide and copper sulfate, are very adherent to the copper surface and impede further corrosion of the copper.

Other architectural metals, such as tin, lead, zinc, and aluminum, which are not noble metals, form an impervious oxide film that protects them from further corrosion in most architectural applications. Steel, however, forms a very porous film that does not offer protection from the continuing electro-chemical reaction.

Exposure to and contact with copper alloys by other architectural metals can be detrimental if the other metal is anodic or chemically positive. Most architectural metals are anodic in respect to copper. An electrochemical re-action occurs under certain conditions. These conditions are the availabil-ity of an electrolyte such as rain or dew, the development of an electric current, which occurs when the metal ions go into solution, and the presence of oxygen. In architectural uses the conditions are readily fulfilled. Humid-ity and pollution, along with salt, will accelerate the effects.

The amount of galvanic corrosion on a less noble metal or more reactive metal is dependent in part on the relative areas of contact between the two metals. In applications using copper or copper alloys, the coupling of large areas of copper with relatively small areas of the more reactive metal is not ad-visable. However, the contrary is usually not a problem. For example, if a cop-per nail is used to attach an aluminum roof to a substrate below, little or no damage will occur. This is because the area of aluminum in relation to the area of copper is so great. In coastal regions this will not be the case because of the added chlorine agents. If, on the other hand, a copper roof is attached with an aluminum nail, the nail will rapidly corrode.

Avoid using combinations of metals in which copper makes up a large area in relation to the other metal. For instance, galvanized nails or clips should never be used with a copper roof. The area of copper is so large in relation to that of the galvanized steel that the nail or clip will rapidly corrode.

The electrolyte that causes staining from copper runoff poses another problem if it is allowed to come in contact with metals more reactive than copper. Controlling the flow of moisture from copper onto another metal is very important to reduce the development of galvanic corrosion. This is a common architectural problem, since copper is frequently used as a moisture diversion material. Copper downspouts damaged by ice often are replaced with galvanized spouts. As the copper guttering or roofing above drains into the galvanized spout, the galvanized surface deteriorates rapidly from im-pingement corrosion. Copper deposits develop into localized corrosion cells that quickly perforate the galvanized downspout. (See Table I-2, page 12.)

Copper in contact with lead or stainless steel of the passive S30000 se-ries is generally not a problem. Copper in contact with anodized aluminum has shown little damage, whereas copper in contact with unfinished alu-minum will cause staining and pitting of the surface. The damage is more su-perficial than structural, because the aluminum will develop an impervious barrier, inhibiting further corrosion.

Copper should never be coupled with galvanized steel or steel. The copper will quickly damage the steel and perforate the surface. The zinc in galvanized steel will develop corrosion cells and help to dissolve the un-derlying steel.

Care should be taken to eliminate the runoff of moisture passing from or over a copper alloy fabrication and onto another metal. The opposite direction of moisture flow is not a concern since copper is, in most cases, the more noble metal.

Painting the metal surfaces that will come in contact is a good idea. However, paint should be applied to both metals. The paint should be thick and applied liberally. If the more anodic metal were to have a breach or scratch in the paint, the galvanic current would be concentrated at this point and corrosion would accelerate at the exposed metal.

Humidity is directly proportional to the corrosion of metals. Once corrosion begins, the metal surface becomes more sensitive to atmospheric humidity and, thus, to corrosive action. This is a problem for metals in storage; once the metal surface becomes wetted and corrosion staining has started, cleaning becomes a temporary measure. The metal will stain quickly at the next exposure.

Wherever possible, use an electrical insulator such as a gasket to eliminate the electrical current from developing between two metals. With copper, the gasketing must not trap moisture or line corrosion can occur. The gasketing should be Teflon or silicone. Neoprene will corrode copper.

There must be three things in combination for corrosion to occur:

Oxygen (air),
An electrolyte (moisture such as rain water), and
The flow of an electrical current.

Prevent any one of these three and corrosion will not occur.

CARE AND MAINTENANCE

Expansion and Contraction Considerations

The thickness, or in the case of copper, the proper weight of the sheet to use for a particular application, depends on a number of variables, the first of which are expansion and contraction. Temperature changes will affect a copper sheet by building up stress within the metal as the metal wants to grow or shrink in size. This stress is cumulative; that is, the longer the dimension, the more expansion/contraction stress will accumulate.

In recent years, high-yield-strength variations of copper have been introduced.[7] The high-yield-strength coppers provide nearly a 15% increase in yield capacity. Because of the improvement in strength, it is available and used in thinner sheet stock, 12-ounce copper (12 ounces per square foot of surface) in lieu of 16-ounce copper for roofing and sheet metal work.

Copper is particularly susceptible to thermal stresses because of its low elastic limit. That is, it does not return back to its original shape once distortion occurs. Compared with other metals used in architecture, copper can be expected to expand and contract more than most. The 150°F (66°C) temperature shown in Table 2-4 is the differential between the minimum expected temperature and the maximum expected temperature. Keep in mind that a roof, particularly a dark copper roof, will get much warmer than the

---

[7]High-yield varieties of copper are available from some domestic copper mill producers. "Tough-12" is a product developed by the Revere Copper and Brass Company specifically for roofing and sheet metal work.

TABLE 2-4. Expansion Characteristics of Copper

METAL	EXPECTED EXPANSION IN 120 INCHES (3,048 MM) FOR A 150°F (65°C) TEMPERATURE VARIATION	
Iron	0.121 inches	3.07 mm
Steel	0.131 inches	3.33 mm
Copper	0.176 inches	4.47 mm
Aluminum	0.230 inches	5.84 mm
Lead	0.291 inches	7.39 mm

air temperature maximum expected for the summer months. High temperatures should be expected to occur on exterior copper with southern exposures. When using copper on exterior surfaces, detail design should take thermal movement into consideration. If copper is installed during the winter months while the metal and building surface are cool, allow for the expected movement resulting from thermal stress development as the metal heats up. The same should be considered when installing metal during the heat of the summer. For example, if a copper roof is installed during the summer and the panels are fitted tightly to the retaining cleats, as the metal cools, it will shrink and open the seams. Good detail should allow for this expected shrinkage.

As another example, consider the expansion and contraction capabilities of a particular design of a copper gutter installed in Kansas City.

The gutter has downspouts at 60-foot spacing with an expansion joint located at the center between the downspouts. The gutter is installed in the fall when temperatures are 50°F (10°C).[8]

The coefficient of elasticity is: 0.0000098 in/in/°F

Minimum design temperature of the copper: 20°F (-7°C)
Maximum design temperature of the copper: 150°F (66°C)

Contraction temperature difference: $50 - (-20) = 70$°F
Expansion temperature difference: $150 - 50 = 100$°F

Amount of maximum contraction this gutter should experience in inches:
$C_n = 30$ feet $\times 0.0000098 \times 70$°F $= 0.021$ feet or 0.245 inches.
Amount of maximum expansion this gutter should experience in inches:
$E_x = 30$ feet $\times 0.0000098 \times 100$°F $= 0.029$ feet or 0.353 inches.

Therefore, the expansion joint between the two runs of gutters should accommodate expansion of a minimum of $0.353 \times 2 = 0.706$ inches, or roughly $3/4$ inches.
This joint must also be able to accommodate contraction that each side will experience of 0.245 inches.

---

[8]*Architectural Sheet Metal Manual*, 5th ed. Chantilly, VA: SMACNA, 1993.

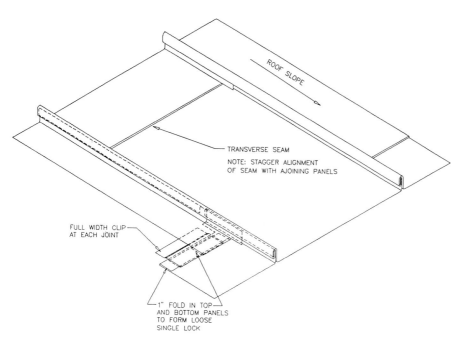

Figure 2-2. Typical transverse seam on copper.

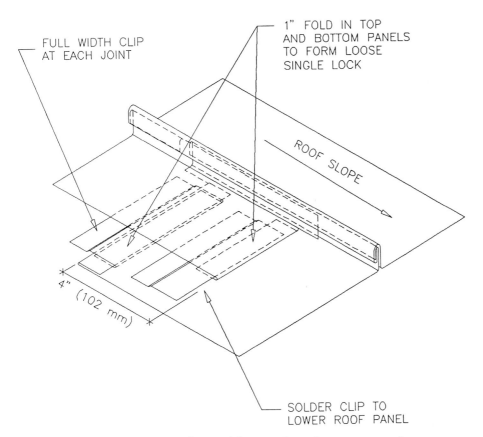

Figure 2-3. Transverse seam designed for use where driving rain can force water into seam.

Surface Deflections—Oil Canning

Good design allows a sheet to expand or contract without accumulating stress around seams or fasteners. Accumulation of stress will cause work hardening and eventual metal fatigue. It will also cause buckling and accentuate the "oil canning" tendencies of the metal skin. Oil canning is not caused exclusively by the accumulation of expansion and contraction stresses; rather, it is a characteristic of thin metal sheets. Often stresses are induced into the material when it is reduced in the cold-rolling process. These stresses are not consistent across the sheet and will show up as waves in the surface of the sheet. As the sheet of metal is being reduced in thickness, the outer edges tend to be held more tightly by the rolls than the center section of the sheet. This differential in pressure will induce stress as the sheet is stretched slightly at the edges.

Stress Considerations

Transverse seams in metal roofs and lap joints in metal flashings, along with cleated edges, eliminate the buildup of cumulative stresses when detailed and installed correctly. Individual sections need only be able to resist friction forces between the support surface and the metal surfaces as the metal fabrication expands. These friction forces develop compression loads onto the copper fabrication as it expands because of increased temperatures. The shape of the copper fabrication then acts as a compression member as it undergoes expansion forces, similar to the way a column resists the load of the structure it supports.[9] As the column gets longer, buckling can occur if the geometry of the column does not provide enough section to adequately pass the stress down its length.

With a copper standing seam panel, for instance, as the length between transverse seams increases so does the friction resisting the expansion of the sheet. This frictional resistance load is developed from the surface the "column" is sitting on and from the side joints where adjoining panels are connected. The standing seam panel must be stiff enough to resist the restraining force and still "push" the cumulative stresses down the sheet. If it is not adequately stiff, then local buckling will occur. This buckling will eventually cause overstressing of the sheet and metal fatigue failure. Failure will appear first where the accumulation of stresses is maximum, usually around bends in the sheet, clips that restrict movement, or seams that have been rigidly joined.

Stiffness can be achieved by increasing the sheet thickness, increasing the tensile strength, or increasing the section of the sheet. Increasing the thickness may be impractical because of cost, availability, or seaming restrictions. Increasing the tensile strength or increasing the temper of the sheet will increase the oil-canning behavior in some applications.

Increasing the unit section is usually the least complicated choice. For instance, with a standing seam roof panel, move the seams closer or make them taller. Through this action, the effective amount of metal is increased and the geometry per unit area is increased. Adding a pencil rib in the flat of the metal pan will also help reduce oil canning of the flat surface. A rib in the center of a roof sheet will, however, make the transverse seam and edge cleat difficult to construct.

The main requisite for a successful and long-lasting copper installation consists of an understanding of gauge, temper, shape, and attachment to the structure. Many of the very old copper installations that are still fulfilling their designer's directive without signs of failure are of a heavier gauge than typically used today. Copper sheets, approximately equivalent to a 32-ounce thickness, were commonly used for metal roofing work prior to the turn of this century. These sheets were fire refined and tougher than the electrolytically refined copper sheets used today. Standing seams were hand folded with blocks and mallets around copper clips. The thicker gauge added stiffness to the sheet, and the hammering of the ribs work hardened the vertical planes of the copper pans.

Today, thinner copper sheets are of the cold rolled temper, which adds stiffness to them, thus balancing with the reduction of material thickness. The copper sheets used on older structures were available only in 8 foot (2438 mm) lengths. The sheets were loose locked at the standing seam, and multiple loose-locked transverse seams were used to build up the longer runs necessary for most roof applications. This reduced the buildup of stress along long runs. Technology has made available the long copper sheet with minimum joints. Thus, the particular copper shape must be able to transfer the stresses that develop over much longer distances.

Guidelines for Design and Fabrication

There are some excellent manuals discussing correct detailing of copper as well as other metal materials. You can find some of these manuals listed in the bibliography. The common problem facing the designer or specifier is inherent to a custom design—it is impossible to find a reference for the precise situation at hand.

There are a few basic guidelines that set forth the basic principals and characteristics of metal fabrications. These guidelines must be incorporated into a design to ensure a successful installation. They are formulated from test results and field observations of copper applications. Buckling and pinches developed in a copper sheet either during installation because of improper handling, or after installation resulting from the buildup of stresses generated by thermal movements of the sheet copper, can be avoided if these guidelines are followed.

Combined with the beauty of the architect's design, good detailing, and a conscientious sheet metal craftsperson, a copper roof and wall cladding will last decades, perhaps even centuries.

The guidelines to follow with all designs using copper are these:

1. *Friction Reduction.* All metal panels must expand and contract. The ideal condition is for a copper panel to be floating in air. Since there are not many installations that achieve this, put a slip sheet under the panel. A slip sheet is a thin, inexpensive building paper that separates the metal surface from the 30 lb. roofing felts that act as an additional moisture barrier on the roof.

This slip sheet allows the copper to expand and contract with minimal friction buildup. Friction buildup can occur if the 30 lb. felts or other waterproofing underlayment adheres to the underside of the copper. Friction can also occur at the clips used to anchor the sheets to the support underlayment. Clips used to attach the copper sheet much be of a similar material to prevent galvanic action. Stainless steel clips of S30400, S31000, or S31600 alloy can be used in areas where additional strength is required. However, exercise caution when using dissimilar metals in coastal or polluted regions. Increase the clip thickness and frequency when uplift pressures are a concern.

Use floating clips on long expanses so that metal fatigue will not occur as the roofing sheet expands and contracts. Keep in mind that the farther a clip is placed from the fixed point on the metal roofing sheet, the more expansion will occur at this clip. The clips should be double fastened so they do not rotate as the sheet tends to expand and contract.

2. *Use of Gravity.* All metal roofing panels must be fixed at some point. Try to place the fix point at the top, to allow the panel to expand down the slope. Place an expansion cleat at the gutter or eave connection to hold the panel edge down and still allow the panel to expand. The fix point can also occur at the center, to direct the expansion both up and down the roof. This effectively reduces the total expansion at extreme ends of the roofing; such placement becomes necessary if the roof has a penetration, such as a vent pipe through the central region.

3. *Dimensional Economics.* Copper is not an inexpensive material. Therefore, it is good practice to maximize the coverage of the sheet between seams or battens. Thin copper sheeting is typically available in 24, 30, and 36-inch (610 mm, 762 mm, and 914 mm) widths; thus, if you allow for the standing seam you can maximize the coverage obtained from the raw flat sheets. If the desired standing seam spacing is 18-inch (457-mm) centers and the seam height is $1^1/_2$ inches (38 mm), then a 24-inch (610-mm) wide sheet will provide the most economical raw stock dimension.

4. *Section.* By *section* is meant the unit cross-sectional area, or, more precisely, the section modulus of the sheet. In areas where the expected wind velocity is minor and oil canning is not a great concern, the maximum width of a sheet can be used, allowing for the seam, of course. In areas where greater uplift pressures may occur and where oil canning is a concern, the seam spacing should be close. Seams at 12 inches (305 mm) on center can provide a relatively flat panel and a stiffer section. Seams at 17 to 18 inches (432 mm to 457 mm) will maximize coverage from a 24-inch (610-mm) coil, but may not give the best flatness or uplift resistance. In addition, the edges will need more closely spaced clips since they now must resist a greater total load. Closely spaced clips may increase friction at the edges and accentuate oil-canning tendencies.

5. *Gauge and Temper.* Specify cold rolled Eighth Hard Temper ASTM B370. Soft copper should not be used in the majority of roofing

applications. Soft copper should be used only if the metal fabrication is to undergo extensive forming, and then only in heavier thicknesses.

For most applications 16-ounce copper should suffice. If durability, flatness, or extensive stress/strain conditions are major criteria, then increase the thickness. Keep in mind there is no substitute for metal thickness; it is the single most important structural component of sheet metal construction. The thinner the gauge, the greater the tendency to buckle, regardless of the tensile strength. A more rigid shape helps resist buckling in any particular gauge, as long as the rigidity is in the direction of the greatest stress.

6. *Attachment.* Attachment is critical if the copper fabrication is to endure for an expected service life. Exposed fasteners, soldered seams that do not allow individual expansion and improperly cleated seams can all cause metal fatigue as a copper sheet expands and contracts. Stresses build up at the areas where the sheet is confined, and eventually the metal around these areas will work harden, buckle, and crack.

There are two types of seaming typically used in joining copper sheets to one another and in joining copper to a structure. A transverse seam is a horizontal lap joint between two fabricated metal sections. Many sheet copper applications are formed from 24-inch or 36-inch (610-mm or 914-mm) wide by 120-inch (3048-mm) long flat sheets. Since most uses require runs of greater than 120 inches (3048 mm), a joint or transverse seam is required.

A transverse seam is perpendicular to the direction of the greatest thermal stress. On a standing-seam roof this joint is attached to the roof with a copper clip and single lock. On a low-pitch roof this seam is cleated. The seam must not allow for the transfer of the thermal stresses to the next sheet down the roof line. Accumulated stresses are released by allowing for thermal movement at the transverse seam. The top of the metal panel is fixed with the cleat, and the bottom of the panel is allowed to expand or slip over the top of the next lower panel while the edge is still cleated to the roof.

On installations occurring during the heat of the summer, the panels should be engaged loosely at the transverse seam. Pulling the seams tight may create buckles when the sheets contract during cooler temperatures.

The other type of joint is the side seam along the length of the fabrication. This joint must allow the copper surface to expand without binding or buckling with the adjacent section. For a roof this joint is a double-lock or single-lock style standing or batten seam. This joint allows for movement of one section over the other while being held down by clips along the edge. A batten seam uses a wood batten between panels. The wood batten is covered with copper and then seamed, similarly to the standing seam.

One of the greatest fatigue problems experienced with a standing seam roof occurs when a person walks on the roof and steps on the seams. This binds the panels together and destroys their ability to expand and contract.

In thin sheet copper applications, of less than 24-ounce copper, soldering of flat seams is an excellent means of joining two sections together. Solder is tin alloyed with lead and has a relatively low melting point. Copper

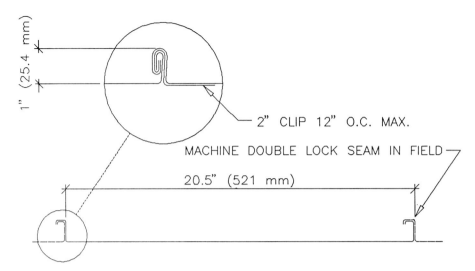

Figure 2-4. Double-lock standing seam geometry.

takes to soldering extremely well because of its ability to absorb heat readily. Solder, when properly fluxed, will run into the joint, filling the void between the two sections of metal. A lock seam, single or double, can be soldered to produce a watertight joint. The joint can be riveted with copper rivets prior to soldering. A soldered joint is intended to produce a watertight continuous condition. Folding and locking the seam will produce a combined sheet as strong as if it were a single sheet. Soft solder alone is weak and tends to creep, so the locking of the joint and riveting are equally important to the joining of the sheets because of the strength they add.

Preventive Measures During Design and Construction

The most important factors in extending the life of copper and copper alloy fabrications are proper installation procedures, good detailing of joints and transitions, and good design. Preventive maintenance techniques, when correctly used, will provide an extremely long-lasting fabrication, regardless of atmospheric conditions.

Choosing the right substrate to attach the metal fabrication and consideration of adjoining materials will reduce the damaging effects of corrosion. When soldering joints, clean and neutralize all acids used to clean and prime the surfaces. Improperly soldered joints will not transfer accumulated stresses and will allow moisture, along with atmospheric corrodents, to collect along seams and eventually pit the copper. Solder should be used to seal seams that have been mechanically joined by loose lock seams or one of the folded and interlocked seams.

Globs of solder apparent on a surface should be removed. If allowed to remain, the rough surface will trap moisture and soot, which can develop into corrosion cells. Free-draining surfaces, sloped surfaces that shed moisture quickly, should not have solder on exposed areas. Try to make such surfaces watertight without soldering by using mechanical connections that interlock yet still allow expansion and contraction. This is part of what is known as "the rain-screen principle." The rain-screen principle involves allowing moisture to flow across an unsealed joint. Pressures within the joint coupled with gravity prevent the moisture from entering the joint.

Well-draining details, the elimination of shelves that collect soot and moisture, should be developed. For copper roofs, maximize slopes to shed moisture rapidly. Flat surfaces and surfaces draining onto the copper, will, over time, work to corrode the copper surface. Erosion occurs when constant draining wears away the copper sulfate that develops to protect the metal surface.

Copper and copper alloys should not be restrained from moving because of temperature changes. A copper fabrication must be properly held onto the substrate, but it must be allowed to contract and expand without building up stresses. Buckling will occur when stresses in sheet copper exceed the yield point of the metal. Stress, which develops in the sheet copper, can lead to metal fatigue. Once buckling occurs or fatigue develops, repairs are impossible and the metal must be replaced.

Large-radius bends at transitions reduce cracking resulting from buckling stresses. Expansion cleats should be used to allow the metal to expand without generating stresses. Design anchorage back to the substrate to eliminate the stressing of the copper from negative wind pressures. Wind buffeting will cause metals to fatigue and can tear or weaken the restraining fasteners.

Storage Stains

Copper roofing sheets are frequently stored in coils. If the coils are not stored in a dry place, the surface will show signs of initial oxidation when the sheets are uncoiled. The oxidation, seen as a purplish or golden color, is induced by the interference film of copper oxide. The oxidation occurs near the center of the sheet and appears as a mottled stain. Initially, a roof or surface of this material will look strange, but it will eventually develop the natural brown copper sulfate as exposure to the environment continues. The stain is an oxide interference film and will eventually blend in and form the characteristic brown color of the oxide. Polished copper in storage will tarnish rapidly, even under a protective plastic film. Air, trapped under the protective film, continues to tarnish the copper surface, sometimes making small brownish spots. After a few months of storage, the surface will require repolishing to its original luster.

If water ever sits on the surface of any of the copper alloys, it will stain and extensive refinishing will be required to restore the surface. Water stains will leave a mottled, almost "burled walnut" appearance on copper. This stain will take months of exposure to blend with other copper. Copper alloys, if allowed to get wet, will develop black spots. These spots are very difficult to remove and will require extensive repolishing.

Copper roofing and architectural metal fabrications require little if any maintenance unless they have been lacquered or antiqued. Foreign materials on the copper surface should be removed. In general, however, copper should be allowed to continue aging naturally.

Removal of Surface Deposits

Surface deposits such as iron oxide, encrusted bird deposits, and other tenacious deposits can be removed by surface blasting with pulverized walnut husks. Walnut husks, applied at 30 psi (207 Pa), can remove surface contamination without damage to the underlying copper.

Removal of Patina

Occasionally, because of unevenness or for aesthetic considerations, the natural patina on copper has to be removed. To remove the patina, you might consider a sponge bath over the surface with a mixture of six parts phosphoric acid, one part nitric acid, and 50% distilled water. The pH should be in the range of 1 to 1.5. Use a binding agent to develop a paste, and allow the paste to remain on the surface for approximately one minute.

Remove the waste by sponging the surface with sponges soaked in a solution of sodium bicarbonate. The solution should have a pH of 10. Follow this sponging with a thorough fresh water rinse.

FASTENING AND JOINING

Soldering

Solder on sheets of thicknesses more than 20 ounces should be considered as waterproofing. Soldering of heavy sheet fabrications over time will require periodic maintenance owing to the creep or flow effect of the alloy.

Solder typically used in the sheet metal industry is 50-50 or 60-40 tin to lead (Table 2-5). The greater the tin content, the lower the melting

Figure 2-5. Soldering copper drain tube onto stainless steel bowl.

TABLE 2-5. Various Solders

SOLDER TYPE	MELTING POINT °F	MELTING POINT °C
50-50 Tin–Lead	361–421	182–216
95-5 Tin–Antimony	450–464	232-240
94-6 Tin–Silver	430-535	221–279
95-5 Tin– Silver	430–473	221–245
96.5-3.5 Tin–Silver	430	221

temperature. The 50-50 alloy will melt at 414°F (212°C), and 60-40 will melt at 370°F (188°C). In addition, there are a number of low-lead solders available. These solders reduce or replace the lead content in the alloy mix. The 95-5 alloy, which contains 95% tin and 5% lead is a common low-lead solder. Other solders are available that use antimony instead of lead. The low-lead solders are more expensive than the old standard 50-50 solders.

When joining flat seam roofs or large fabrications, expansion joints are necessary to allow for the release of accumulated stresses.

With the standing seam and batten seam panels, expansion and contraction of the skin is taken up in each individual panel at the transverse seam and at the standing seam.

Welding

Copper can be welded with the use of arc welding techniques.

OAW
SMAW
GTAW
GMAW (reduces distortion)

Argon gas is used for thin sheet or plate sections to be welded, and helium for thicker members. The high thermal conductivity of copper poses some problems with joint fusion. It is recommended to use wide root gaps and tack the parts together in more frequent locations. Copper does not weld effectively with hydrogen and oxygen gases. Adequate welds are difficult to produce.

3

The Major Copper Alloys— Brasses, Bronzes, and Nickel Silver

INTRODUCTION

A few centuries after the discovery of copper by Stone Age people, either by chance or by design, one of them added another metal to the mix. This other metal had some interesting effects on copper. It caused copper to melt at a lower temperature and to flow more easily into a casting mold. This metal was tin, and the name given the new copper-tin alloy was bronze. These newfound characteristics of bronze were responsible for bringing the metal, and humankind with it, out of the Stone Age and into modern civilization. Bronze could now be shaped and cast into utensils, which would forever change human behavior. The start of the Bronze Age, with the first purposeful alloying of metals, was dawning.

This new copper-tin alloy was much harder than pure copper, but still malleable enough to hammer into shapes. The sharpened edge of a forged ax or sword would remain sharp longer than the copper instrument, and, with the improved hardness, real tools could be fabricated. Bronze was used throughout antiquity to make armor and weapons for warriors to take into battle and statues of those warriors when they did not return.

Bronze is not used for weapons anymore, but we still make a statue or two, to commemorate the missing warrior. Technically, bronze is the name of the copper alloy that contains copper and tin. Used today in referring to architectural metals, bronze is actually a description of color, rather than of element content. From a metallurgical standpoint, the term bronze is applied to copper alloys that contain other elements, besides zinc, as the main alloying metals. Architectural Bronze and Commercial Bronze are actually brass alloys that contain no tin in their alloying metal constituents whatsoever. Statuary Bronze is the name given to describe a range of bronze-color tones produced by chemical enhancement on different copper alloys, both brass and bronze alloys. Persons who work with the metal and architects who design with it have, over the years, referred to the copper alloys as bronze if their color ranges are in the darker brown tones.

Brass is the name given to those alloys of copper that contain zinc as their main alloying constituent. The term *brass*, however, is often used in reference to the color of some of the copper alloys. Brass nautical fittings, polished to a lustrous golden yellow, exhibit the color most associate with the term *brass*. The levels of zinc and other metals, such as nickel and aluminum, give the copper alloys a range of available colors, unmatched by any other metal.

The copper alloys are unique among architectural metals because of the natural color and tone change from one alloy to the next. There are distinct color differences exhibited by the copper alloys, which depend on the alloying elements mixed with the copper. Table 3-1 describes various alloys and the colors they possess.

TABLE 3-1. Various Colors of the Copper Alloys

ALLOY	DESCRIPTION	ALLOYING CONSTITUENT	COLOR—NEW POLISHED STATE
C11000	Commercially Pure Copper	None	Soft red
C21000	Gilding Metal	Zinc	Red to reddish brown
C22000	Commercial Bronze	Zinc	Golden red
C22600	Jewelry Bronze	Zinc	Color of 14k gold
C23000	Red Brass	Zinc	Reddish gold
C24000	Low Brass	Zinc	Golden color with red tint
C26000	Cartridge Brass	Zinc	Yellow with red tint
C27000	Yellow Brass	Zinc	Yellow
C28000	Muntz Metal	Zinc	Golden yellow
C61400	Aluminum Bronze	Aluminum	Dark gold
C65500	Silicon Bronze	Silicon	Reddish brown
C74500	Nickel Silver	Zinc and Nickel	Silvery gold to white

HISTORY

At about 3000 B.C. the transition from the Copper Age to the Bronze Age took place. This is the time when the first indications of deliberate alloying of copper with other materials were demonstrated. Copper was first alloyed with materials such as tin and arsenic to develop bronze. Early tin mines in the Caucacus region and in the Cornwall region of England's southeast coast were valuable sources of the alloying metal 2,500 years ago. Today, less than 2% of the world's supply comes from these regions.

Ancient sources for tin were recently located at archeological sites in southern Turkey. This site, near Tarsus, may well have been one of the major sources for tin in the Mediterranean. The region just north of Italy was the home of the Aunjetitz culture in Bohemia. These people were renown for intricate bronze work.

Bronze acquired its name, it is believed, from the ancient city in Italy, Brundisium. One of the oldest seaports in Italy, Brundisium probably operated as a gateway for tin ores brought in from Cornwall and other tin-mining regions surrounding the Mediterranean. Extensive metalwork, dating back to the second and third centuries B.C., attests to the proliferation of copper-tin alloying in this area.

Bronze exhibited better casting properties and greater durability than pure copper. Indications of an early knowledge of copper alloys are shown in Babylonian tablets dating from about 2500 B.C. The tablets include records specifying early tin-copper ratios for bronze alloys. Early metalworkers found that by adding tin in quantities in the 5% to 10% range, better swords, daggers, and other armaments could be fashioned.

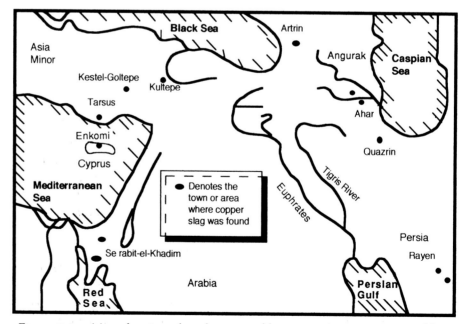

Figure 3-1. Map of region where bronze and brass constituents were mined by early civilizations.

Currency and ornamentation were also common uses for the bronze material. Because of its color and relatively heavy weight, it was cherished by early civilizations and became one of the first currency forms. Sometime during the third century B.C., the Colossus of Rhodes was erected. This bronze statue of Apollo, considered one of the Seven Wonders of the World, was more than 100 feet (32 meters) high. The statue spanned the harbor, and ships would pass beneath it as they entered. This enormous statue has since disappeared, sometime during the decline of the Roman Empire. But its existence attests to the availability of bronze and the extensive metalworking knowledge of the people of this region and time. Artifacts uncovered from the region in what is now southwestern Iran show an understanding of casting that would rival the craftsmanship of today. A statue of Queen Napir-Asu of the Kingdom of Elam, who reigned in this region during the fourteenth century B.C., was cast of bronze and then covered with another shell of copper. The sculpture weighed in at 3,750 pounds (1701 kg). The copper outer shell was intricately detailed and was cast in two halves, then cleanly pinned to the bronze core.

PRODUCTION AND PROCESSING

The copper alloys are produced by converting essentially pure copper into specific alloys. The pure copper is provided as cake or billets of 99.95% purity by producers of cathode or electrolytically refined copper. Additional copper is obtained from recycled scrap and provided directly to the brass mill. At the mill, alloying constituents are added to molten copper in quantities to meet alloy design specifications. The molten alloy is cast into a thick, large rectangular block. While hot, the cast block is rolled through high-pressure rolls to create a ribbon of sheet material. The cast block can also be cut to smaller sections to allow for extrusion shaping or drawn shaping.

ENVIRONMENTAL CONCERNS

The environmental issues concerning copper alloys are essentially the same as those for copper, which are discussed in Chapter 2, with one additional concern. Some brasses contain lead as an alloying constituent. Various alloys and their lead constituent are indicated in Table 3-2. When lead is added, it does not dissolve into the brass alloy mix. Instead, the lead disperses throughout the metal and remains as a separate solid trapped within the brass. Lead gives brass a "free cutting" characteristic. This basically means that during certain fabrication operations, the brass will come off in small shards, improving tool life and wear.

In the United States, federal law considers brass alloys containing less than 8% lead as "lead free." Some mechanisms used to pump well water in the United States were manufactured using leaded brasses. The testing of water from these pumps showed high levels of lead, apparently leaching from the brass. The lead could also have come from the solder used to seal the parts of the pump. Most of the brass alloys used in architecture contain less than 8% lead. These alloys under normal working conditions and normal

TABLE 3-2. Various Alloys and Their Lead Content	
COPPER ALLOY	LEAD CONTENT
C 31400 Leaded Red Brass	2%
C 36000 Free Cutting Brass	3%
C 38500 Architectural Bronze	3%
C 79600 Leaded Nickel Silver	1%
C 83600 Composition Bronze	5%
C 84400 Cast Semi-Red Brass	7%
C 85200 Cast Brass	3%
C 85400 Cast Yellow Brass	3%
C 97300 Cast Nickel Silver (56% Cu)	10%
C 97600 Cast Nickel Silver (64% Cu)	4%
C 97800 Cast Nickel Silver (66% Cu)	2%
Cast Gold Bronze	3%
Cast Bronze	2%
Tin Bronze 83-7	7%

exposures should pose no health concerns. It is, however, recommended to avoid drinking or eating from fabrications made from these alloys.

THE ALLOY GROUPS OF COPPER

Copper has an extensive array of alloys developed for different aspects of industry. Only a few are commonly used in architecture. The rest of the alloys were developed for specific forming or structural characteristics. The various families of alloys are listed in Table 3-3.

THE ARCHITECTURAL ALLOYS

The Brasses

Brass is the name given to those copper alloys that contain zinc as the major alloying constituent. The brasses are broken down into a series of groups classified by the crystal makeup of the copper alloy. As more zinc is added the crystal structure of the copper alloy changes, and this change affects the color and the forming ability of the metal.

Basic Metallurgy

The brasses undergo three metallurgical phases as zinc is added: alpha, alpha plus beta, and beta. These phases represent three distinct changes in the crystal structure of a metal. All metals are made of crystal structures at the molecular level. These crystals group together into the grains that make up the substructure of the metal.

Copper alloys have a distinct crystal structure, the simplest form, the face-centered cube. The face-centered cube at the atomic level is made of a

TABLE 3-3. The Alloy Groups of Copper

	DESCRIPTION	UNS ALLOY
Wrought Alloys	Brass Alloys (Cu-Zn)	C20500–C28500
	Leaded Brasses	C31200–C38590
	Copper-Zinc-Tin Alloys	C40400–C49080
	Phosphor Bronzes	C50100–C55284
	Aluminum Bronzes	C60800–C64210
	Silicon Bronzes	C64700–C66100
	Manganese Bronzes	C66400–C69710
	Copper-Nickel Alloys	C70100–C72950
	Nickel Silver Alloys	C73500–C79800
Cast Alloys	High Copper Alloys	C81400–C82800
	Red and Leaded Brass	C83300–C83810
	Semi-Red and Leaded Brass	C84200–C84800
	Yellow and Leaded Brass	C85200–C85800
	Manganese Bronze	C86100–C86800
	Silicon Bronze	C87300–C87900
	Tin Bronze	C90200–C91700
	Leaded Tin Bronze	C92200–C92900
	High-Leaded Bronze	C93100–C94500
	Nickel Tin Bronze	C94700–C94900
	Aluminum Bronze	C95200–C95900
	Copper Nickel Bronze	C96200–C96900
	Nickel Silver	C97300–C97800
	Copper Lead	C98200–C98840
	Special Cast Alloys	C99300–C99750

cube lattice crystal with an atom at each corner of the cubic space and an atom at the center of each of these faces.

As zinc is added, the cubic crystal structure undergoes changes that add certain characteristics to the original alloying metals. Zinc is basically dissolved into the cube by substituting zinc atoms into the crystal's lattice structure. Other elements are also used to alloy with copper, but zinc affords the most dominant properties to the copper alloys.

The alpha phase is the first transformation the copper alloy crystal goes through when zinc is added. The alpha brasses contain from 2% to 38% zinc. These brasses are not as corrosion resistant as the other copper alloys and often are used as decorative elements in architecture. The alpha brasses are well suited for cold working with press brakes and roll forming, as well as deep drawing, spinning, and other more severe forming processes. The alpha brasses cannot be hardened by heat treatment, but they may require annealing to reduce brittleness when deep-drawing operations are repeated on the same fabrication. "Orange peel" surface conditions will occur if specifications for time and temperature of annealing operations are not closely followed.

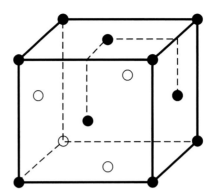

Figure 3-2. Cubic form of the copper crystal.

As the zinc content increases to about 28% to 38%, the alloy becomes less and less corrosion resistant and prone to dezincification corrosion. Dezincification appears in the form of pitting as local corrosion occurs along the grain boundaries.

Common alpha brasses are commercial bronze, red brass, and cartridge brass. One characteristic of alpha brasses is that their color will become more yellow as zinc is added; cartridge brass is the most yellow of the alpha brass alloys.

The copper alloys in the alpha phase become more malleable, at least until the 30% zinc level is reached. As the zinc content increases beyond this concentration, hardness and strength continue to increase but ductility is reduced as the alpha grain becomes saturated with zinc.

Once the copper alloys reach a level of 37% zinc, the crystal structure moves into a transition phase, known as the alpha-plus-beta phase. The alpha-plus-beta brasses are the copper alloys that contain from 37% to 50% zinc. As zinc is added in the alpha phase to the 38% level, beta grains become more apparent. The resulting alloys are stronger but less malleable than the alpha brasses.

In alloys that are about 62% to 58% copper, both alpha and beta grains are present. Muntz brass and architectural bronze are the most common of the alpha-beta brasses. Used as architectural fabrications such as handrails, ornamental plates, and panels, such brasses take to polishing well. These copper alloys are hard, yet they can take moderate cold working well.

Severe forming operations, such as spinning and deep drawing, are not easily performed on these alloys because they work harden rapidly and will split along the grain boundaries. Welding or brazing also becomes more difficult because of localized changes that make the metal surrounding the weld brittle.

Beyond the range of 42% zinc, the copper alloy enters the beta phase. In the beta phase, the copper alloy is brittle and difficult to cold form. Few architectural applications use the beta-phase brass alloys.

The addition of other alloying elements come into play in the alpha-plus-beta brasses. These elements replace a portion of the zinc content without affecting the basic metallurgy of the alloy. Small percentages of lead, up to 3.5%, are alloyed to improve the machinability of the alpha-plus-beta-phase brasses. The lead makes the alloy slightly softer and less resistant to impact. Extruded architectural bronze, which is widely used as ornamental handrailing and other extruded shapes, contains small percentages of lead to aid in the extruding operation. Lead also enables the copper alloy to be "free cutting." The small portion of lead causes the brass to come off in small shards when cut, making it easier to mill and process the metal fabrication.

Tin is added to improve corrosion resistance. Tin helps to resist dezincification, a corrosion behavior of brasses that robs zinc from the alloy. Naval brass, an alloy used on and around nautical structures, contains 1% tin to achieve improved corrosion resistance. This alloy is similar to Muntz Metal in color and alloying constituents, although Muntz does not contain tin. Cast Yellow Brass 405.2 also contains 1% tin.

Iron, aluminum, silicon, and manganese are all added in small amounts to improve corrosion resistance and formability. Architecturally, these additions are nonconsequential.

The Color Effects of Alloying

As the zinc content of the brass alloys increases, the color of the metal becomes progressively more yellow (see Figure 3-3). Brass alloys in the alpha phase reach a maximum of zinc saturation at 35%. This maximum saturation of the alpha crystal structure also represents the extent of the yellow color. Beyond this point, the alloys enter the alpha-plus-beta phase and, as zinc is added, the brasses progressively show less and less of the yellow color. Muntz Metal, which has 40% zinc, is less yellow than Yellow Brass at 35%. At 42% zinc, the color of the copper alloy is comparable to Red Brass at 15% zinc. Reference Table 3-4 for the various colors depicted by the brasses.

TABLE 3-4. The Architectural Brass Alloys*

GROUP	ALLOY NUMBER	ALLOY NAME	ZINC %	OTHER %	FORM	COLOR
Alpha	C21000	Gilding Metal	5%		Wrought	Copper color
	C83600	Composition Bronze	5%	5% Lead	Cast	Red copper color
	C84400	Cast Semi-Red Brass	9%	7% Lead	Cast	Red with golden tones
	C22000	Commercial Bronze	10%		Wrought	Golden copper color
	C22600	Jewelry Bronze	12.5%		Wrought	Close to 14k gold color
	C23000	Red Brass	15%		Wrought	Golden red
	C24000	Low Brass	20%		Wrought	Reddish yellow
	C85200	Cast Brass	24%	3% Lead	Cast	Yellow
	C85400	Cast Yellow Brass	29%	3% Lead	Cast	Yellow
	C26000	Cartridge Brass	30%		Wrought	Yellow
	C27000	Yellow Brass	35%		Wrought	Yellow
Alpha-	C85700	Cast Yellow Brass 405.2	35%	1% Lead	Cast	Golden tone
Plus-	C28000	Muntz Metal	40%		Wrought	Golden tone
Beta	C38500	Architectural Bronze	40%	3% Lead	Wrought	Reddish gold to golden tone

*Zinc is the main alloying constituent.

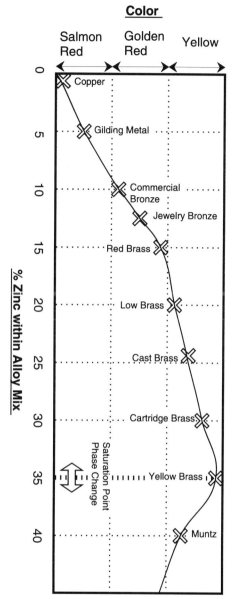

Figure 3-3. Graph of percentage of zinc content versus color.

Alloy C21000

Common Names: Gilding
95-5 Brass

Alloy Mix—Nominal Composition
Copper 95%
Zinc 5%

Wrought Alloy—Available Forms
Strip ASTM B36
Tubing ASTM B587

TABLE 3-5. Common Tempers—Architectural Uses

TEMPER	ASTM NUMBER	YIELD STRENGTH (ksi)	YIELD STRENGTH (MPa)	ROCKWELL Hardness b
Quarter Hard	H01	32	221	32
Half Hard	H02	40	276	34

Natural Polished Color	Reddish bronze, copper color
Oxidized Color	Brown tones, some interference colors of yelow, reds, and purples
Weathered Color	Gray-green patina

Gilding is very malleable; it can be easily formed using conventional cold forming equipment. This copper alloy is very soft. Even at Half Hard Tempering, the Rockwell Hardness is only 34. Sheet width is limited, because it is customary to produce in strip dimensions rather than sheet.

Gilding is not in common architectural use because of its similarity in appearance to copper. Copper is more readily available and has similar properties. Alloy C21000 has high-corrosion-resistant properties. This alloy is used as a base for gold plating or leafing, thus its name. It is also used to manufacture the jackets of bullets. Gilding can be finished in any of the common mechanical and chemical finishes. It will receive a statuary finish well. Table 3-5 lists various tempers available.

Alloy C83600

Common Names:	Composition Bronze
	85-5-5-5 Brass
	Ounce Metal
	Leaded Red Brass

Alloy Mix—Nominal Composition

Copper	85%
Zinc	5%
Tin	5%
Lead	5%

Cast Alloy Form Only

Natural Polished Color	Reddish copper color
Oxidized Color	Brown tones
Weathered Color	Gray-green patina

This casting alloy has been used for brass castings for centuries. Composition Bronze is still the preferred cast architectural and ornamental alloy for casting plaques, statuary, and other ornamentation. The lead content makes alloy C83600 suspect for handling potable water, but other uses should not pose an environmental or handling hazard.

This alloy has low-strength characteristics and lacks hardness.

Yield Strength   17 ksi   117 MPa

Composition Bronze can be cast using the following methods:

Sand	ASTM B584
Continuous	ASTM B271
Investment	AMS 4855

Alloy C84400

Common Name: Leaded Semi-Red Brass

Alloy Mix—Nominal Composition
Copper	81%
Zinc	9%
Tin	3%
Lead	7%

Cast Alloy Form Only
Natural Polished Color	Reddish orange
Oxidized Color	Brown tones
Weathered Color	Gray-green patina—slightly mottled

Alloy C84400 is a common cast architectural alloy used for general purpose ornamental castings. The additional tin aids in corrosion resistance in exterior environments. Because of its lead content, this alloy is not recommended for transfer of drinking water. Other uses, not involving food or drink, should pose little hazard.

This alloy has low-strength characteristics and lacks hardness.

Yield Strength   15 ksi   103 MPa

Alloy C84400 can be cast using the following methods:

Sand	ASTM B584
Continuous	ASTM B505
Centrifugal	ASTM B271

Alloy C22000

Common Names: Commercial Bronze
                         90-10 Brass

Alloy Mix—Nominal Composition
Copper	90%
Zinc	10%

Wrought Alloy—Available Forms
Strip	ASTM B36
Sheet	ASTM B36
Plate	ASTM B36
Tubing	ASTM B587

TABLE 3-6. Common Tempers, Alloy C22000 Commercial Bronze

TEMPER	ASTM NUMBER	YIELD STRENGTH (ksi)	YIELD STRENGTH (MPa)	ROCKWELL Hardness b
Quarter Hard	H01	35	241	42
Half Hard	H02	45	310	58
Hard	H04	54	372	70

Natural Polished Color	Reddish bronze with golden tone
Oxidized Color	Brown tones with orange tone
Weathered Color	Gray-green patina

Commercial Bronze is an ornamental alloy sought for its rich bronze hue. This alloy is frequently used for mirror- or satin-polished ornamental panels and trim. The surface will tarnish if not protected with clear lacquer or coated with a statuary finish and oiled. Commercial Bronze can be cold formed with little difficulty.

Various cold-rolled tempers are available. Quarter and Half Hard Tempers are preferred for architectural uses. These tempers correspond to the small grain size, which allows for a smoother, more reflective surface when polished. See Table 3-6 for the standard tempers available.

Alloy C22600

Common Name: Jewelry Bronze

Alloy Mix—Nominal Composition
Copper   87.5%
Zinc   12.5%

Wrought Alloy—Available Forms
Strip   ASTM B36
Drawn Shapes

Natural Polished Color	Golden red—matches closely to 14 k gold
Oxidized Color	Orange tone
Weathered Color	Gray-green patina—mottled

This copper alloy is not in common architectural use. The available dimensions of the flat product is limited to strip widths. Alloy C22600 is more commonly used in jewelry applications and plaques. The color of polished Jewelry Bronze matches closely to that of 14 karat gold.

Alloy C22600 can be formed and shaped similarly to Commercial Bronze and Gilding. It is available in all cold-rolled tempers, with grain size improvement occurring with the Quarter and Half Hard Tempers. See Table 3-7 for the standard tempers available.

TABLE 3-7. Common Tempers of Alloy C22600, Jewelry Bronze

TEMPER	ASTM NUMBER	YIELD STRENGTH (ksi)	(MPa)	ROCKWELL Hardness b
Eighth Hard	H00			
Quarter Hard	H01	37	255	47
Half Hard	H02	47	324	61
Hard	H04	56	386	73

Alloy C23000

Common Names: Red Brass
  Rich-Low Brass

Alloy Mix—Nominal Composition
Copper   85%
Zinc   15%

Wrought Alloy—Available Forms
Strip   ASTM B36
Sheet   ASTM B36
Tubes   ASTM B135

Natural Polished Color   Golden red
Oxidized Color   Reddish-orange tone
Weathered Color   Gray-green patina—mottled

Red Brass matches closely in color to Architectural Bronze, particularly in the polished satin finish. This alloy can create a subtle color tone difference when used in conjunction with Commercial Bronze or the Cartridge Brass alloy.

This alloy is slightly harder than C22000 and C22600; however, its formability is still very good. Red Brass will take a polish very well and can be used to develop a rich statuary finish. The full range of tempers are available on Red Brass sheet and strip forms. See Table 3-8 for standard tempers available.

TABLE 3-8. Common Tempers of Alloy C23000, Red Brass

TEMPER	ASTM NUMBER	YIELD STRENGTH (ksi)	(MPa)	ROCKWELL Hardness b
Quarter Hard	H01	39	269	55
Half Hard	H02	49	338	65
Hard	H04	57	393	77

Alloy C24000

Common Name: Low Brass

Alloy Mix—Nominal Composition
Copper   80%
Zinc   20%

Wrought Alloy—Available Forms
Strip   ASTM B36
Tubes   ASTM B135

Natural Polished Color   Reddish yellow to orange
Oxidized Color   Orange-brown tone
Weathered Color   Dark gray-green patina—mottled with
                            dark streaks

This alloy is similar to Red Brass in most characteristics. The dark orange-red color of satin-polished Low Brass ranges between the color of the yellow brasses and the bronze tones. This copper alloy has good formability characteristics. Quarter and Half Hard Tempers are used for architectural applications. See Table 3-9 for a list of these tempers.

Alloy	**Common Names**
C85200	Cast Brass
C85300	70-30 Yellow Brass
C85400	Commercial #1 Yellow Brass
C85700	Cast Yellow Brass 405.2

Alloy Mix—Nominal Composition

Alloy C85200		**Alloy C85300**		**Alloy C85400**		**Alloy C85700**	
Copper	72%	Copper	70%	Copper	67%	Copper	63%
Zinc	24%	Zinc	30%	Zinc	29%	Zinc	35%
Tin	1%	Tin	0%	Tin	1%	Tin	1%
Lead	3%	Lead	0%	Lead	3%	Lead	1%

Cast Form Only
Natural Polished Color   Yellow with slight red tone
Oxidized Color   Orange-yellow tones
Weathered Color   Dirty yellow—mottled with dark streaks

These alloys are from a family of ornamental cast brasses. They will polish well, but weather poorly if not protected. Like many of the high-zinc alloys,

TABLE 3-9. Common Tempers of Alloy C24000, Low Brass

TEMPER	ASTM NUMBER	YIELD STRENGTH (ksi)	(MPa)	ROCKWELL Hardness b
Quarter Hard	H01	40	276	55
Half Hard	H02	50	345	70

weathering will appear mottled and dark. These alloys are used as ornamental grilles, light fixture parts, door handles, and other ornamental features for which the rich yellow brass appearance is sought. These alloys will receive statuary finishes and other coloration processes well.

Each of these alloys has a yellow, brassy color; alloy C85700 appears most yellow. These alloys have similar strength characteristics and hardness. The addition of tin helps in corrosion resistance. The melting point of each alloy is near the same temperature, approximately 1,725°F (940°C).

Yield Strength   12 ksi   83 MPa

Alloy C85700 has a slightly higher yield strength.

Casting Techniques Available

Centrifugal ASTM B271	Alloy C85200
	Alloy C85300
	Alloy C85400
	Alloy C85700
Continuous	Alloy C85200
	Alloy C85400
Sand ASTM B584	Alloy C85300
	Alloy C85400
	Alloy C85700

Alloys C85300, C85400, and C85700 can also be cast with plaster and permanent mold processes.

Alloy C26000

Common Name: Cartridge Brass

Alloy Mix—Nominal Composition
Copper   70%
Zinc   30%

Wrought Alloy—Available Forms
Strip	ASTM B36
Sheet	
Tubes	ASTM B135
Shapes	ASTM B129

Natural Polished Color	Golden yellow
Oxidized Color	Dirty gold color
Weathered Color	Brown color with dark streaks

This alloy is not recommended for exterior use unless coated with a clear protective film. The high-zinc alloy is subject to pitting when exposed to severe environments such as the seacoast or urban industrial areas.

Cartridge Brass acquired its name from the munitions industry, where it is used for ammunition components such as shell cartridges. This alloy is

TABLE 3-10. Common Tempers of Alloy C26000, Cartridge Brass

TEMPER	ASTM NUMBER	YIELD STRENGTH (ksi)	YIELD STRENGTH (MPa)	ROCKWELL Hardness b
Quarter Hard	H01	40	276	55
Half Hard	H02	52	358	70
Hard	H04	63	434	82

very ductile and can be used where elaborate forming is required. It will take polishing and statuary finishing well.

Cartridge Brass is available in most of the standard cold worked tempers, reference Table 3-10.

Alloy C27000

Common Name: Yellow Brass

Alloy Mix—Nominal Composition
Copper   65%
Zinc   35%

Wrought Alloy—Available Forms
Strip   ASTM B36
Sheet
Plate
Tubes   ASTM B135

Natural Polished Color	Yellow
Oxidized Color	Dirty golden yellow
Weathered Color	Dark brownish gold with dark streaks

This alloy is not recommended for exterior use unless coated with a clear protective film. The high-zinc alloy is subject to pitting when exposed to severe environments such as the seacoast or urban industrial areas.

This alloy represents the yellow color limit of the copper alloys. It is very ductile and can be used where elaborate forming is required; it will also take polishing and statuary finishing well.

Yellow Brass is available in all of the standard cold-worked tempers, reference Table 3-11.

TABLE 3-11. Common Tempers of Alloy C27000, Yellow Brass

TEMPER	ASTM NUMBER	YIELD STRENGTH (ksi)	YIELD STRENGTH (MPa)	ROCKWELL Hardness b
Eighth Hard	H00	35	241	50
Quarter Hard	H01	40	276	55
Half Hard	H02	50	345	70
Hard	H04	60	414	80

Alloy C28000

Common Names: Muntz Metal
60-40 Brass

Alloy Mix—Nominal Composition
Copper   60%
Zinc   40%

Wrought Alloy—Available Forms

Strip   ASTM B36
Sheet
Plate   QQ-B-613
Tubes   ASTM B135

Natural Polished Color	Golden yellow
Oxidized Color	Reddish gold
Weathered Color	Dark reddish brown with mottled gray-green

Muntz Metal is used frequently by designers seeking an alloy that exhibits the elegant soft golden tone. Muntz Metal acquired its name from its inventor, George F. Muntz, who developed the alloy for the lining of boat hulls in 1832.

Muntz Metal is very durable. This alloy falls at the start of the alpha-beta transition, and thus its behavior under stresses varies from that of the ductile Yellow Brass and Cartridge Brass. Cold working is possible but difficult. The alloy will crack under severe cold-working operations. Welding and brazing this alloy is not an easy task. Discoloration around the weld and cracks will occur if the proper welding techniques are not used. Spinning and deep drawing are difficult because the alloy work hardens. Interstitial annealing is required to form severe shapes. The annealing generates oxides on the surface, which are difficult to remove. This alloy should be used for only limited cold-forming operations.

Under hot-forming operations this alloy performs well. Hot-forming processes are not commonly used for shaping architectural metal features. Hot forming will build up oxides on the surface, and discoloration of the surface will require postpolishing.

Using lower cold-rolled tempers does little to soften this hard alloy. Muntz Metal can be polished to a mirror finish. Satin polishes and custom polishing are also possible with this striking architectural alloy. It will take statuary finishes in all the available ranges. When used on the exterior of a

TABLE 3-12. Common Tempers of Alloy C28000, Muntz Metal				
TEMPER	ASTM NUMBER	YIELD STRENGTH (ksi)	(MPa)	ROCKWELL Hardness b
Soft Annealed	O60	21	145	80 (F-scale)
Hot Rolled	M20	21	145	85 (F-scale)
Eighth Hard	H00	35	241	55
Half Hard	H02	50	345	75

building, Muntz Metal should be protected with a clear coating. There are only a few available cold-working tempers. Muntz will rapidly work harden, and thus the range of tempers is limited, reference Table 3-12.

Alloy C38500

Common Names: Architectural Bronze
                      Leaded Muntz Metal

Alloy Mix—Nominal Composition

Copper   57%
Zinc   40%
Lead   3%

Wrought Alloy—Available Form

Extrusion

Natural Polished Color   Golden yellow
Oxidized Color   Reddish gold
Weathered Color   Dark reddish brown with mottled gray-green

The makeup of Architectural Bronze is similar to that of Muntz Metal, except for a small amount of lead. The lead is necessary to facilitate extruding. Architectural Bronze is available only in extrusion form.

In the natural unweathered state this alloy matches the color of Red Brass; it is also a close match to Muntz Metal. It will weather fairly rapidly to a dark-brown mottled color on unprotected exterior exposures. Architectural Bronze is commonly used for handrails, door frames, door thresholds, and other ornamentation that can be produced easily from extrusion processes. It can be mirror polished, satin polished, or given statuary bronze tones. Generally, this alloy is used for interior applications.

Architectural Bronze has good machining characteristics. The lead develops the free-cutting characteristic, which aids in machining. This alloy is available in only one temper. Reference Table 3-13 for yield strength and hardness.

The Bronzes

The copper alloys discussed in this section are those considered as bronze, based on their metallurgy. The bronzes contain alloying elements other than zinc. In architecture, the use of these alloys is limited. Not all the alloys of

TABLE 3-13. Common Temper of Alloy C38500, Architectural Bronze

TEMPER	ASTM NUMBER	YIELD STRENGTH (ksi)	(MPa)	ROCKWELL Hardness b
As Extruded	M30	20	138	65

the group are included here, only those that are manufactured in forms typically used in building construction.

The Tin Bronzes (Including the Tin-Zinc Bronzes)

The tin bronzes are perhaps the only true bronzes, or at least the only copper alloys that relate to antiquity, when the first purposeful alloying of copper occurred. These alloys are not common architectural alloys. If they are used, the cast forms of the tin bronzes are the predominate materials.

When copper is alloyed with tin up to the 10% level, it retains ductility and appears copper red in color. Beyond the 10% level, the alloys become very brittle and take on an orange-yellow color. The higher-tin-content bronzes are available only in the cast form.

These alloys weather to a gray-green patina over time, as attested by the rich green of cast statues. (However, many of these statues are of cast brass.)

The tin bronzes are usually cast for initial color. Many foundries have custom recipes for developing particular color and casting properties. Some of these recipes fall outside of current alloy designations. For example, a popular tin bronze known as Art Bronze contains 97% copper, 2% tin, and 1% zinc. This casting alloy has a deep, rich, dark red color when cleaned and polished. Another tin bronze used in architecture is Gold Bronze. This alloy is also a custom cast mixture containing 89.5% copper, 2% tin, 5.5% zinc, and 3% lead.

Alloy C91300, known as Tin Bronze 81-19, contains 81% copper and 19% tin and is used to cast bells. Not a common architectural alloy, this brittle metal is very hard and will not cold form.

Another popular tin bronze alloy is known as Cast Bronze. This alloy contains 86% copper, 7% tin, 5% zinc, and 2% lead. The appearance after cleanup and polishing is dark orange with a slight golden hue. Reference Table 3-14 for the various chemical makeups of these alloys.

TABLE 3-14. Cast Tin Bronze Alloys Used in Architecture

ALLOY	NAME	COPPER %	TIN %	ZINC %	OTHER %	COLOR
None	Gold Bronze	89.5	2.0	5.5	3% lead	Dark gold
None	Cast Bronze	86.0	7.0	5.0	2% lead	Dark orange
C93200	Tin Bronze	83.0	7.0	3.0	7% lead	Dark orange
C91300	Tin Bronze 81-19	81.0	19.0	0.0	0	Dark bronze

The Aluminum Bronzes

The aluminum bronzes are not common architectural alloys; however, they do possess certain properties that may be of interest to the designer. First, their color is a brown-gold, unlike that of any of the other copper alloys. They will weather to a dirty brown with streaks of white.

These alloys are among those with the highest strength of all the copper alloys. They have excellent corrosion resistance and are lighter in weight than the other copper alloys. They can be cold worked; however, they are about the only copper alloys that do not solder. Brazing is also difficult.

TABLE 3-15. Aluminum Bronze Alloys for Architectural Considerations

ALLOY	NAME	COPPER %	ALUMINUM %	OTHER	FORM
C61400	Aluminum Bronze	91	7	2% Iron	Sheet
C61500	Aluminum Bronze	89	10	1% Iron	Sheet
C95300	Aluminum Bronze	89	10	1% Iron	Cast

Aluminum bronzes are available in sheet, plate, pipe, tubing, and some custom drawn shapes. These alloys are available in two tempers, soft and hard. Recommended alloys for consideration in architectural applications are listed in Table 3-15.

The Silicon Bronzes

Silicon bronzes are alloys of copper that contain silicon as the main alloying constituent. These alloys are very strong and workable in the cold state. Initially, they resemble copper or the high copper brasses in color. There are many varieties and variations of silicon bronze; however, alloy C65500 is recommended for architectural use because of its economy and the available forms it can take.

Alloy C65500—Silicon Bronze. This copper alloy, C65500, consists of 97% copper and 3% silicon. Its natural, unweathered color is a reddish brown similar to that of commercial bronze, but with a slight rose tone. This alloy could be considered closest to true bronze, since it contains no zinc and its color is similar to that of the copper-tin alloys used by the ancients. Silicon bronze will weather to a russet brown color, eventually developing a mottled gray-green patina.

Silicon bronze has high strength as compared with other copper alloys. In the plate and sheet forms, these alloys can be cold worked without imbrittlement. They can be polished to a reflective specular color or given a soft, satin finish. The surface polish will tarnish rapidly if not protected. Silicon bronze is available in sheet, plate, pipe, tube, and drawn shapes.

Alloy C87200—Silicon Bronze. The casting alloy C87200 contains 91% copper, 4.5% zinc, and 4.5% silicon. Used for creating large cast sculptures by the lost wax method, this alloy has a low melting point and a good fluidity, strength, and corrosion resistance. Its color is a deep copper bronze, which takes to statuary finishing well. Unprotected, this alloy will weather to a gray-green patina.

The Nickel Silvers

The nickel silver alloys of copper are Old World metals with an elegant golden-silver color. These are the alloys of copper that contain zinc and nickel as the major alloying elements. Early alloys actually contained small measures of silver; however, their name derives from their color. (Other names given this metal are "German Silver," "White Copper," "White Bronze," "Paris Metal," and more currently, "Dairy Bronze."

This type of copper alloy was very popular during the early 1900s when the Art Deco style was in prominence. It was used with other copper alloys and with surface polish treatments to develop artistic designs in building entryways and spandrels. The Art Deco techniques used layers or inlays of other metals to develop soft transitions of color. Intricate castings of nickel silver would be selectively given mirror or satin polishes to enhance the appearance of depth.

Nickel silver contains copper, zinc, and nickel in various proportions. The more nickel, the whiter or more silvery the alloy appears. When polished, these metals take on a bright silver tone with a golden tint. As it weathers, a golden-brown hue initially develops. Eventually, the alloy will develop a soft gray-green patina with a mottling of brown tones.

Nickel silver has good corrosion-resistant properties. The alloy can be cold worked using conventional sheet metal equipment and is available in sheet and plate forms. Simple extruded shapes are possible with the leaded alloy C79600. Nickel silver can be cast using investment, sand, and permanent mold casting techniques. Cast alloys of nickel silver are C97300, C97600, and C97800. The C97600 alloy has a high fluidity when molten, permitting very fine detail in casting. Alloy C97800 has a high nickel content and is white-silver in color.

All forms of the alloy take polishing very well. Bright, lustrous finishes can be applied to the surfaces of the nickel silvers. There are more than 30 published alloy forms of nickel silver. For architectural and building construction use, the varieties listed in Table 3-16 predominate.

Figure 3-4. Cast nickel silver ornamental grille.

TABLE 3-16. The Nickel Silver Alloys Used in Architecture

ALLOY	NAME	COPPER %	ZINC %	NICKEL %	LEAD %	OTHER %	COLOR	FORM
C74500	Nickel Silver 65-10	65	25	10	0	0	Golden silver	Sheet
C75200	Nickel Silver 65-18	65	17	18	0	0	Silver white	Sheet
C79600	Leaded Nickel Silver	45	42	10	1	2% Mn	Golden silver	Extrusion
C97300	Cast Nickel Silver	56	20	12	10	2% Sn	Silver white	Cast
C97600	Cast Nickel Silver	64	8	20	4	4% Sn	Silver white	Cast
C97800	Cast Nickel Silver	66	2	25	2	0	White	Cast

AVAILABLE FORMS

Each of the copper alloys, because of its particular crystal makeup, has certain characteristics that make it better suited to a certain fabrication. Color, formability, and availability in the industry limit certain alloys to particular uses. Table 3-17 lists the various forms of the copper alloys. The following is a discussion of the common architectural forms of the different copper alloys and the limitations of those materials.

Strip

Strip is a cold-rolled, wrought form of the copper alloys. It is the name given to describe the form of the metal when it is in narrow-width coils. Basically, *strip* is a term for a long, narrow sheet. The maximum width of the strip of metal is 24 inches (610 mm). Lengths can vary, depending on the thickness of the copper alloy provided in strip form. Thicknesses available in strip form range from 0.005 inches (0.127 mm) to 0.188 inches (4.78 mm).

Strip forms of the metal are used in manufacturing facilities that stamp or punch shapes on continuous lines. Strips are also used by manufacturing facilities that roll form trim or shapes from a particular copper alloy.

The finish surface of strip material is characterized by fine grains developed through the cold-rolling process and the annealing process. The surface can receive polishing and buffing processes. Polishing very thin strip to mirror and specular finishes is difficult. The thin strip tends to warp.

Sheet

Sheet and plate are the most common forms of copper alloys used in architecture today. From the metal roof of commercially pure copper sheet to the Muntz Metal column cover fabricated from plate, this wrought form of the alloy has a preponderance of possible uses.

The term *sheet* is given to the wrought form of the copper alloy available in cold-rolled, thin panels. The panels are cut to dimension and provided in a stack of similar-sized sheets. The thickness of any given sheet ranges from 0.005 inches (0.127 mm) to 0.188 inches (4.78 mm). This is similar to strip thicknesses. What distinguishes sheet from strip is the length

TABLE 3-17. Available Forms of the Copper Alloys

ALLOY	NAME	STRIP	SHEET	PLATE	TUBE	EXTRUSION	SHAPE	CASTING
C21000	Gilding Metal	✓		✓	✓			
C22000	Commercial Bronze	✓	✓	✓	✓	✓	✓	
C22600	Jewelry Bronze	✓						
C23000	Red Brass	✓	✓	✓	✓			
C24000	Low Brass	✓						
C26000	Cartridge Brass	✓	✓	✓	✓		✓	
C27000	Yellow Brass	✓	✓	✓	✓			
C28000	Muntz Metal	✓	✓	✓	✓		✓	
C38500	Architectural Bronze					✓	✓	
C61400	Aluminum Bronze 91-7		✓	✓	✓		✓	
C61500	Aluminum Bronze 89-10	✓	✓					
C65500	Silicon Bronze	✓	✓	✓	✓			
C74500	Nickel Silver 65-10	✓						
C75200	Nickel Silver 65-18	✓		✓				
C79600	Leaded Nickel Silver					✓	✓	
C83600	Composition Bronze							✓
C84400	Cast Semi-Red Brass							✓
C85200	Cast Brass							✓
C85400	Cast Yellow Brass							✓
C85700	Cast Yellow Brass 405.2							✓
C93200	Tin Bronze 81-19							✓
C95300	Aluminum Bronze							✓
C97300	Cast Nickel Silver							✓
C97600	Cast Nickel Silver							✓
C97800	Cast Nickel Silver							✓
None	Gold Bronze							✓
None	Cast Bronze							✓
None	Tin Bronze							✓

limit and the available width. Sheets are available in lengths up to 12 feet (3,658 mm). Greater lengths are possible but not easily achieved. This length limitation is a variable set by handling and polishing limitations. It is difficult to skid and package great lengths, and polishing equipment has predetermined bed sizes.

Available width varies from one copper alloy to the next. The maximum widths of the different wrought copper alloys are listed in Table 3-18. Some of the maximum widths are currently available only from overseas sources, because domestic sources of the sheet form have limited facilities.

The sheet form of the metal can be polished and buffed similarly to the strip form of the copper alloy. Sheets are available in various cold-rolled tempers.

TABLE 3-18. Available Widths of Wrought Copper Alloy

ALLOY	NAME	MAXIMUM AVAILABLE WIDTH (INCHES)	MAXIMUM AVAILABLE WIDTH (MM)
C21000	Gilding Metal	49.2	1250
C22000	Commercial Bronze	49.2	1250
C22600	Jeweler's Bronze	24.0	610
C23000	Red Brass	49.2	1250
C24000	Low Brass	49.2	1250
C26000	Cartridge Brass	49.2	1250
C27000	Yellow Brass	49.2	1250
C28000	Muntz Metal	49.2	1250
C61400	Aluminum Bronze 91-7	49.2	1250
C61500	Aluminum Bronze 89-10	49.2	1250
C65500	Silicon Bronze	49.2	1250
C74500	Nickel Silver 65-10	29.5	749
C75200	Nickel Silver 65-18	29.5	749

Temper

Temper is produced in sheet and strip forms of the copper alloys by degrees of cold rolling and thermal treatments through annealing processes. Copper alloys in sheet and strip form are produced in large coils of metal at various thicknesses. These coils have undergone final annealing, so the grain size is fairly large and the metal is relatively soft. The coils are unwound and passed through rolls that reduce the thickness by applying pressure to the metal ribbon. The thickness of the metal decreases by degrees as each of the cold-rolled tempers are achieved. For instance, the Quarter Hard Temper requires a reduction in sheet thickness of 10%. The Extra Spring Temper requires a thickness reduction of more than 68%.

Temper correlates to a degree of hardness or softness, but the hardness will vary from one metal to the next for the same temper. Hardness is measured by the Rockwell or Brinell test. Reference Table 3-19 for the temper designations used on copper alloys.

Temper also correlates to the grain size produced by the cold-rolling operation. The grain size of a fully annealed sheet is approximately 0.120 mm, whereas the grain size for the Eighth Hard Temper is in the range of 0.015 mm to 0.025 mm. Grain size decreases as the level of temper increases.

Temper increases   →   Grain size decreases
                                       Ductility decreases
                                       Hardness increases
                                       Surface smoothness increases
                                       "Oil Canning" tendency increases

TABLE 3-19. Available Cold Rolled Tempers	
ASTM TEMPER DESIGNATION	BRASS MILL DESCRIPTION
H00	Eighth Hard
H01	Quarter Hard
H02	Half Hard
H03	Three-Quarter Hard
H04	Hard
H06	Extra Hard
H08	Spring
H10	Extra Spring

Plate

Plate is the term used to describe the hot-rolled, thick form of copper alloys. Before the sheet or strip form of the metal is reached through cold rolling and reduction, the cast block of metal undergoes plate hot rolling. Plates are large panels of the metal with a minimum thickness of 0.188 inches (4.78 mm). Maximum widths of copper alloys in the plate form vary, depending on the alloy. Most alloys are available from United States mills to a width of 36 inches (915 mm). European and Asian mills can supply plate in widths of 49 inches (1,250 mm). Nickel silver alloys are the exception. They are available in narrower widths, maximum 36 inches (915 mm).

Because plates are created by a hot-rolling process, the surface finish is not as good as cold-rolled sheet or strip. Surface streaking, pits, and surface roughness are acceptable characteristics of hot-rolled plates. The hot block of copper alloy is reduced in other rolls by forcing and stretching the metal through subsequently narrower gaps. As the material is spread out to the required thickness, all the oxides, pits, and other nonhomogenous materials are also spread throughout the sheet. Subsequent polishing of the sheet does not remove these imperfections. The surface of plates can be ground and polished, but the resulting quality and reflectivity does not reach the levels sheet or strip can achieve.

The temper available on plates is the "As Hot Rolled," which basically means that the temper is not a characteristic created or controlled by the hot-rolling process.

Table 3-20 lists the various copper alloys available in plate form and the corresponding ASTM specification.

Tubes

Tubing is available in round, rectangular, and some special polygonal shapes. Standard lengths are 20 feet (6,096 mm). Drawn shapes are thin-walled hollow cross sections that can be polished and used in railings, light fixtures, and other ornamental shapes that require reduced weight or less severe fabrication.

TABLE 3-20. Copper Alloys Available in Plate Form

ALLOY	NAME	ASTM SPEC	FEDERAL SPEC
C21000	Gilding Metal	ASTM B36	QQ-C-576
C22000	Commercial Bronze	ASTM B36	
C23000	Red Brass	ASTM B36	
C26000	Cartridge Brass	ASTM B36	QQ-B-613
C28000	Muntz Brass		QQ-B-613
C51100	Phosphor Bronze	ASTM B100	QQ-B-750
		ASTM B103	
C61400	Aluminum Bronze	ASTM B169	QQ-C-450
		ASTM B171	
C65500	Silicon Bronze	ASTM B96	QQ-C-591
		ASTM B100	
C75200	Nickel Silver	ASTM B122	QQ-C-585

Tubes are generally cold drawn. This means that the block of metal is pushed through a die at high pressure. For tubing, the die has a mandrel to support the inside wall of the tube. The metal is at room temperature for cold drawing, whereas hot drawing uses heated material to modify certain structural characteristics. Most copper alloy tubing used in architectural and ornamental applications is cold drawn.

The architect should understand that standard dimensional variations governing wall thickness, squareness, and straightness are allowed in ASTM B251 for copper alloy tubing. This standard allows for curvature of up to .5 inches in 10 feet (12.7 mm in 3.05 meters) and radius corners on rectangular sections. Closer tolerances are possible, but at a higher cost.

Standard commercial twist tolerances are also allowed in ASTM B251. One degree of twist per foot is allowed. If closer tolerances are required for a particular fabrication, it is wise to verify its availability.

Extrusions

Extruded shapes are an economical means of achieving particular complex shapes in aluminum. Certain copper alloys can also be extruded into complex configurations.

Extrusions are small cross-sectional shapes that are "pushed" through a shaping die. A billet of metal is heated and pushed under great pressure through a hardened steel die that has a cross section of the desired shape cut out of the die face. The shape is drawn out of the die in lengths in excess of 20 feet (6,096 mm). The extruded metal is stretched or tension leveled to straighten and eliminate any twisting of the shape. The hot shape is then allowed to cool.

The finish initially provided on extruded shapes is called "as extruded." This finish is characterized by small die lines running the length of the shape. Such wrought forms of the copper alloys can be polished and buffed. They are hot-formed products, but extruding leaves a fine grain structure running the length of the parts. Streaks and pits in the hot form are stretched and

Figure 3-5. Various brass extrusion shapes.

elongated as the metal is extruded, which tends to blend the imperfection down the length. Grinding of the surface is not advised for most extruded shapes; this is an operation best suited for flat surfaces. Many extruded surfaces are curved or shaped and do not facilitate grinding operations.

Handrails, angles, channel shapes, curtain wall frames, tubing, thresholds, and other kinds of hardware are common forms of copper alloy extrusions. Limitations of copper alloy extrusions are cross-sectional geometry, maximum dimension, quantity, and alloy type. Cross-sectional geometry is an inherent limitation of extrusions. Shapes do not necessarily have to be symmetrical, but must be balanced. That is, the volume of metal passing through the die extremities must be similar.

Another limitation of extrusions is overall size. A metal part must be able to fit within a circle of a certain size. The metal billet that is used to push a particular shape is of a maximum diameter. Typically, the maximum allowable circumscribed shape of copper alloys must fit within a 6-inch (152.4 mm) circle. The larger the shape being pushed, the thicker it must be. If larger parts are required, they can be built up from a series of extrusions.

Quantity is the most common restriction of extruded shapes. If an insufficient quantity of a custom shape is ordered, a metal extruder will not fabricate the extrusions or will charge a high price to make it worth the effort and setup costs. Extrusions are economy-of-scale fabrications. Extruded parts are consistent from one section to another but require that the entire billet of metal be pushed; otherwise it has to be recycled to form another part. Custom die charges are expensive. In fabricating large quantities of a particular part, the die charge is spread out over this quantity. In fabricating small quantities, the die charge, as well as the excess billet charge and setup charge, is concentrated in just a few feet of the section. Once the die is made, then the cross section of the part is established—not in stone, but in metal.

To change the configuration usually requires an entirely new setup and die charge; therefore, it is very important to review the die configuration prior to fabrication. Multiple assemblies mean multiple die charges.

Most extrusion manufacturers have standard shapes. Considerable cost can be saved if one of the standard stock shapes can be used for a particular fabrication.

There are various alloys of copper developed for the extrusion process. These alloys are as follows:

C31400 Leaded Commercial Bronze
C38500 Architectural Bronze
C46400 Naval Brass
C51000 Phosphor Bronze
C61400 Aluminum Bronze
C63000 Aluminum Bronze
C67500 Manganese Bronze
C79600 Leaded Nickel Silver

Each of these alloys contains small portions of lead. Additions of small amounts of lead are necessary when elaborate shapes are to be formed. Complex shapes require that the alloys be malleable when hot, allowing the metal to flow with reduced resistance into the extreme flanges and around grooves. These alloys are very strong and durable when cooled.

Shape

The term *shape* designates the bar, angle, channel, and other shapes produced by the cold or hot drawing method. Bar stock is available in round, square, rectangular, octagonal, and half-round cross sections. Some alloys are available in minor structural shapes, such as angles and channels. The shapes are manufactured by cold forming through repeated dies. The cross section of these shapes is small, no greater than 4 inches (101 mm).

Castings

Bronze and brass castings are commonly used in architectural and ornamental environments. Casting is the technique of forming by pouring molten metal into a form and then allowing the molten metal to cool and solidify. Copper alloys are some of the simplest metals to cast; whereas it is difficult to achieve good casting quality with pure copper. The surface of pure copper castings can be rough and coarse, and shrinkage cavities within the castings are prone to develop. Additions of alloying elements, even in minute amounts, will improve copper's castability.

Color matching with adjoining parts, shaping, hardness, and workability are limitations of castings with copper alloys. When casting a particular item, be sure that the caster is aware of how it will be finished and what areas will be exposed to view. Sample colors of the alloy can be obtained to verify the resulting color of the casting. If polished surfaces are desired, elimination of pits and air pockets on exposed surfaces is critical. In casting, the metal is fed into the mold from feeder arms. If the shape is relatively large, multiple feeders are necessary to prevent intermittent cooling of the part. Often, when such feeders are used, the molten metal is more turbulent and air or gases are trapped in the rapidly cooling part. These pockets of gas appear as pits in the metal when the fabrication is polished, which may require filling with metal filler to conceal them. These air pockets often occur throughout the part, so additional polishing and grinding will not eliminate them.

Color matching with other brass parts is very difficult since the alloys are so different. Statuary finishes help blend cast parts with fabricated sheet or extruded parts, but the variations of alloy makeup will cause the chemical reactions of the statuary finish process to create different shades. Matching a flat sheet of one alloy to that of another by applying the statuary finishes will also result in different hues. Statuary finishes do not conceal major blemishes, but tone down the color and reflectivity.

CASTING TECHNIQUES

In considering a copper alloy for casting, castability and fluidity are important characteristics. Castability is the ease of casting using general means, such as sand casting. Fluidity is the ability of a particular molten material to fill a mold cavity completely. If an alloy is difficult to cast, then sources and available casting techniques will be limited. If an alloy does not have good fluidity, then casting intricate forms and shapes may be impossible. The skill and practices of a foundry, however, are as important to the final result of a cast product as the fluidity and castability of the alloy. How the molten metal is fed to the mold, the care taken in mold design, and the control of the alloying constituents are variables good foundries understand. Reference Table 3-21 for a general listing of the castability of various copper alloys.

Figure 3-6. Pouring molden silicon bronze into dies.

TABLE 3-21. Castability and Fluidity of Cast Copper Alloys

ALLOY	NAME	CASTABILITY	FLUIDITY
C83600	Composition Bronze	Excellent	Fair
C84400	Semi-Red Brass	Excellent	Fair
C85200	Cast Brass	Good	Good
C85400	Cast Yellow Brass	Good	Good
C85700	Yellow Brass 405.2	Good	Good
C87200	Silicon Bronze	Poor	Good
	Gold Bronze	Good	Fair
	Cast Bronze	Excellent	Fair
	Tin Bronze	Excellent	Fair
C93200	Tin Bronze 81-19	Excellent	Fair
C95300	Aluminum Bronze	Poor	Fair
C97300	Cast Nickel Silver	Fair	Poor
C97600	Cast Nickel Silver	Fair	Poor
C97800	Cast Nickel Silver	Fair	Poor

The most widespread means of casting copper alloys today is the technique known as investment casting. Investment casting is a very old process that has been used since before the days of great pyramids. The Shang Dynasty in China (1766–1122 B.C.) used this technique to cast statues and sculptures from bronze alloys.

The investment casting technique is commonly known as "the lost wax method" of casting, or precision casting. There are two basic procedures used: the first is investment flask casting, which is seldom used today, and the second is the more common method, known as ceramic shell casting, or permanent mold casting.

In *permanent mold casting* a ceramic shell is formed around a wax model or pattern. This can be done by coating the wax model with a stucco slurry and letting it dry and harden into the mold. The wax is then removed by heating, to leave a hollow shell. Often this hollow shell is sectioned for ease of removal from the cast shape and to allow duplicate pours to be made. The molten metal is then poured into the mold and allowed to cool. Sometimes a rubber mold is made around the wax model, then coated with the ceramic material, thus making a permanent mold. The end result is a very close likeness of the wax model. Metal molds are often made for production parts to allow for ease of reuse of the molds. The metal can also be heated for better flow of the molten casting material into place.

Sand casting is also a common technique used for parts that do not require a smooth surface. The mold for this technique is made in a sand-filled box. The sand is a slurry containing a binder that hardens around the shape used for a model. The molten metal is poured from feeders into the sand mold and allowed to cool. The part or fabrication is removed from the box, and the sand is broken away. The resulting surface usually requires additional cleanup to smooth the coarse surface. This technique is the most economical and flexible. The molds cannot be reused, but are relatively inexpensive. Sand

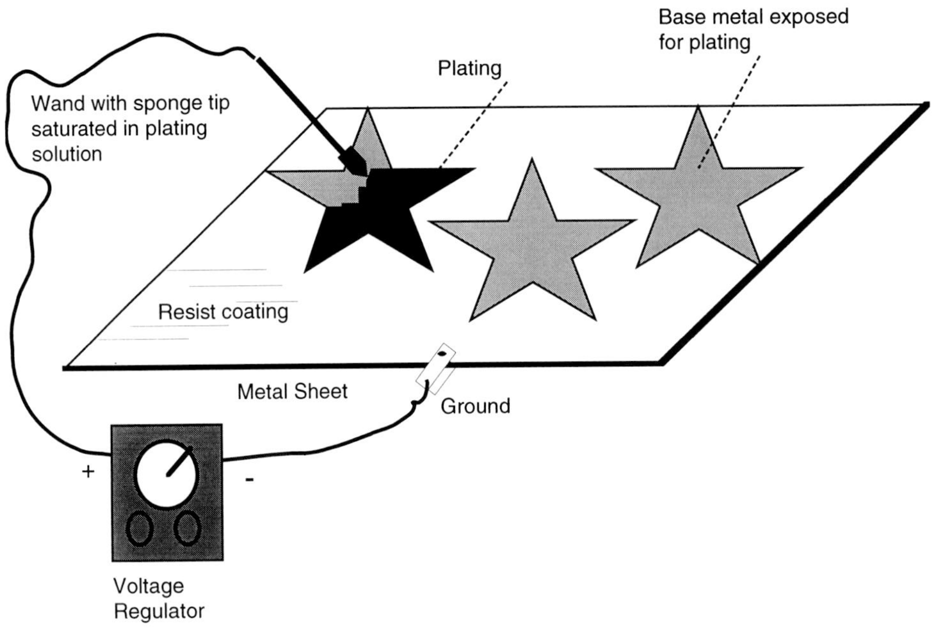

Figure 3-7. Selective plating technique.

cast molds can be made from old parts or previously cast panels. This allows for quick and simple matching and small-run potential.

Centrifugal casting, another technique for casting copper alloys, uses a revolving drum. The copper alloy is poured into a mold in the drum. As the drum revolves, centrifugal force takes the metal into the recesses of the mold. This technique has been used to produce small architectural features such as hardware and fixtures.

Continuous casting is used to form long shapes similar to extrusions. The cast shape is continuously removed as it solidifies. Length is not limited by a mold. Typically used for producing billets, slabs of alloy, and rectangular shapes, this technique has advanced to fabrication of more elaborate forms.

Casting is limited to certain alloys and shapes. Questions as to shrinkage and cost of the mold must be asked of the foundry before choosing casting as a technique for fabrication. Intricate designs can be accomplished by casting at far less cost than fabrication, but the cost of the initial model may prohibit small-quantity runs. The foundry that pours the metal must have knowledge of how the casting can be poured from different directions to eliminate air pockets and improper cooling.

CASTING OF BRONZE SCULPTURES

In ornamental application, copper alloys are best known for their use in cast statuary. A bronze sculpture is a huge mass of material. Yet the detail can be very intricate, down to the texture of hair and skin.

In casting a large sculpture, the first step is to construct a supporting frame. A modeling material is then applied to this frame to develop the features of the sculpture. The modeling material is usually clay. Once the shape and detail are established in the modeling material, a rubber compound is painted onto the surface. A thick layer or number of layers of rubber are applied to the sculpture surface. The rubber will harden and capture all the intricacies of the sculpted clay.

After the rubber has cured, it is removed from the original clay model and reassembled with an external frame or support. Molten wax is poured into this new rubber mold, producing a copy of the original clay construct. Changes can be made to this wax model, and duplicates can be produced. Wax bars are added to develop a channeling path for the copper alloy to enter the mold. This is known as a sprue system. Each wax section with sprues attached is coated with a ceramic slurry to develop a hard shell on the outside of the wax.

The wax is removed by steam, and the ceramic mold is heated and cured in an oven. The ceramic mold is then preheated, and the copper alloy is fed into the mold via the sprue system. Once the copper alloy casting has cooled sufficiently, the ceramic mold is chipped away and the sprue system is cut off the shape. For large castings separate sections are cast and welded together.

The resulting piece has rough edges and sprue marks, which are removed by grinding and filing. Artistic detail is refined into the copper alloy at the places where sections were joined and at recesses that may not have cast properly.

Next, the surface is cleaned and finished with the statuary coloring process. The ceramic mold leaves a smooth surface, but oxides must be removed before the statuary finish can be applied. Finally, a coating or series of coatings of Incralac are applied, followed by a coating of wax.

FINISHING

Good architecture often calls attention to some particular surface used in such a way that a subtle elegance is achieved. The natural tones of granite and wood can, when used effectively, produce this effect. The copper alloys, among all the architectural metals, are perhaps the best suited for achieving a natural elegance. Their various surface tones offer the designer natural colors to work with—colors that only copper and gold, among all the metals, possess.

The copper alloys can be left to weather naturally, or they can be enhanced and protected to sustain an intermediate appearance. An enhanced surface is maintained by application of a barrier coating to temporarily resist further oxidation of the copper surface. Without the barrier, further oxide development will occur.

There are mechanical finishes that can change the general reflectivity, and thus the tone of the surface, of the copper alloys. Chemical treatments can be applied, which accelerate the weathering process within controlled restraints to obtain a particular color and tone. This tone can range from the deeply oiled, rich statuary finishes to the deep verde antiques.

Additional chemical conversion coatings can be applied to the surface of the copper or copper alloy to develop various rich colors ranging from orange to black.

Mechanical Finishes

Mill Finishes—The As-Fabricated Finishes

Like other architectural metals, the copper alloys produced at the mill in sheet, plate, bar, or rod are provided in an "as-fabricated" or "mill" state. The mill state is an unpolished form, often rough and spotted with mill oil or surface oxide. The mill state is the surface and texture finish induced on the metal by the rolling and reduction of the sheet. This as-fabricated condition also denotes the surface of the metal when it is extruded or drawn, prior to any secondary polishing. This is the most economical finish, since it leaves the metal in an initial, unprocessed condition. It is not usually considered an architectural finish.

The surface condition of the mill finish varies, depending on how the metal is refined. Hot-rolled products, such as heavy plate and bar shapes, have relatively dull surfaces in the mill form. These heavy shapes are slightly oxidized and appear darker or discolored. Extruded products are usually clean of dark staining, but they can have die marks running down the parts. Die marks are induced by the fine imperfections in the die used to extrude the shape. In extrusion and plate form, the mill finishes are delivered unprotected to the fabricator or secondary processor. Sometimes they are interleaved with a protective paper to prevent scratching or gouging. Sheet and plate stock are furnished flat and straight and well crated against damage. All mill finish parts should be free of major scratches, wear marks, and water stains.

There are three variations of the as-fabricated or mill finish surface that can be ordered from the metal producer:

Unspecified
Specular
Matte

The unspecified mill finish is left up to the particular mill, which provides the metal with the characteristics induced by the hot or cold rolling processes. On cold-rolled products this can be a shiny, reflective surface or a somewhat dull surface if the steel rolls are dull. On hot-rolled products mill oil will be present on a dull, discolored sheet. Cast products will be rough and coarse in texture. This is the most economical finish for a copper alloy. Architectural products are rarely ordered in this finish.

A mill can produce a reflective finish by passing the sheet or plate material at room temperature, or in the cold state, through highly polished steel rolls. This finish may be on one or both sides and is known as a specular finish. On tubing, bar stock, and other shapes, this finish is achieved by grinding and minor polishing operations; it is not available on cast products.

A matte finish produces a dull surface on a copper alloy sheet, plate, extrusion, or casting. It can be achieved by hot or cold rolling, followed by annealing of the metal. On castings, this finish is produced by blasting with sand or other media. On extrusions, this is the finish left on the surface by the steel die. The finish is smooth and lacks reflectivity. Further protection or polishing is necessary to keep the finish from oxidizing. For extrusions, this is the standard finish provided to a fabrication or polishing plant.

The copper alloys typically are used with secondary finishes produced by either mechanical or chemical means or a combination of both. If major work is to be performed, such as brazing or spinning, then the fabricator usually purchases the material in the raw as-fabricated condition from the mill and produces the final finish after shaping, brazing, and forming of the piece. In this case the recommended finish to accept from the mill is either the specular or the matte.

Secondary Processes—The Polished Finishes

Generally, a mill product is scheduled to go to a processor prior to going to a fabrication plant. A processor is a facility that enhances the metal by first cleaning and degreasing the surface and then applying the final finish. This surface can be polished to produce a directional texture such as the satin finish or buffed and polished to produce a mirrorlike surface. The sheets or bar stock is protected in special PVC wrap or interleaved with paper, then packaged to order. The secondary processor does not stock a lot of polished copper alloys, but polishes a specific mill product to order. Copper alloys will tarnish during storage, even with the protective PVC wrap, so to keep a stock of polished copper alloy would require repolishing at some point.

The final fabricator often takes the mill product form if severe forming or welding is necessary. Once the material is fabricated to the specified configuration, the fabricator induces the final finish onto the surface using hand- or belt-sanding equipment. If a particular sheet metal fabrication has large areas untouched by the forming processes, prepolished sheets will be obtained and the surfaces marred by the forming process will be repolished to blend with or match the untouched areas.

Satin Finishes. Satin finishes can be either directional or nondirectional. These finishes are characterized by tiny scratches induced into the surface of the metal, or as minute indentations.

Directional Satin Finishes. The directional satin finishes are similar to the familiar No. 3 and No. 4 finishes available on stainless steel. The directional satin finishes are distinguished by tiny parallel lines or scratches on the surface of the metal. These scratches are applied by passing an abrasive roll over the metal surface or passing the sheet under a fixed abrasive roll. The edges of the sheet are clamped as this coarse-grained abrasive wheel is passed over the surface to produce a series of parallel grit lines.

Directional finishes are typically applied along the length of a sheet or strip. They can, however, be applied perpendicular to the length of the sheet or selectively applied to portions of mirror-finish or as-fabricated sheets. This custom application of a directional satin finish is more expensive because of the special handling and setup necessary to produce it. Selectively brushing requires the application of a plastic resist material. The resist material must withstand the abrasive action of the brush.

The types of directional finishes are as follows:

1. *Fine satin.* This finish is produced by wheel or belt polishing on sheet or finished fabricated parts. A 150- to 180-grit abrasive is used, sometimes

preceded with 100-grit belts. The finish is also produced by passing a Scotch-Brite[1] pad over the metal in parallel strokes. A fine satin finish is characterized by very fine parallel scratch lines and reduced surface reflectivity.

2. *Medium satin.* This finish is produced by wheel or belt polishing with a 100-grit to 150-grit abrasive. Characteristic of this finish are a more diffuse reflectivity and deeper parallel scratches than seen in fine satin.

3. *Coarse satin.* This finish is produced by wheel or belt polishing using an 80-grit abrasive. This finish has deep scratches and is rough to the touch. This finish is used initially to take down welds prior to application of medium and fine polishing.

4. *Uniform.* This finish is produced by a single pass of an 80-grit abrasive belt moving at 5,500 to 6,000 feet per minute. It is typically used on long shapes such as tubes, bars, and extrusions. The coarse grain produced works well on small cross sections with narrow exposure planes. On sheets, the coarse satin finish is often uneven in appearance. The light reflecting off the deep scratches makes certain regions appear darker than others, whereas on the small surface planes of extrusions and bar stock, the reflection is more consistent.

Nondirectional Satin Finishes. These finishes, applied in the fabrication or custom metal finishing shop, can produce an artistic effect on a metal surface, offering the designer unlimited options:

1. *Hand rubbed.* This finish is produced by pumice and solvent applied to the surface of a copper alloy with a fine brass wire brush. The pumice takes the sheen off the surface and produces minute scratches. However, the pumice slurry does not induce contamination in the surface. The finish achieved is dull and free of oxides. This finish can precede fine satin finish applications.

2. *Brushed.* A brushed finish is applied with nylon discs or abrasive cloth wheels of different levels of coarseness. This finish is used as a means of blending or joining other satin finish surfaces together at a weld, or it can be a custom finish on the metal.

Custom Brushed Finishes

• *Swirl.* This finish creates an interesting play of light by abrading the metal surface with a "band" of swirling grit lines. The different curves and the variable-angle grit lines break up the light and produce a three-dimensional effect. The swirl finish is produced by passing a grinder over the surface of the sheet or fabrication in a series of tight arcs. There are many variations of this hand-crafted custom finish. One interesting result requires the grinder to remain constantly on the metal surface, creating a continuous blend of abrasive swirls.[2]

• *Engine Turn.* An engine-turn finish is characterized by a pattern of overlapping circular grit lines. This custom finish is produced by a rotating

---

[1]"Scotch-Brite" pads are nonmetallic abrasive pads manufactured by the 3M Company, St. Paul, Minnesota.
[2]This finish was created in the A. Zahner Company shop by the architect, John Albright for the Sears Tower 2000 renovation project.

abrasive block or blocks moving over the surface of the sheet or fabrication. The blocks can be round or square and of different cross sections to produce various sizes of circles. A computer controls the placement of the circles by moving the sheet or the rotating block to a predetermined coordinate.

- *Distressed.* This custom finish is characterized by grind marks applied to the surface at various angles. To achieve this effect, a circular grinder with 80- to 100-grit abrasives is used. The edge of the grinder is brought down onto the surface at alternating angles to produce a texture of overlapping abrasive rectangles. Variations on this finish are innumerable. The additions of "whips" and swirls can create artistic effects.

- *Glass Bead.* This finish is produced by spraying fine particles of round glass balls against the metal, with the use of special blasting equipment. Varying degrees of texture with different matte lusters are obtainable. The finish is usually slightly mottled or irregular. Distortion of thin metal sheets can result from the impact of the abrasive media and the transfer of energy into the sheet. Castings often receive this treatment, beginning with a coarse texture of clean sand followed by a finer abrasive blast of glass bead.

The finish is best applied after a part is fabricated. The part should be of considerable thickness to resist the impact energy of the blast particles. All oxides, fingerprints, and grease must be removed prior to application of these finishes. Blasting over oily parts will permanently contaminate the surface.

Interesting effects can be obtained by masking certain sections of a mirror- or satin-polished part and blasting the exposed areas with glass bead. Removal of the masking material leaves a contrasting surface of matte and specular finish or matte and directional satin finish.

Using various abrasives on different sections of a part can produce subtle differences of tone. Experimental and control samples should be obtained when working with these custom finishes. Moreover, it is important to use caution when touching these finishes after an abrasive application, as they fingerprint easily. Use cotton gloves and apply a clear coating of lacquer, wax, or silicone rapidly after abrasive blasting.

Specular (Mirror) Finishes. The specular or "mirror" finish is characterized by a bright metallic luster of high reflectivity. Such finishes are induced on the surface by first polishing with successively finer abrasive wheels or belts, sometimes of leather, saturated with abrasive materials, followed by final polishing with 320-grit abrasive. The reflective surface is then buffed with rouge and large cloth wheels to bring out the high, reflective luster of the metal. There are two levels of specular finish, the smooth specular and the specular.

The *smooth specular* finish is the most expensive and the most reflective surface available. This attractive finish is acquired by buffing the surface repeatedly with fine abrasive and finishing with a large-diameter cloth wheel and jeweler's rouge. The smooth specular treatment removes most evidence of surface scratches and imperfections. However, pits and imperfections in plate and other hot-rolled products, as well as feeder air bubbles in cast products, cannot be polished out.

A *specular finish* is distinguished by a degree of minor surface imperfections and scratches left after abrasion. The finish is still very reflective, but in direct light or when viewed from an angle, small scratches are visible. This finish is less expensive than the smooth specular finish and can be more easily repaired in the field.

Another mirrorlike finish is the *buffed finish*. The buffed finish is achieved by successive polishing and grinding operations followed by polishing and buffing, as for the smooth specular finish, but with more extensive buffing. This finish is applied to hardware and fixtures. More expensive than the other finishes mentioned and characterized by a bright, lustrous tone, it is limited to small fabricated parts.

Polishing and buffing the copper alloy surface to mirror reflective levels actually causes the metal surface to flow. Friction heat at the surface takes the high points down and smooths the metal.

Electropolishing

Copper alloys can be electropolished to brighten the surface. Electropolishing uses electrochemical reactions on the metal surface to remove microscopic irregularities. This creates a clean, more consistent surface and, thus, a higher luster. Electropolishing uses phosphoric acid in an agitated heated bath. Nitric and sulfuric acid are also used to brighten the surface. An electrical current is pasted through the part or sheet. Electrochemical action removes microscopic surface irregularities to the order of 10,000 ü. This level is of the same magnitude as visible light. So as the surface irregularities are selectively dissolved into the electrolyte, the roughness of the metal decreases and the surface reflectivity increases.

Embossed and Coined Finishes

Patterned or embossed finishes are available on flat, thin sheets. These finishes are induced into the metal surface by passing the sheet, usually still in the coil form, between patterning rolls. The rolls have a design imprinted in them, one being the negative of the other. As the metal passes through the rolls, the design is imprinted into the metal. Designs vary, but the pattern always repeats or is continuous along the length of the coil. Sometimes the design occurs only on one roll, the other roll being smooth. This technique is known as "coining" of the metal surface. Coining produces a shallow design pattern that is pronounced on one side of the sheet.

Through the embossing of the sheet, the surface of the metal is stressed and thus additional stiffness is imparted to the sheet. Therefore, thinner sheets can be used. By breaking up the surface into patterns, distortions and surface imperfections can be concealed. This finish is available only on flat sheet metal.

Mechanical Finish Designations

Frequently the finish textures of copper alloy parts are mistakenly referenced to the American Institute of Steel designations for stainless steel finishes, such as a No. 3 or a No. 8 finish. The processes used to achieve the finishes on stainless and copper alloys are similar, and most users of the materials understand the requirements. However, it is recommended to qualify sample requirements to ensure the end results without confusion.

The Copper Development Association has adopted the Aluminum Association method of finish and texture designations because many of the same processes are used on all nonferrous metals. These finish designations will reduce the possibility of confusion and offer the designer a greater palette to choose from. A control sample is still recommended, since many of the finish designations vary slightly from one fabricator to the next. The list of finishes in Table 3-22 is not complete. The custom finishes are not listed, because they are considered handcrafted applications and are limited to custom-finishing operations.

CHEMICAL FINISHES

Chemical finishes are commonly used as architectural enhancements to the copper alloys. Artificial coloring of metal objects dates back to the third millennium B.C. Bronze statues, prevalent in ancient times, were thought to be colored by either chemical means, paints, or gilding. The Romans used colored bronze for decorative columns. The columns, as depicted in drawings and paintings, were oxidized to produce dark brown colors or gilded with bright gold leaf.[3] Japanese metalworkers have a tradition in chemically coloring copper alloys that dates back to the 1400s. The Japanese cast copper alloys by pouring the molten metal into molds submerged in water. The result was a rose-colored metal, the surface of which was enhanced with statuary colors or coated with enamels.

TABLE 3-22. Mechanical Finish Designations for Copper Alloys—NAAMM* Specifications

AS FABRICATED	MIRROR POLISHED	DIRECTIONAL	NONDIRECTIONAL
M10-Unspecified	M20-Unspecified	M30-Unspecified	M40-Unspecified
M11-Specular	M21-Smooth Specular	M31-Fine Satin	M41-(unassigned)
M12-Matte	M22-Specular	M32-Medium Satin	M42-Fine Matte
M1x-(to be specified)	M2x-(to be specified)	M33-Coarse Satin	M43-Medium Matte
		M34-Hand Rubbed	M44-Coarse Matte
		M35-Brushed	M45-Fine Shot Blast
		M36-Uniform	M46-Medium Shot Blast
		M3x-(to be specified)	M47-Coarse Shot Blast
			M4x-(to be specified)

*NAAMM: National Association of Architectural Metal Manufacturers.

[3]James Fergusson. *History of Architecture in All Countries*, Vol. 1. New York: Dodd, Mead and Company, 1883, p. 372.

The copper alloys, like pure copper sheet itself, continually change in color with time and exposure to the natural environment. Often, ornamental uses require the metal to achieve a particular level of oxidation and, then, for this level to be maintained. The desired level could be an intermittent oxidation stage known as a statuary finish or the green-gray patina. Picasso, it is said, used to bury his bronze sculptures below the pens of farm animals in order to take advantage of the natural oxidizing agents to generate a strong, stable patina. Today there are successful chemical treatments available for copper alloys to generate stable patinas and the brown statuary finishes of the intermediate color levels.

The single most important step in chemical coloration of a copper alloy is thorough cleaning of the metal surface. This initial step is a chemical finish in itself and consists of washing, dipping, or spraying the particular object in a chemical solution. Many of the pretreatment processes are proprietary, but the results are to clean the metal surface of oxide film, grease, foreign material such as fingerprints, and polishing compounds. This cleaning process should not etch the metal surface, but should remove only foreign particles adhering to the surface.

The cleansing process is made up of two steps; the first is degreasing. Degreasing by exposing the metal to vapors from solvents is known as vapor degreasing. However, vapor degreasing with chlorinated solvents such as trichloroethane is damaging to the environment and should be eliminated from consideration. Aqueous cleaning systems have been developed to replace the ozone-depleting solvent degreasing methods. These systems use agitation to remove films and foreign particles from the surface of metal parts. If the parts are small, fine fiber brushes and detergents should be used, followed by thorough rinsing in clean water. Dilute oxalic acid and finely ground pumice will remove stubborn oxides from the surface of copper alloys, and alcohol can be used to remove fingerprints and mild grease.

The metal can be immersed or sprayed with a heated solution of cleansers, followed by thorough rinsing with water. Alkaline cleansers such as trisodium phosphate or sodium metasilicate (4 to 8 ounces per gallon of water) should work in most cases to leave the metal surface clean and free of contaminants. A light tarnish may be apparent on the copper alloy surface; to remove this tarnish use a very weak acid and follow with a thorough rinse with fresh water.

The second step in the cleaning process is the removal of oxides. Heavy oxides on copper require acid for removal. Pickling in dilute sulfuric or hydrochloric acid will remove scale and excess lubricants. Oxalic and phosphoric acids in dilute proportions will also work effectively. An acid can etch the surface by removing portions of the metal. The thickness of the metal will be reduced slightly in the etching process, and the metal surface will not be as bright. Brightness can be restored by using mechanical means, such as rubbing with nonmetallic pads or by dipping in a nitric acid bath.

Observe caution when using acids. Proper eye care, clothing, and ventilation are essential. Have proper disposal plans in place for rinsing and removal of the waste material.

Try removal of oxides and foreign particles using mechanical means. It may save handling and disposal problems generated by the chemical approach. Abrasive blasting with glass beads and hand rubbing with a fine abrasive pad, such as a Scotch-Brite pad, with a pumice slurry can remove most oxides and prepare the surface for chemical coloring.

There are a number of methods available to put color into the copper alloys. All coloring techniques require the acceleration of the conversion coating or oxide film on the metal surface. The objective is to duplicate by chemical means the natural process of weathering, then to stabilize the finish so that a continuation of the process does not occur.

Statuary finishes are achieved by the formation of copper oxide, sometimes with a mixture of sulfides, on the metal surface. The film developed is similar to the natural oxidation process, with tones ranging from black to light gold. There are many methods to produce the statuary tones on copper and brass, and the best work is done by craftspersons familiar with the processes.

To produce a statuary finish on copper alloys, three primary methods are used:

1. Polysulfide compositions
2. Selenious acid compositions
3. Alkaline oxidizing compositions

The polysulfide compositions are the oldest techniques used to produce the statuary finishes on copper alloys. These compositions are dilute solutions of ammonium sulfide, potassium sulfide, or sodium sulfide. Potassium sulfide is also known as "livers of sulfur" and is commonly used to generate black to light brown tones on statues and other sculptures made of copper alloys.

The technique involved with the livers of sulfur and the other sulfides is to develop a solution containing approximately 5% to 10%, by volume, of the sulfides with clean, preferably distilled, water. Apply the solution over the entire surface of the copper alloy sheet or fabrication. Allow the solution to react with the copper surface for a few minutes. A darkening effect will occur. Rinse the solution from the surface using clean water. Once the surface has been rinsed of the sulfide, apply a solution of two ounces of oxalic acid in a gallon of water over the entire surface. Rinse the surface of the acid solution and repeat the application of the livers-of-sulfur solution. The surface will become dark as the sulfide reacts. The surface should be dark brown to black in color; a blotchy, inconsistent film is a sign of poor precleaning. Try rubbing any blotchy region with Scotch-Brite and reapply the solutions as indicated. Coloring experts will "cut the finish back" by rubbing the surface with Scotch-Brite pads in the direction of the grain. Cutting the finish back will lighten the surface by removing some of the oxide created by the darkening technique. The sulfide and acid solution treatment is repeated and then cut back to produce deep, rich statuary colors. Depending on the level desired, the colors can range from light browns to dark matte black. Application of ferric nitrate solution and heat after the blackening process will create dark reddish tones.

Selenious acid solutions are available from different chemical supply houses. These solutions work rapidly at room temperature to produce different statuary tones. Light brown to black colors can be produced by using mixtures of selenious acid, cupric sulfate, phosphoric acid, and zinc sulfate.

These premixtures are relatively expensive; however, they will create statuary finishes on clean copper alloy surfaces quickly and easily. As with the livers-of-sulfur mixture, the color produced on the surface should be cut back using Scotch-Brite pads to the desired level of statuary color.

Multiple applications of the chemical coloring compounds are required to achieve deep, rich statuary finishes. Cutting back of the surface film by using abrasive pads or pumice, followed by repeated applications of the chemical treatment, highlights the statuary finish and develops the rich bronze colors. The technique is to darken the metal surface beyond the desired color and cut back the surface to desired levels with an abrasive pad.

The alkaline oxidizing compositions consist of immersing copper alloy in heated baths of caustic soda, usually with admixtures of copper sulfide or copper acetate. These treatments can be performed on sheet and strip material or on tube and extruded shapes. The color produced can be a deep matte black or a brown tone, depending on the time of immersion. These hazardous strong oxidizing solutions develop cupric oxide or cuprous oxide films across the surface. The films are black matte in color and very adherent. The surface can be highlighted by rubbing with Scotch-Brite pads to tone the color.

The color of the statuary finish can be adjusted by controlling the amount of time the copper alloy is exposed to the various processes, the temperature of the processes, and the concentration of the compositions. Because of the number of variables involved, in situ applications of statuary finishes on large objects are difficult. In many cases, the results are unpredictable. Quite often, the finishes obtained require maintenance or additional barrier coatings to sustain any successful coloration. The oxide film will age on exposure if not maintained regularly. As a statuary finish weathers, streaking and surface blotching will develop if water is allowed to stand on unprotected surfaces.

A clear lacquer or oiling of the surface is required to stabilize a statuary finish and resist further natural oxidation. Lacquering will wear down over a period of time and will require removal and reinstallation. If oiling is used, application of linseed oil should be done monthly.

Because of the many variables involved, color samples should be prepared as control samples for matching purposes. It should be understood that variations from part to part will occur when chemical coloring techniques are used. The color samples must be made on the same alloy as the final fabrication. The natural color of the base metal, surface texture of the metal, thickness of the oxide film developed, and the physical homogeneity of the surface are all important variables that affect color. Cleaning the surface, however, is still most critical for a successful overall finish.

Statuary finishes are used on the copper alloys rather than on commercially pure copper. Copper will develop a brown oxide instead of a statuary oxide color. The smaller the surface area being treated, the better the uniformity achieved. This is because of the manual process used to apply the color. The best approach is to work on one area or sheet to achieve the desired color, and then match adjoining areas or sheets to the successful color. Use small samples as control coupons that can be positioned over the part being oxidized.

Variations in color produced will occur when different alloys receive the same treatment. For instance, Commercial Bronze alloys will color differently from Red Brass alloys even when the same treatment is applied to both metals. The oxidizing process is alloy sensitive. It is the variations in alloy constituents that cause variations in the color tones produced.

Patinas

The brasses will develop the green patina over time and exposure to sulfates and chlorides. The patina on these copper alloys is often mottled and contains dark spots where localized corrosion processes, such as dezincification, occurs. Most brass or bronze statues will develop a rich green patina, but it is almost always accompanied by the dark gray or black spots of these other corrosion processes.

Brass hardware on ships and boats or used on buildings along the seacoast will develop a green and brown patina rapidly. Surface pitting will occur if the brass is not cleaned frequently. Eventually, the entire outer surface of the brass will be coated with a rich dark green patina.

Prepatination Processes on Brasses and Bronzes

Like copper, the brass and bronze alloys can have an accelerated patina developed over their surfaces. Some of the artificial patinas can be very lustrous, unlike the dull gray-greens that occur naturally. They involve heating the object and applying pastes and slurries to cause the chemicals to react with the base metal. Successful patinas on brass and bronze alloys use heated copper nitrate solutions or pastes made from copper sulfate and copper acetate. The processes should be performed by professionals experienced with working with chemicals and corrosives.

Because many of the treatments involve heating the metal fabrication with a torch, large objects or objects with flat, thin planes are difficult to color successfully. Cast forms and smaller fabricated shapes will color adequately if cleaning is performed correctly.

The following technique can be used to develop a deep olive green patina on a dark background. The formulation is used on Yellow Brass, Cartridge Brass, and Muntz Metal.

1. Develop a solution of copper nitrate by dissolving 200 grams in approximately 1 liter of water.
2. Develop a solution of potassium polysulphide by dissolving 50 grams in approximately 1 liter of water.
3. Heat the fabrication or sheet using a torch.
4. While wearing a face shield and working in a well-vented area, brush the copper nitrate solution onto the surface. The surface should be hot enough to cause the solution to evaporate in approximately 20 seconds. Heat until the rapidly developing blue-green patina begins to turn black. Brush all excess solids from the surface.
5. Continue heating and reapply the copper nitrate until the blue-green patina appears again.

6. While the surface is hot, apply the polysulphide solution. As the color deepens, slow the heating and allow to cool once the color of choice is achieved.
7. Once cool and dry, apply a layer of wax.

Additional Coloring Possibilities

In addition to the statuary and patina colors, very lustrous tones of gray, green, yellow, orange, red, and purple can be developed on copper alloys. These alloys can achieve more attractive finishes than any other metal. The processes of developing such colors is similar to the statuary coloring process in that the surface must be clean and the process involves developing chemical conversion coatings on the alloy surfaces. There are many techniques, and the various recipes can be found in the extensive study on the subject by Richard Hughes and Michael Rowe.[4]

For the copper alloys, immersion of the part in a boiling solution of copper sulfate will produce colors in the red, orange, and purple range. A deep, rich black color can be obtained by immersion of a heated casting into a hot bath of copper sulfide. Caustic soda (sodium hydroxide) solutions will produce lustrous colors on the surface of copper alloys.

In all cases, these colors should be applied by professionals knowledgeable about working with chemical treatments and the hazards associated with such chemicals.

Producing consistent colors on large sheets and repeating a color from sheet to sheet are very difficult with all but the black and brown tones. Producing these colors on small objects is easier, and consistent results are more controllable.

For all the coloring processes on copper alloys it is necessary to apply a barrier coating over the conversion oxide or sulfide coating in order to prevent further oxidation of the surface. Even the mechanical finishes must be so coated. There are various clear coatings available for the copper alloys. Each has advantages and disadvantages, depending on environment and cost. The next section of this chapter discusses a few of the more popular coatings.

Specifications for Various Chemical Finishes Available for Copper Alloys

For specifying the chemical and clear coatings applied to sheet, plate, and extruded copper alloys, a system, similar to that defined for mechanical finishes, was developed. The designation for any of the specific finishes is a capital letter indicating the process and a two-digit number identifying the degree, type, and method of chemical treatment.

The four capital letter designations are:

M—Mechanical Finishes
C—Chemical Finishes

---

[4]Richard Hughes and Michael Rowe. *The Coloring, Bronzing and Patination of Metals.* New York: Watson-Guptill Publications, 1991, Reference plate 5. 117.

O—Clear Organic Coatings
L—Laminated Coatings

To use the system the specifier would need to identify the complete finish designation. For instance, suppose the finish desired is a medium statuary finish. The designation would be M32-C55.

- M32 specifies a medium satin directional texture.
- C55 specifies a chemical finish using the conversion coating produced by the sulfide method. The final color would be cut back to a medium statuary finish.

Reference Tables 3-22 and 3-23 for a listing of various standard mechanical and chemical finishes.

Hardware Finishes

Frequently, an architect or designer specifies a US number for a particular brass or bronze finish. Such a designation is from the United States Finishes Designation System. The Department of Commerce, Commercial Standard CS22-40, defined what a US standard finish was. The National Bureau of Standards stopped using the US designations for copper alloy finishes in the 1950s. Manufacturers of brass and bronze hardware, however, continued to use the US finish designations and developed samples of the finishes as they perceived them. Yet it was not a certainty that one US finish from one manufacturer would match that of another.

In an attempt to clear up some of the confusion, the Builders Hardware Manufacturers' Association (BHMA) has defined new classifications and developed representative samples of the different finish designations. These new hardware classifications can be compared with the old hardware classifications, as indicated in Table 3-24. The American National Standard for

TABLE 3-23. NAAMM Specification Codes for Chemical Finishes

NAAMM DESIGNATION	DESCRIPTION
C10	Cleaning—unspecified; at the option of the finisher
C11	Degreasing—organic solvent
C12	Chemically cleaned
C1X	Other—to be specified
C50	Patina (Ammonium Chloride Process)
C51	Patina (Cuprous Chloride Process)
C52	Patina (Ammonium Sulfate Process)
C53	Patina (Carbonate)
C54	Statuary Finish (Oxide Method)
C55	Statuary Finish (Sulfide Method)
C56	Statuary Finish (Selenide Method)
C5X	Other—to be specified

TABLE 3-24. Hardware Finishes on Copper Alloys

US NUMBER	BHMA SPEC NUMBER	CLOSEST SHEET/PLATE SPECIFICATION	DESCRIPTON
3	605	M22 O7x	Specular polished brass, clear coated
4	606	M31 O7x	Satin polished brass, clear coated
—	607	M31 C54	Satin polished brass, oxidized and oil rubbed
—	608	M31 C5x O7x	Satin polished brass, oxidized, relieved and clear coated
5	609	M31 C5x O7x	Satin polished brass, blackened, satin relieved, clear coated
7	610	M31 C5x O7x	Satin polished brass, blackened, bright relieved, clear
9	611	M22 O7x	Specular polished bronze, clear coated
10	612	M31 O7x	Satin polished bronze, clear coated
10B	613	M31 C54	Satin polished bronze, dark oxidized, oil rubbed
—	614	M31 C54 O7x	Satin polished bronze, dark oxidized, relieved, clear coated
—	615	M31 C5x	Satin polished bronze, dark oxidized, relieved, waxed
11	616	M31 C5x	Satin polished bronze, blackened, satin relieved, clear
13	617	M31 C5x	Satin polished bronze, dark oxidized, bright relieved, clear
19	622	C5x O7x	Flat black oxidized, clear coated
20	623	C5x O7x	Light statuary bronze, clear coated
20A	624	C5x O7x	Dark statuary bronze, clear coated

Materials and Finishes, ANSI A156.18, has incorporated the BHMA complete listing of finishes for hardware.

For example, a designer may have chosen a hardware for doors and hinges with a satin brass color finish to match other brass ornamentation, such as light fixtures or door panels made from brass sheet. The architect may designate the finish for all items as a "US 4." The sheet brass material used to manufacture the light fixture and door panel should be specified as a satin finish, M32, with a clear coating of lacquer or urethane.

Samples should be required for approval prior to ordering any part or sheet. The architect, hardware supplier, and fabricator of the light fixture and door panel must see the same resulting finish and agree to the acceptable variations prior to having all the parts arrive at the project.

CORROSION CHARACTERISTICS

Galvanic action will create many different types of corrosion that are electro-chemical in nature. Dezincification corrosion, deposit and impingement corrosion, crevice corrosion, and stress corrosion are all types of localized galvanic corrosion.

Dezincification

A broad type of corrosion, dezincification, affects the brass alloys. Dezincification is generally localized to small areas on the brass surface. The brass is dissolved as an alloy and stripped of its zinc constituent. The copper is then redeposited onto the surface of the metal.

In very acidic exposures the entire surface of the brass may be attacked and undergo what is called "layer-type" dezincification. Brasses with less than 15% zinc are resistant to dezincification corrosion. Copper alloys such as Muntz Metal, Yellow Brass, and Architectural Bronze are subject to dezincification corrosion because of their high zinc content.

Severe dezincification will make it impossible to restore the surface of the metal. Extensive pitting and discoloration caused by dezincification on the metal surface render the appearance less than acceptable. Costly grinding and repolishing are required to restore the surface to a close equivalent of the original finish.

Pitting Corrosion (Deposit and Impingement Corrosion)

Pitting corrosion is a type of galvanic attack whereby localized cells of electrical polarity are set up. The attack of the metal in this instance is characterized by cavities or pits in its surface. The metal surrounding the pit is relatively untouched.

One form of pitting corrosion is *deposit corrosion*, which occurs when detrimental foreign matter is allowed to come in contact with the brass surface. The foreign material is porous and allows a corrosion cell to develop between itself and the copper portion of the alloy. This type of corrosion can be prevented by periodic maintenance of the copper alloy surface. The damage done by deposit corrosion will necessitate cleaning and repolishing of the surface.

Impingement corrosion is another type of pitting corrosion occurring when a solution is allowed to flow over a metal. The protective natural film is prevented from developing on the surface of the metal, and the base metal is allowed to corrode continuously. Impingement corrosion attacks all copper alloys and is also known as erosion corrosion.

Line corrosion, a type of impingement corrosion, occurs when moisture is held against a surface of a metal, for example, by capillary action in a seam or when moisture stands in a vessel at a constant level. The corrosion is characterized by a line at the moisture-level point.

Stress Corrosion Cracking

Stress corrosion cracking is the most irreversible type of corrosion, because it involves failure of the metal's structural stability. Stress corrosion cracking occurs when copper alloys are subjected to stress, moisture, air, and ammonia.

Stress is usually internal rather than generated by applied loads. Internal stresses may be produced by cold working a metal into a particular shape or configuration or by welding and polishing. The ammonia is atmospheric, and the amount required is very small. It could come from building materials or decaying matter.

Stress corrosion cracking can occur along the intergranular paths or cut right across the grains of the metal. This type of corrosion occurs in the brasses containing more than 15% zinc.

Incorrect assemblies of copper alloy fabrications, without regard to expansion and contraction from temperature variations, will build up stress and weaken a metal. This weakening can lead to corrosion fatigue failure, which appears as cracks and splits in the metal skin.

CARE AND MAINTENANCE

Clear Coatings

Quite often, a design using copper alloys requires the color, either the natural polished finish of the statuary or patina coloring, to be maintained at some desired level. The dark rich brown statuary finish on a cast statue or the bright reflective gold color of a Muntz Metal column cover are examples of copper alloys that are intended to remain as first installed. The installed color and finish are to remain, without developing into the green patina or dark brown oxides that nature constantly seeks.

Many factors affect the performance and appearance of the clear coatings used to protect such finishes: Some of handling, dust, abrasion, humidity, ultraviolet radiation, and pollution. It may be preferred under certain circumstances to use a coating that can easily be removed and refinished at a later date. Other more durable coatings will, as deterioration occurs, create a more unsightly condition, with spotty decay intermixed with glossy areas. In such cases removal and repair of the surface are difficult and costly.

Clear coating of the statuary finishes requires some basic conditions. First, the surface must be clean. For highly polished brass surfaces, this means all residual polishing compounds must be removed. Polishing compounds can become engrained in the surface; these must be removed for a successful coating application.

All clear coatings should be accompanied by oxidation inhibitors. Benzotriazole, an excellent inhibitor of copper oxide development, forms a very thin layer, one molecule thick, over the surface of copper. Corrosion protection using benzotriazole is extended to the copper alloy surface and to the conversion coatings developed in the chemical coloring processes. Benzotriazole is insoluble and will inhibit corrosion of copper alloys in many atmospheres, even in water, for a short period of time. It can be added to lacquer coatings and wax coatings to prevent corrosion under the clear film. Benzotriazole is simple to apply and should be incorporated into any coating for ornamental copper or copper alloys. To apply benzotriazole, dissolve about 40 grams in a gallon of water. This will produce approximately a 0.25% solution. Wipe the solution over the surface at room temperature. This should be the final cleaning step used on copper alloys. For more lasting protection, consider heating the solution to approximately 140°F (60°C) and letting it sit on the sheet or surface of the part for up to two minutes.

Once the brass or bronze surfaces are cleaned and awaiting the final clear coat application, protect the sheets with interleave paper saturated in benzotriazole. This form of protection will work for months under most normal exposures. The interleave paper should be soaked in a solution of approximately 1% benzotriazole. Let the paper dry and interleave it between the copper alloy sheets.

Benzotriazole should also be added to lacquers or waxes used to protect a copper alloy sheet. Often with the use of lacquers and waxes, small brown spots will develop under the surface within a few days or weeks of applying the coating. Initially, the coating will seem to be sound and the highly polished surface will appear clean, but soon small brown specks will become apparent just beneath the clear coating. Adding benzotriazole, up to 10%, to waxes and lacquers used to coat the metal will prevent this deterioration.

Incralac, a lacquer that contains benzotriazole, was developed by the International Copper Research Association for protecting copper and copper alloys exposed to exterior environments. This lacquer will work extremely well on interior applications as well. Incralac is a wet-applied acrylic lacquer with more than 25 years of successful results. Like all clear coatings, it eventually will degrade and require removal and reapplication. Depending on the severity of the environment, Incralac can be applied in a series of coats. For statues in urban environments, often up to eight coats of Incralac are applied to the surface. After Incralac application, a final protective coating of wax is used. The wax brings out the highlights in the bronze sculpture and aids in initial corrosion resistance.

Among the variables involved with clear coatings is the environment of the application. In controlled environments, such as a shop or paint booth, adequate time and preparation can assist in the application of the clear coatings. However, so many applications, particularly the second and subsequent applications, must occur in place. For those applications, air-drying coatings are the only option.

The thermoset or baked coatings perform well on small fabrications or castings. On sheet or large flat applications, these coatings are difficult to apply. Warpage of the sheet resulting from the temperature needed to catalyze the solvents may destroy the fabricated part.

There are waterborne top coats for developing thick clear coatings on hardware and small copper alloy fabrications. The coatings can be as great as 1 mil thick and are composed of acrylic urethane. These coatings are very durable and chemical resistant. The waterborne coatings replace the high-solvent lacquer coatings and require a cure temperature of 325°F (163°C).

With the statuary surface finishes, imperfections in the clear coatings are not as obvious or as critical. Minor surface tarnish will not be visible, since these finishes are accelerated oxidation processes. It is important for such surfaces to be passivated by the coatings so that further oxidation does not occur or that abrasive wear and tear is borne by the clear coating rather than by the colored coating.

Many architectural applications require the shiny, untarnished brass or copper surfaces to remain. These lustrous metallic surfaces reflect all imperfections in the coating. Imperfections such as "orange peel," surface tarnishing, and fingerprints are usually deemed inappropriate and require cleaning and reapplication of the clear coating. Coating highly polished surfaces of a metal with any of the clear coatings can be a nightmare for the unfamiliar shop. Clear coatings can go on smoothly and evenly, only to congeal and orange peel as they dry. Brown tarnish spots that were not there prior to the application of the clear coating can pop out under the coating as degradation of the copper surface is caused by reactions of the solvents remaining in the paint. Dust particles in the air can settle on the still-tacky clear coating during the time necessary to dry. The particles become imbedded in the surface as the film dries. Clean-room filtration systems, necessary to minimize dust contamination, can be a costly requirement in the application of clear films.

Tarnish inhibitors should be applied as soon as possible after a polished surface is brought to the required luster. Dust should be kept to a minimum both during and after application of the clear coating. When a satin polish is desired, use only silicone carbide pads (Scotch-Brite) or stainless steel pads. Steel pads, steel wool, and other wool pads may contain chemicals that affect the surface or leave steel in the surface to react with the copper. Pumice stone in slurry of 5% oxalic acid solution, rubbed on with a clean cotton cloth, is a good means of achieving a finish surface polish that will not generate a corrosive reaction, but will remove surface tarnish prior to the application of the protective coating.

All polish residue must be removed prior to application of tarnish inhibitors and final clear coatings. Moderate temperature and low humidity must be sustained during the coating process. High humidity will adversely affect the results.

For statuary finishes, and even some patinas, oiling of the surface is recommended. Mixtures of 5% to 10% lemon oil USP or Lemon Glass Oil blended with a high grade of paraffin oil may be rubbed into the metal surface. This should be repeated until the surface darkens evenly. For good results, this coating should be maintained by periodic waxing and a schedule established for complete removal and reapplication of the coating.

Reference Table 3-25 for various clear coatings and their relative performance.

Precautions—Chemical Coloring and Clear Coatings

Because of the variables involved in chemical coloring, matching precisely a sheet or plate fabrication to a hardware part can be difficult. The alloys may not be of precisely the same composition. The sources of the parts may be domestic or foreign. Plate and extruded parts, even of the same alloy, will color differently owing in part to the nature of their grain structure and

TABLE 3-25. Various Clear Coatings Available for the Copper Alloys

CLEAR COATING	RELATIVE COST	APPLICATION	ENVIRONMENT	UV RESISTANCE	REMOVAL AND MAINTENANCE
Microcrystaline wax	Low cost	Air drying	Interior	Poor	Not difficult
Nitrocellulose	Low cost	Air drying	Interior	Poor	Not difficult
Acrylic	Moderate cost	Air drying	Exterior	Good	Not difficult
Epoxy	Expensive	Thermoset	Interior	Poor	Difficult
Silicone	Expensive	Air drying	Exterior	Good	Not difficult
Alkyd	Low cost	Thermoset	Exterior	Good	Difficult
Acrylic Urethane	Moderate cost	Thermoset	Exterior	Good	Difficult
Cellulose Acetate Butyrate	Moderate cost	Air drying	Interior	Poor	Not difficult

distortion during fabrication and processing. A range of colors should be submitted and approved prior to fabrication, and the potential problems of exact color matching should be understood by the designer and the fabricator.

Clear coating copper alloys can be very difficult, as mentioned previously. Some of the maladies associated with clear coatings are listed in Table 3-26.

TABLE 3-26. Clear Coating—Troubleshooting

PROBLEM	PROBABLE CAUSE(S)	SOLUTION
Dark spots under coating	Polishing compound left on surface. Surface not cleaned. Curing of coating oxidation on sheet.	Remove coating, reclean, and apply a tarnish inhibitor such as benzotriazole before recoating.
"Fish eyes" on coating (minute gaps in coating)	Surface of sheet not completely cleaned. Possible presence of oil.	Remove coating and reclean. Use a cheleating additive to even coating.
Dust in coating	Dust particles in air deposit on the surface as coating cures.	Remove and reapply. May need a cleanroom environment.
Orange peeling	Coating thickness.	Sand and apply a thin second coat, or remove and apply a thinner coating.
Premature weathering	UV inhibitor not used.	Strip the coating and clean surface. Reapply using Incralac or other coating with UV inhibitor.

Aesthetic Problems Associated with Internal Stress

Directional properties develop in the copper alloys as they do in other cold-rolled sheet and strip materials. These directional properties occur at the grain level of the metal. The grains of cold-rolled metals break up and elongate in the direction of the cold-rolling process. These grains undergo recrystalization by subsequent annealing. This causes the grains to enlarge, but does not eliminate the directional nature of the grains.

The results of this directional tendency become apparent when the metal is shaped and formed. For instance, in deep drawing of copper alloy parts, a malady called "earing" occurs. This is a condition evident when the effort of shaping the metal into a dish or cup causes an uneven relationship around the rim. Such a development results from the uneven distribution of stress through the sheet.

Another malady that develops when deep-drawn processes are performed on copper alloys is a surface defect known as "orange peel." This condition is characterized by a coarse or rough surface similar to the surface of an orange peel. This condition can be rough to the degree that it will not polish out. Orange peeling can be eliminated or reduced by using a finer grain size.

A condition that becomes apparent when a copper alloy surface is mirror polished is known as "flow lines." This condition appears as waves when viewed at a slight angle. The waves are not apparent to the touch, but only when viewed as a reflection.

Oil Canning

A further surface irregularity common to metal sheet or plate is the tendency of a large flat section to "oil can." This problem is not unique to the copper alloys, but because of their highly reflective surfaces, it appears more pronounced in copper than in like thicknesses of other less-reflective surfaces. Bright, flat surfaces will show minor distortions more intensely owing to the reflection of light being concentrated and directed back to the eye at different angles. There is no substitute for metal thickness when flatness is a major requirement. The thicker the sheet metal, the flatter the appearance will typically be. Thicknesses less than 0.10 inches (2.5 mm), when specular polished, may show surface distortions or oil canning. Laminating a rigid backing will help, but probably not completely eliminate, the deflection of the surface. Using a satin texture will reduce the oil-canning appearance by causing the light to be more diffuse as it is reflected back to the eye. Texturing, embossing, adding shape, or stiffening will actually reduce or eliminate the surface deflection. If flatness is necessary, then keep the thickness greater than 0.10 inches (2.54 mm) on a specular polished sheet.

One reason for oil canning is the uneven distribution of internal stress throughout a sheet. Uneven stress is generated when the sheet or strip undergoes cold working. The process of cold working develops tension at the surface of the sheet and compression at the center. Thus, the stress distribution across the sheet is not consistent; the center of the sheet may have different stress distributions than its edges. Such uneven stress can cause localized bending of a thin sheet.

Storage Stains

Copper and copper alloys can develop stains during periods of storage. A humid environment will cause staining to occur in a relatively short period of time. Copper sheets, even when stored in coils, can develop a purplish-to-yellow mottling across their centers. The stain is an oxide interference film that will eventually blend in and form the characteristic brown color of the oxide. Polished brass in storage will tarnish rapidly, even under a protective plastic film. Air, trapped under the protective film, continues to tarnish the brass surface, sometimes making small brownish spots. After a few months of storage, the surface will require repolishing to the original luster.

If water sits on the surface of any of the copper alloys, it will stain and necessitate extensive refinishing. Water stains will leave a mottled, almost burled walnut, appearance on copper alloys. This stain will take months of exposure to blend with the rest of the copper alloys. Copper alloys, if allowed to get wet, will develop black spots, which are very difficult to remove and will make extensive repolishing necessary.

Maintenance of Brass and Bronze Fabrications

Brass and other copper alloys should be periodically maintained. Lacquers should be removed when they show signs of yellowing and cracking. Waiting for the copper alloys to develop the blotching stains will necessitate major, often expensive, cleaning and polishing. Relacquering should be performed when the surfaces have been properly neutralized of all cleaning and polishing agents.

A periodic maintenance program should be established in order to sustain the surface appearance of quality copper alloy work. The quality of clear coating applied over a copper alloy surface will dictate the frequency of necessary maintenance. Maintenance usually requires removal of the clear coating and repolishing of the surface, followed by a field-applied lacquer. For best results, a heated lacquer application system can be used. Exterior lacquer coatings should be maintained at least once every five years. For heavy traffic areas, consider a program of more frequent maintenance.

Bronze statues are not often exposed to handling or severe environments. They usually receive a conversion coating, followed by multiple coats of lacquers. These surfaces should be expected to last as long as 10 years before removal, cleaning, and repolishing.

Removal of Patina

Occasionally, because of unevenness or aesthetic considerations, the natural patina on a copper alloy has to be removed. To remove the patina, consider a sponge bath over the surface with a mixture of six parts phosphoric acid, one part nitric acid, and 50% distilled water. The pH should be in the range of 1.0 to 1.5. Use a binding agent to develop a paste, and allow the paste to remain on the surface for approximately one minute.

Remove the waste by sponging the surface with sponges soaked in a solution of sodium bicarbonate. The solution should have a pH of 10. Follow this sponging with a thorough fresh water rinse. This should remove the green patina. Repeat the process for stubborn areas.

For bronze and brass sculptures or large, rigid objects, use high-pressure water and detergent spray. Follow with pulverized walnut shells to remove encrusted oxides and corrosion particles from the surface. On thin sheets, the high-pressure spray may distort the shape of the panel or object. On sculptures and other thick fabrications, the pressure is not of consequence. The walnut shells are applied with an abrasive-blast machine at a pressure of 40 to 80 psi (0.27 MPa to 0.55 MPa units). This should result in a metallic brass or bronze appearance. To resist rapid oxidation, coat the fabrication quickly with the statuary finish or spray it with a dilute solution of benzotriazole.

WELDING AND JOINING

Many of the copper alloys can be joined using welding and soldering techniques. Welding techniques used on the copper alloys are:

OAW
SMAW
GTAW
GMAW

Use argon gas for thinner sections and helium gas for thicker members. The GMAW method is recommended to reduce distortion because of the reduced heat involved. Repair castings using the GTAW method of welding, with a mix of 3 to 1 helium to argon.

For the copper alloys, the color of the weld is most critical. Brasses require lower heat and power in welding; otherwise, they will crack after cooling. Use silicon bronze filler rod or phosphor bronze rod for strength. Naval brass alloys are also good fillers. Experiment to obtain the correct color.

For the Yellow Brasses, use GTAW and GMAW welding methods with copper silicon fillers and argon shielding. Zinc fuming will damage the weld.

Lead in alloys will cause the weld to crack; welding of leaded alloys is difficult at best.

4

Iron, Steel, and the Stainless Steels

The cast irons and wrought irons, along with the steels and stainless steels, have the element iron as the main alloying metal. In architecture, iron and steel are superior materials with unmatched strength and hardness. At the beginning of this century, large amounts of chromium were alloyed with the steel alloys to create a metal that had the structural strength and stability of iron and steel but also had the ability to resist the corroding effect of the environment: stainless steel.

This chapter will discuss first the history and use of iron and steel and finish with the history and use of stainless steel, the metal that has moved into prominence as an ornamental architectural metal.

IRON AND ITS ALLOYS

INTRODUCTION

Iron, particularly in the ubiquitous form of carbon steel, is a common architectural metal. Protective paints and coatings of other metals conceal the iron alloys' true appearance under oxygen-inhibiting barriers. This workhorse of metals, with all the strength of Superman, has little tolerance for nature.

Steel is the name given to metals whose principal element is iron. Steel is commonly used as a decorative thin metal skin or as enhanced ornamental structural material, but in all instances, only when painted or coated. Coatings of zinc and paints provide steel with the ability to withstand the environment while giving strength to a particular ornamental shape or form.

Iron and steel have a specific gravity of 7.87. This puts the metal in an average category, heavier and denser than aluminum but not as heavy as copper or lead. Iron is not found naturally occurring in its pure state, but its oxide and sulfate forms make up one twentieth of the earth's crust. Iron, usually combined with nickel, is found within meteorites.

Iron is one of the Second Transition Metals in the Periodic Table of Elements. There are nine of these metals; nickel, cobalt, platinum, and palladium are a few of the others in this category. These elements often occur in the pure form in nature, with the exception of iron. Iron, however, along with the other Second Transition Metals, are found in levels of high purity within meteorites. Meteorites most likely accounted for the initial availability of pure iron used by early civilizations. Early Egyptian tools made of iron contained nickel, implying a possible meteorite source. The British still refer to the study of iron and steel metallurgy as "siderurgy" from the Greek term for iron, *sideros*, which means "from the heavens."

HISTORY

Implements made of iron date back to about 2000 B.C. These implements were manufactured from iron recovered from meteorites. It is unlikely that humans had the skill or knowledge of iron refinement from mineral sources at this time. The art of refining iron, or smelting, developed from enhanced techniques used by the ancient metalsmiths to refine copper and bronze. Some of the first indications of iron smelting are found in the area along the southern shores of the Black Sea, an area known as Anotolia, now Turkey. The sands around the Black Sea contained rich deposits of magnetite, or ferrous oxide. The Chalybes, a tribal people in the region, experimented with and developed techniques for iron smelting. The Chalybes were under the influence of the Hittite empire. The Hittites are credited with introducing iron to the regions around modern-day Turkey and Mesopotamia, mainly by conquering other tribes. The Hittites were known for their warring nature and, for a while, probably experienced an advantage over their bronze-wielding adversaries.

Early processes of ore refinement entailed layering charcoal in a furnace, with iron ore intermixed. As the heat built up within the furnace, carbon monoxide passing over the hot ore acted to refine the mineral. As temperatures reached the 1,200°C to 1600°C levels, iron released from the oxide absorbed some of the carbon. Oxygen was kept from coming in contact with the molten mass by the layers of red-hot charcoal. The mass was removed and hot worked into layers by hammering. Slag waste and silicates were intermixed with the forged mass of metal as it was hammered and folded.

The absorption of carbon into the iron mass became a critical feature in early ore refinement. The Chalbyes were experts in this technique, developing different ways of adding carbon and rapid quenching to increase the hardness of the metal.

As the people in the region became more experienced with the working of iron and steel, special hard blades and weapons were developed. Damascene steels were forged blades of steel that had exceptional hardness and durability. This almost mythological steel was well known for its hardness and decorative surface of wavy lines.

The ancient Egyptians were highly skilled in the smelting of copper but showed little understanding of the workings of iron. Just south of the Egyptian Empire was the Nubian Kingdom of Africa. Here, practice of iron smelting and forging was being perfected by the people of Meroë. Techniques of wire making were perfected by certain members of the tribe who passed the techniques down within a secret and sacred order of early metalsmiths. They collected the ore and placed it in crude furnaces constructed of mud and charcoal. The charcoal was produced by burning a special wood gathered in this region. The wood was burned down to embers of carbon-rich charcoal. The iron ore was layered in the charcoal and air was pumped into the crude kiln by tribespeople working around the clock at animal skin bellows. The iron-rich slag was then removed and hammered with special stone tools. They made wire by pulling molten metal through specially constructed wood and stone dies wedged in a bough of a tree. The wire was pulled through an orifice and kept suspended from the ground as it cooled.[1]

In India, just south of Delhi, is a pillar of almost pure iron. The level of purity is so high that corrosive elements in the atmosphere have had little effect on the pillar's surface. Apparently, this 6.2-meter high pillar has stood since the fourth century A.D., which can only attest to the knowledge of refining and casting gained by some regions of the world.

Modern uses of the metal centered on wrought iron and cast iron. Until the mid-1900s, wrought iron was the major form of the iron alloys in use. Sheet iron, a form of wrought iron, was first rolled in the form of sheets in a factory in Trenton, New Jersey, in the early 1800s. This new metal form was used to roof the White House in 1804, replacing the original slate roof. Some of the sheets had pressed designs, which added decoration and strength to the flat surfaces. Sheet iron enabled the early sheet metal fabricator to create strong and durable ornamentation and wall cladding.

Corrugated sheet iron was patented in England in 1829. The corrugated form provided strength and rigidity to the flat sheet. This added rigidity allowed the thin metal to span across supports, thus eliminating the necessity for solid wood surfaces.

Cast iron was developed from the early refinement techniques and was used for columns in building construction in the late 1800s and early 1900s. Cast iron was also widely used for decorative building facades in the early 1900s. Window mullions and spandrel panels were cast from this form of steel and assembled into architectural facades. Cast iron is a very brittle and inelastic form of steel. It cannot take bending stresses very effectively. There

---

[1] The New York American Museum of Natural History has an interesting montage on this tribe of people south of Egypt. Some of the crude but ingenious tools are on display.

were a couple of unfortunate buildings designed and constructed using cast iron beams to go with the cast iron columns. Once loaded, the cast beams could not take the bending stresses, and the buildings collapsed before they were ever completed.

Steel is produced commercially through one of four processes. The oldest method used today is the Bessemer process, the invention of which is generally credited to Sir Henry Bessemer[2] of Great Britain. The process utilizes a blast of air through molten pig iron to remove the impurities. The air combines with excess carbon and other elements and leaves a superior refined steel alloy.

Another process, known as the basic oxygen process, removes impurities by rapidly adding pure oxygen into the molten mass of pig iron and scrap iron. The molten mixture boils as the oxygen enters and oxidizes the impurities.

A third process, the open-hearth process, not as common today, uses a furnace with an open hearth for viewing by the operator. Pig iron and scrap are put into the furnace, and impurities are gradually burned off.

The electric arc furnace is the modern-day approach to steel making, owing mainly to the increased demand for low-alloy steels and stainless steels, as well as the advent of specialty steel mills. Invented by K.W. Siemens in 1878, the process uses electric current.

PRODUCTION AND PROCESSING

Current world sources of iron ore, other than the great recycling market, are Russia, Brazil, Australia, India, and the United States. The ores most often used for refinement are magnetite and hematite, both of which contain a high percentage of iron.

Today iron ore refinement is performed commercially in blast furnaces. The iron ore is mixed with coke and fed through the top as superheated gases are passed through the mixture from below. The iron ore is removed by chemical reactions of the impurities with the hot gases. The gases combine with burning coke, forming carbon monoxide. The carbon monoxide reacts with the iron oxide, removing the oxygen and leaving iron to be removed as pig iron. Pig iron, which contains approximately 4% carbon along with other impurities, is the initial source used in developing wrought iron, cast iron, and the steels. Ferrous scrap is another important source used in production. Recycling of ferrous scrap provides the majority of the raw material used in steel manufacturing today. When the scrap is refined, certain elements and waste materials are removed from the molten metal.

In the electric arc furnace technique of steel refinement, a current is passed between two electrodes through the mass of pig and scrap metal. The mass is heated by the resistance of the electric charge as it passes through the metal from one electrode to the other. The impurities are removed by passing controlled amounts of oxygen across the molten mass.

[2]The Bessemer process is said to have been invented independently by Sir Henry Bessemer in England and by William Kelly of Kentucky. William Kelly built five experimental blast furnace systems, used to decarbonize steel, from 1851 to 1856. He first started working with experimental models in 1847. Unfortunately, he did not apply for a patent until 1866, whereas Bessemer filed for his patent in 1855. The process took on the name of Bessemer and Sir Henry has received most of the recognition for the process.

As the molten steel is poured into a large ladle, additive elements such as aluminum, manganese, and chromium are introduced in controlled amounts. Some of these elements were previously removed in the scrap refinement but are now returned, also in controlled amounts, either to add certain alloying effects or to reduce oxygen levels.

The steel, in the controlled alloyed state, is cast. For sheet metal, this operation is performed in a continuous-casting or strand-casting process, producing large slabs. The large slabs then undergo further hot rolling to the desired thickness and to establish grain direction. As steel solidifies during casting, many internal structural variations occur. These variations are modified and enhanced by alloying different metals into the casting.

ENVIRONMENTAL CONCERNS

Much of the steel manufactured today comes from recycled scrap. Manufacturing steel from recycled waste uses less than half the energy of manufacturing steel from ore. Unfortunately, galvanized steel and painted steel scrap have very low values. The zinc and paint coatings contaminate the melted scrap and require removal. Removal of paint produces vapors which must be collected to avoid causing air quality hazards.

Iron and steel are not considered toxic substances. Iron oxide fumes and dust generated by welding and grinding operations can cause benign pneumoconiosis. Relatively high concentrations of iron oxide are necessary for the development of the illness.

ARCHITECTURAL ALLOYS

There are three principle architectural forms of iron alloys:

Wrought Iron
Cast Iron
Steel

Wrought Iron

Wrought iron is a low carbon form of the iron alloys. The carbon content of wrought iron is less than 0.035%. Formed by hammering the heated metal on anvils or large machine forges, wrought iron is very strong and malleable, particularly in the hot state. Wrought iron has good tensile strength and can be shaped into many intricate forms because of its high elasticity. The more wrought iron is worked, the stronger it becomes. Often strength is enhanced by hammering layers of the hot metal together. Wrought iron is heated, hammered, bent, and twisted to form elaborate railings and the artistic wrought iron fencing.

Cast Iron

Cast iron is a form of steel containing large amounts of carbon. Additions of carbon, to the levels of 2% to 4% by volume, creates a castable alloy of iron.

Figure 4-1. Wrought iron work by Gaudi.

Cast iron is very hard and has excellent compressive strength, but poor elasticity. This form of iron has excellent casting characteristics. It is highly fluid and can be cast into very intricate shapes. The melting temperature of cast iron is approximately 2,102°F (1,150°C).

THE STEELS

Steel alloys are numerous and range from almost pure iron alloys to extreme variations containing many components of other metals and nonmetals. In architectural metals, carbon is a major additive needed to achieve particular properties. Therefore, many of the architectural steel alloys, other than the stainless steels, are known as carbon steel. Carbon steel is the base metal used for sheet metals such as black iron, galvanized steel, prefinished galvanized steel, terne-coated steel, tin plate, and numerous other common metals.

For all steels, carbon is the principal hardening element. As the amount of carbon increases, the hardness and tensile strength increase. This is because carbon molecules fill in the interstitial spaces between the grain boundaries that form as the steel solidifies. By filling the interstitial spaces, or voids, slippage between boundaries is resisted. The filling of voids provides formability to the metal when it is cold worked. Carbon content will, however, inversely affect the ductility of the steel alloy and the weldability of the metal. As carbon content increases beyond a certain point, the metal will not form adequately and welding will cause carbon to precipitate to the surface.

Other elements commonly added to the steel alloy are manganese, which contributes to strength and hardness and benefits surface quality. Phosphorous, when added in controlled amounts, also increases hardness and strength. Sulfur is added to lower the ductility of the metal, and silicon to reduce oxygen in the steel.

To a lesser degree, other elements are added, such as copper to improve corrosion resistance, nitrogen to raise the yield and tensile strength, and boron as another hardening element.

Hot-Rolled Steel and Cold-Rolled Steel

Steel for architectural uses has two main classifications: hot rolled and cold rolled. Hot-rolled steel sheet is rolled hot out of the casting to a specified thickness. The temperature is sufficient to develop a thin, often adherent, layer of oxide on the surface of the metal. The oxide can be removed in subsequent pickling operations.

Cold-rolled sheet steel, one of the most common architectural metals, is produced from the hot-rolled steel product. After pickling the hot-rolled steel to remove the scale, the sheet metal is rolled through cold rolls to further reduce its thickness. As the metal is cold rolled, it work hardens and must be annealed to recrystallize the grain structure. Cold working stretches

Figure 4-2. Ornamental steel frames.

the grains, making the metal brittle and unworkable for subsequent forming operations. In the annealed state the metal is soft and the surface is not of finish quality. The metal can be passed through additional cold-rolling processes to develop a particular temper. Cold-rolling processes also smooth the surface for subsequent finishing and coating operations.

Cold-rolled steel is available in two classes. Class 1 is required for architectural applications in which surface quality and flatness are of prime importance. Class 2 is not used architecturally. Sheet steel of this class is subject to surface imperfections not amenable to architectural applications.

The Weathering Steels

All forms of metal, both ferrous and nonferrous, develop oxide films when exposed to the atmosphere. Oxide development is a natural occurrence by which a metal tends toward its natural density. All metals have this entropy. When exposed to the atmosphere, they all seek to return to the more natural, energy-balanced state.

In most common carbon steels, the oxide develops to the point that total decay may occur, as signaled by the powdering of red rust. The oxide film on most steels is porous and allows moisture and oxygen to continue corroding the surface. Some high-strength steel and copper-bearing steels develop dense adherent oxide films that slow the rate of corrosion, similarly to the oxides on copper and aluminum. These steels are the "weathering" steels. Cor-Ten is the trade name for one of the more common of these weathering steels developed by US Steel Corporation. Bethlehem Steel had another form, called Mayari R.

These copper-bearing weathering steels are characterized by a reddish brown, coarse rust film that rapidly develops on the surface. Initially the surface is a bluish black color. As moisture, even slight condensation, collects on the surface, streaks of reddish brown rust develop. Further exposure leads to a coarse fine layer of rust eventually covering the entire surface of the metal. The runoff of the rust film quickly stains adjacent materials such as concrete wainscots and sidewalks below the weathering steel installation. Designing with this material requires special attention to runoff and moisture control.

The rust film on a metal surface is very adherent. The impervious iron oxide film will work to resist further corrosion of the base metal. Eventually, the runoff slows down to a point where further staining is minimal.

Copper-bearing weathering steel is a high-strength, low-carbon steel alloy containing small amounts of chromium and copper. The nominal range of alloying elements is as follows:

Carbon	0.12% to 0.16%
Chromium	0.50% to 1.00%
Copper	0.30% to 0.49%

The alloys are very hard and have a yield strength of 50 ksi (345 MPa). The addition of chromium provides corrosion resistance characteristic of a hard, impervious surface film.

Cor-Ten was used as an architectural metal in the 1960s and 1970s. Major uses of Cor-Ten continue today in structural members, left exposed to weather naturally, requiring little if any maintenance. Eero Saarinen used Cor-Ten for the John Deere headquarters building facade. Saarinen noticed the interesting dark brown steel surfaces on railway cars used to transport grain. An unfinished weathering steel was used for grain cars because of the abrasive nature of the grain kernels on painted steel cars.

Cor-Ten and other weathering steels have since been discontinued as an architectural metal in sheet form because of premature perforation of the sheet steel. Problems developed when the corrosion of the inside surface of the steel occurred at accelerated rates in comparison with the outer surface. When sheet metal thicknesses less than 0.032 inches (0.813 mm) were used, some areas experienced premature perforation of the metal. In addition, areas where water was allowed to erode the surface, such as on leaders and scuppers, the protective layer of rust was removed. Corrosion of the surface continued until portions of the metal dissolved completely.

There are, nonetheless, many successful installations of the metal. These are typified by projects in which the inner side of the sheet metal was protected with a thick paint coating. Moreover, the sheets used in these instances were of thicknesses greater than 0.032 inches (0.813 mm).

Enameling Steel

The use of vitreous enamel coatings on steel dates back to late 1940s. Vitreous or porcelain enamel coatings are fused glass coatings applied over steel. Because of the high temperature required to fuse the glass to the steel, in the range of 1,356°F to 1,598°F (730°C to 870°C), special low-carbon steel is required. If low-carbon steel is not used, imperfections in the finish porcelain coating will develop. As the temperature is raised, carbon may "boil" from the steel surface and show as pockmarks in the porcelain coating. Fish scale imperfections may also appear, which are caused by a hydrogen reaction on the steel surface. The steel can also sag from its own weight, as it becomes slightly plastic at the high temperatures.

There are five types of enameling steel in common use:

1. Cold-rolled aluminum-killed steel. Low-carbon steel used for deep-drawing operations.
2. Cold-rolled rimmed steel. Low-quality steel. Used for concealed regions.
3. Enameling iron. Premium grade of commercially pure iron. This sheet material has a very low carbon content with a small amount of manganese added to improve formability. The manganese strengthens the sheet during firing. Enameling iron must be fired twice: first with a ground coat, then with a finish coat. The ground coat helps the final finish coat adhere to the metal surface.
4. Decarburized enameling steel. This cold-rolled sheet contains very low levels of carbon. Produced by annealing in an open-coil process to remove carbon as the annealing gases pass over the surface, this enameling steel is suitable for direct coating.

5. Interstitial-free enameling steel. This type of enameling steel has a moderately low carbon content. It is used when forming conditions require a stronger base metal.

For architectural products, the most commonly used of these materials are enameling iron and decarburized enameling steel. Different companies involved with porcelain coatings on steel may have their own suggestions.

FINISHES

The iron alloys are not commonly used in the uncoated state. The natural color of the iron alloys, when cleaned of scale and mill oil, is a dark blue-gray. The steels have a dark blue-gray color, while the heavier carbon alloys of cast iron are almost black. Cast iron has a dark carbon-gray surface, which will weather to black intermixed with rust. The rust, if allowed to continue, will cover the entire cast section and eventually corrode the entire part.

The wrought iron forms are dark blue-gray in tone. Heating blackens the surface, and subsequent shaping drives the dark tones into the surface. Wrought iron is almost always coated with a thick paint to inhibit corrosion.

Depending on the surface quality, steels appear blue-gray in color. The hot-rolled surfaces are grainy and darker, almost blue-black. The cold-rolled surfaces are smoother and have a dark blue color. On exposure to humid air, the steels develop rust over the surface in an very short time. The rust is porous and will continue to consume the steel sheet if allowed.

The iron alloys used in architectural applications, with the exception of the weathering steels, are always protected from the atmosphere with a barrier coating. The most common coating is zinc and is known as galvanized steel or galvanized iron.

For all the coatings, the surface quality of the iron alloy is of critical importance. Cast iron is typically supplied sand blasted. Sand blasting removes the surface scale and leaves a porous, coarse finish. Wrought iron is either sand blasted or pickled in acids to remove surface scale. The steels have specific surface levels requirements, depending on their final use and form.

Both hot-rolled and cold-rolled steel are available in four surface quality classifications:

Commercial Quality. Cold-rolled commercial quality is the common classification of steel used in architectural applications. This steel has a matte finish suitable for further finishing and coating with zinc (galvanizing), terne, paints, and porcelain. Commercial quality steel is produced from low-carbon grades of steel without close control of chemical uniformity within the steel alloy.

The other grades are used for specific applications requiring conditional control of the alloying elements for purposes of fabrication.

Drawing Quality. Both hot-rolled and cold-rolled steels of drawing quality are used in applications in which severe forming is required.

Drawing Quality and Special Killed. The special killed, drawing quality steels are used in particularly severe forming and drawing operations.

Structural Quality. The structural grades are used when specific strength characteristics are required. Strength characteristics can be influenced by the alloying elements introduced into the alloy composition.

TEMPER AND HARDNESS CHARACTERISTICS

Control of the temper imparted to cold-rolled steel is critical for subsequent fabrication steps. A list of standard tempers can be found in Table 4-1. The *full hard temper* steels are also known as spring steels. They have a high yield strength and rigidity and are called spring steels because of their tendency to resist forming. Dead soft or fully annealed versions of commercial quality steel have maximum ductility and softness.

When steel is stored for any length of time, aging will occur. Aging is a process by which steel loses some of its ductility and experiences an increase in hardness. Yield and tensile strength will also increase with time of storage.

For commercial quality cold-rolled steel, three basic mechanical finishes are available:

Matte finish
Bright finish
Luster finish

Mechanical finishes on steel are usually precursory to finish coating with paints or other metallic coatings such as zinc or terne. Matte finish, produced by cold rolling on roughened textured rolls, is a dull surface requiring surface preparation prior to painting. Bright finish, also produced by cold rolling, is a reflective, shiny finish. Minor surface finishing will render this steel ready for plating or paint applications. Luster finish is a smooth, bright finish produced by cold rolling on polished rolls. Final painting or plating can be readily performed on this surface.

Prior to paint finishing of steel sheet, the surface must be free of all mill scale, rust, and oil. To remove mill scale, corrosion products, or general dirt, localized areas can be hand cleaned with wire brushes or abrasive papers. For larger areas, power equipment can be utilized. Power grinding, abrasive flap wheels, and power sanders effect adequate removal of difficult scale or corrosion products. Sand blasting and shot peening are best for large regions requiring extensive cleaning. The energy of this cleaning method will warp and distort the surface of many metal parts. Shot peening and sand blasting with clean sand will leave the metal surface roughened and free of all oxides for a short period of time. Abrasive blast finishing is a versatile process, providing a tremendous variety of media for cleaning the surface of a metal. Abrasive blasting uses compressed air to propel glass beads, steel shot, clean

TABLE 4-1. Standard Tempers of Cold Rolled Steel

TEMPER DESCRIPTION	ROCKWELL Hardness b
Fully Annealed/Dead Soft Temper	55 maximum
Cold-Rolled Commercial Quality	60
Quarter Hard Temper	60 to 75
Half Hard Temper	75–85
Full Hard Temper	85 minimum

sand, or any number of particulate materials to remove coatings, adhesives, rust, and other surface contamination. The roughened surface provides good adhesion for subsequent paint applications.

MECHANICAL AND CHEMICAL FINISHES

The steels can be given mechanical finishes similar to those applied to other metals. The surface will fingerprint and develop rust rapidly if not immediately protected. For this reason, mechanical finishes are uncommon on steel except to prepare for application of other barrier coatings.

Grinding, sandblasting, and wire brushing are all precursors to subsequent applications of barrier coatings. Steels can be given chemical treatments to develop different darkening or color tones prior to the application of clear coatings or protective oils. Iron oxide or rust can be selectively induced onto the surface of steel, then protected with a clear paint coating to inhibit further rust. The chemical treatments for decorative coloration of steels should be limited to interior applications and will require periodic maintenance.

Steel Bluing

Bluing of steel is an ancient practice. Steel and wrought iron will develop a deep blue color when heated and then cooled rapidly in water. A thin oxide coating with a blue color develops on scale-free steel surfaces when they are heated. The blue color is a play of the iridescent light interference film and the blue tone of the steel. The color will remain fixed to the surface if protected from further oxide development.

Steel Blackening

Steel can be blackened or given colors ranging from dark brown to black by dipping it in a heated sodium hydroxide (caustic soda) bath. This process can develop a deep black oxide over the surface of the steel. Oxides produced by this alkali method are weak and prone to abrasion if not protected with waxes or oils. Protection against corrosion is enhanced slightly.

Clear Protective Coatings

The clear coatings used on steel are similar to those used on other metals. The application of such coatings is limited to interior locations which are not exposed to the rigors of weathering. The steel surface must be neutralized and clean of all foreign particles prior to the application of clear barrier coatings.

Oiling can be used for small features not intended to be touched. This will require periodic maintenance and can be messy. Light oiling leaves a matte tone as it is absorbed into the pores of the steel surface. Oils used are mineral forms. A silicone spray with an oil-carrier base will also work.

Clear coatings of urethane enamels are effective protective barriers for steel and iron alloys. These coatings are more durable, but recoating is more difficult. The clean, porous steel surface will allow the enamel to adhere tightly. In areas where contact and abuse is minimal, these coatings can last for a long time.

CARE AND MAINTENANCE

There are additional processes that can be performed on cold-rolled steel sheet to improve the quality of the finished product. These are mill processes performed to specification to improve final surface appearances.

To improve flatness, sheet steel can undergo temper rolling. Also known as skin pass or skin rolling, temper rolling is a light cold-rolling operation that improves the quality of the surface and imparts an increase in temper to the sheet.

Another method used to improve surface flatness is roller leveling. The sheet steel is passed through offset trains of small rolls, which bend and flex the sheet up an down. The bending results in a flatter sheet with some increase in temper.

For superior flatness, stretcher leveling should be performed. Stretcher leveling, or tension leveling, is performed on coil sheet steel. The continuous sheet is passed through a mechanism that stretches the metal beyond its yield point. The result is a very flat and consistent sheet. The internal stresses that were induced in the sheet during cold-work operations are overcome during the stretching process. This method reduces the occurrence of "oil canning" in steel sheet products.

Coils have their own set of particular surface imperfections. Coils tend to contain more surface blemishes than cut sheets; inferior sheets can easily be sorted out from a stack, but with coils this is not possible. Coil breaks, which are surface imperfections running perpendicular to the coil length, are typical imperfections imparted to coil sheet steel. Coil breaks occur at different locations across a coil. Coil breaks are distortions induced into the sheet metal surface during uncoiling and sheeting processes.

CORROSION AND COMPATIBILITY CONCERNS

Unprotected iron alloys will develop corrosion particles of iron oxides quickly. Condensation, allowed to develop on iron alloys will cause surface rust to develop. When storing iron alloys, consider applying a thin layer of oil over all surfaces. If rust does occur, clean the surface and oil it immediately. Once the steel surface has developed rust, it becomes sensitized to further corrosion.

Water on the surfaces of steels or iron will generate corrosion cells. Polarity is set up across the steel surface, developing cathodic an anodic regions that feed on themselves. The anodic regions join with oxygen and form rust. The rust is pervious, and additional moisture is allowed to reach unaffected metal, keeping the process going. Dew allowed to collect on steel surfaces is very detrimental. Dew collects pollutants from the air and deposits them in a concentrate on the metal surface. The pH of dew collected in urban regions can be very acidic and cause severe deterioration in a very short time.

Iron alloys will corrode because of galvanic contact with other metals, particularly copper. Copper oxides allowed to deposit on steels will create rapid corrosion of the steel part. Steel in contact with other metals, such as aluminum and zinc, will not corrode as rapidly. Zinc will sacrifice itself as the iron alloy is subjected to the electromotive forces of galvanic contact. Iron rust will stain other surfaces. The reddish brown corrosion particles are tenacious and will adhere and stain both porous and nonporous surfaces.

Steel is usually protected from the environment and from contact with other materials by a barrier coating, either of another metal or by organic and inorganic coatings. Therefore, the compatibility of the iron alloys with other building construction materials is determined by the compatibility of the coatings with these materials.

THE STAINLESS STEELS

INTRODUCTION

By adding chromium to steel, one of the more durable and corrosion resistant metals is obtained. This specialized steel alloy retains, even after prolonged atmospheric exposure, a very lustrous metallic surface. The alloys of steel containing chromium in excess of 10% are known as the stainless steels.

In the early 1900s, particularly 1910 to 1915, alloys of iron-chromium were developed for industrial and commercial uses. Scientists wished to find a steel that would have improved corrosion resistance, improved strength, and better properties when subjected to sustained temperatures. New techniques of oil refining, chemical processing, and other manufacturing processes developed in industrial revolution of the late 1800s demanded more durable metals. Existing steels corroded quickly or developed weaknesses when exposed to high temperatures. This new alloy of steel exhibited properties that would make it a cornerstone of the modern industrial age.

Chromium is the element that provides stainless steels with their superior performance characteristics. Chromium is one of the First Transition Metals. There are fourteen members of this family, which includes such metals as titanium, vanadium, and tungsten. Each of these, when added to steel, develops a stronger, more durable material.

The specific gravity of chromium is 6.92, just below that of zinc. This metal does not occur in the pure form in nature. Usually, it is found in combination with iron and magnesium or aluminum. Chromium is characterized by hardness. Known generally as "chrome," this metal is responsible for the hard, metallic, mirrorlike luster stainless steel develops when polished. A surface that can be intensely polished and is characteristicly tarnish resistant makes chromium a superior architectural material. In subsequent pages this incredible surface will be shown to possess more decorative potential through different mechanical and chemical processes.

HISTORY

The earliest published discussion of iron-chromium alloys and their corrosion-resistance characteristics was authored by Stodard and Faraday in England in 1820. The report discussed the corrosion resistance of iron when chromium was added; however, the amount of chromium added in this case was below that required for passivity of the alloy. Pierre Berthier of France soon afterward discovered that adding certain quantities of chromium to steel created an alloy more resistant to acids than any previously known. These new, higher-chromium alloys were brittle and hard; however, they showed a strong resistance to acidic deterioration. Berthier is credited with the discovery of the stainless steel alloys and the effect of chromium on corrosion resistance.

By adding chromium to low-carbon steel, a special alloy was developed that would produce a microscopically thin, transparent chromium oxide film on the surface. This film resisted staining and would repair itself when scratched or damaged. The phenomena of self-healing and corrosion resistance were overlooked as having any industrial value until many years later. (The corrosion-resistant behavior was of interest, but only as a novelty.) The practical applications of the metal were not apparent to industry until the early 1900s. The benefits and potential of iron-chromium alloys became clear to the industrial world as needs developed for longer-lasting steel parts.

The development of the stainless steels occurred at approximately the same time in different parts of the world. In France and England, the metallurgical principles of stainless steel were developed by the Frenchman Guillet in 1904. Guillet produced low-carbon chromium alloys that retained the corrosive-resistant properties exceeding those of steel and iron. In Germany, Monnartz studied the property of passivity imparted by these new iron-chromium alloys. In Europe and in the United States, the industrial importance of stainless steel was considered and developed commercially almost simultaneously. Frederick Becket of the United States improved the method of reducing chromium ore, thus making the ore more available. Harry Brearley of Great Britain, while searching for a better alloy for making gun barrels, developed the alloy known as Type 420 stainless steel. It was Brearley who coined the name "stainless steel" with his commercial development of the metal for cutlery. Eduard Maurer and Benno Strauss of Germany, while studying iron-nickel alloys for thermocouple protection tubes, arrived at the alloy known today as Type 302 stainless steel. Architectural uses of the new alloy of steel became readily apparent when it went into production. This new, environmentally stable metal, with the ability to sustain a polished luster, provided the designer with a material that would soon become a signature for the modern age of construction.

The architectural appeal of stainless steel is evident in any major urban cityscape. Desired for its reflectivity and unchanging surface characteristics, stainless steel provides the designer with an elegant and durable material. Many major architectural structures across the world are clad with thin stainless steel surfaces. These thin skins are used as major design features and to hold back the effects of time and nature. The Chrysler Building in New York was one of the first major architectural projects designed and constructed with stainless steel as a prime architectural surface. Designed by William Van Allen, the Chrysler Building was constructed in the late 1920s, using stainless steel to enhance a very intricate art deco pinnacle roof. The stainless steel pinnacle roof, furnished with a 2B mill finish, was completed in 1929. The alloy chosen for this ornamental feature was 302 stainless steel. Alloy 302, consisting of approximately 18% chromium and 10% nickel, provides excellent corrosion resistance when exposed to a harsh urban environment. In the early 1980s, the building's roof covering was cleaned and inspected. The stainless surface was found to be in excellent condition, and the surface still retained an incredibly bright metallic luster.

The first stainless steel roof was installed over the Pittsburgh Plate Glass Company Plant in 1924. This roof has been the subject of metallurgic evaluation, since it is still in service many years later. On examination it was found to be in excellent condition. The Gateway Arch in St. Louis, constructed of large 0.25 inch (6.35 mm) thick stainless steel panels, is one of the most fascinating stainless steel shapes ever constructed. Designed by Eero Saarinen and completed in 1966, this structure is fabricated from Type 304 stainless. The surface has a directional satin finish, known as a #4 surface. This structure shows virtually no signs of atmospheric deterioration.

Recently there have been other major architectural structures utilizing stainless steel as a major design surface. Among them are the intricate roof of the Temple of the Reformed Latter Day Saints in Independence, Missouri, designed by Gio Obata of Hellmuth, Obata, Kassabaum, Inc. This structure used more than 4,000 satin-finish #4 stainless steel panels, each individually tapered in two directions. The Fresno City Hall, designed by Arthur Anderson and Associates with Lew and Patnaude Architects, has a massive dull-finished (#2D) stainless steel roof and interior panels and staircase of stainless steel. The Memphis Pyramid, designed by Rosser-Fabrap International Architects, is a pyramid sheathed in a stucco-embossed stainless steel sandwich panel. At the Museum of Science and Technology in Paris, France, rests a magnificent, 36-meter diameter sphere. This sphere is fabricated from 6,433 triangles, 1.5 mm thick and provided with a reflective mirror finish. The alloy chosen by the architect, Adrien Fainsilber, was 316L. Alloy 316 is commonly used in Europe for exterior architectural fabrications because of its superior corrosion resistance.

All these projects used stainless steel as a design material with unchanging durability and a metallic appearance afforded by the reflective surface. No other architectural metal can provide a surface capable of maintaining high reflectivity over time without periodic maintenance or secondary protection.

PRODUCTION AND PROCESSING

At the mill, stainless alloys are melted and cast to particular specifications of alloy content. For architectural uses, higher nickel content is sometimes specified to achieve good surface finish in later processes. Once a slab is cast and initial cooling occurs, the slab of metal is put through a grinding process. High-speed grinding improves the surface of the slab of cast stainless steel prior to initial rolling. The first rolling operation the slab encounters is hot planetary rolling, which passes the hot slab of stainless steel under pressurized rolls. The high pressure of the rolls, 585 tons, reduces the slab thickness to about 0.225 inches (5.7 mm). The appearance at this point is very rough and not flat. The thinner slab of stainless steel then is annealed and pickled. During the initial slab reduction process the stainless steel work hardens slightly and becomes brittle. The initial annealing of the stainless steel removes the hardness that has developed from cold working the metal through the rolls. Annealing also dissolves chromium carbides that form at the grain boundaries.

For the austenitic stainless steels, which are the iron-chromium-nickel alloys used in architectural metal work, the metal is heated at temperatures between 797°F and 1,499°F (425°C and 815°C) to remove the hardness. The initial pickling of the stainless steel slab removes the surface scale developed by the heat treatment of annealing. The pickling process removes the scale through treatment with strong oxidizing agents, such as mixtures of hydrofluoric acid, nitric acid, and water. Scale is a thick, loose, and uneven deposit of metal oxides that develop on the surface of a metal when rapid cooling occurs.

The edge of the slab is trimmed and evened to remove excess material prior to initial cold rolling. During cold rolling, the metal slab is rolled through successive rolls and reduced in thickness to 0.180 inches (4.57 mm). The reduced stainless steel sheet is surface polished by belts on both sides of the sheet to provide uniformity and to eliminate surface inclusions. Surface inclusions develop on hot rolled metal surfaces and are characterized by a pit or surface gouge. Polishing causes the metal to flow at the surface, effectively filling in minor surface inclusions. After the polishing the sheet is again annealed and pickled to soften the work-hardened metal and to remove any further scale that develops.

The sheet of stainless steel goes through additional cold-rolling operations to thin it to a 0.125 inch (3.17 mm) thickness. Stresses develop in the sheet during the cold-reduction process, requiring another annealing and pickling operation. For architectural projects, this step in the process is of critical importance to quality. The stainless steel is in a long ribbon that must be recoiled before entering the pickling bath. The annealing stage that occurs at this point will affect the end color tone of the stainless surface. For architectural projects, it is recommended that all coils be annealed and pickled in one single batch. This will achieve final color uniformity on the stainless steel surface by ensuring that the stainless coils undergo exactly the same conditions of time, temperature, and atmosphere during the annealing process. For large projects, anneal and pickle stainless coils grouped according to their location on the building. This will put those coils of stainless annealed together on the same elevation and help to ensure color consistency.

After the final annealing and pickling step the coil of stainless steel is passed again through a final set of cold rolls. These rolls produce the final surface finish to be provided by the mill. Following the finishing step the stainless steel is degreased and bright annealed if required. The stainless may be protected with a plastic film adhered to the surface by adhesive or electrostatic charge, or left unprotected. The stainless is decoiled, sheared, and skidded into sheets, or stored in coils for later shipping and handling.

Other finishes are produced by a secondary processor who specializes in the finishing of stainless steels. Such secondary finishes greatly affect the final appearance of the metal surface. These are the polished finishes, which are introduced later in this chapter.

ENVIRONMENTAL CONCERNS

Stainless steels have no toxicity. The material is considered hygienic and is used for most food service and preparation operations, dairy operations, and medical surfaces. Stainless steel does not develop any hazardous salts or oxides.

The manufacturing of stainless steels involves the use of chromium. Chromium is a hazardous and toxic metal, but when alloyed in steel it is "fixed" in metallic solution.

Fumes generated from welding operations and dust from grinding operations should be avoided. Welding and grinding should be performed using adequate ventilation and/or respiration equipment.

Mining and recycling operations are similar to those used in steel and iron production. Nickel alloying material is mined similarly to copper. Large amounts of waste material is generated during the mining and ore reduction process. Chromium ore is refined by electrolysis, as in copper and nickel refining.

THE ARCHITECTURAL ALLOYS

At first glance, the number of stainless steel alloys available is staggering. A total of 57 stainless alloys are recognized by the American Iron and Steel Institute. In addition to these, there are innumerable special proprietary alloys considered as part of the family of stainless steels. Further investigation, however, leaves only a few alloys associated with architectural metal work. Reference Table 4-2 for a list of the standard architectural alloys.

TABLE 4-2. Common and Not So Common Architectural Stainless Steels

UNS NUMBER	AISI NUMBER		RANGE OF CONSTITUENTS % IN EACH ALLOY TYPE		
		Chromium	Nickel	Molybdenum	Comment
S20100	201	16.0–18.0	3.5– 5.5	0	Contains 5.5%–7.5% manganese
S20200	202	17.0–19.0	4.0– 6.0	0	Contains 7.5%–10.0% manganese
S30100	301	16.0–18.0	6.0– 8.0	0	
S30200	302	17.0–19.0	8.0–10.0	0	
S30400	304	18.0–20.0	8.0–12.0	0	
S30403	304 L	18.0–20.0	8.0–12.0	0	Carbon reduced to 0.03 max.
S31000	310	24.0–26.0	19.0–22.0	0	
S31600	316	16.0–18.0	10.0–14.0	2.0–3.0	
S31603	316 L	16.0–18.0	10.0–14.0	2.0–3.0	Carbon reduced to 0.03 max.
S41000	410	11.5–13.5	0.50 max	0	Martensitic group
S43000	430	14.0–18.0	0.50 max	0	Ferritic group

Criteria commonly used by the specifier to determine which alloy is best suited for a particular architectural application fall within certain parameters, most important of which are cost and corrosive exposure. The environment to which the metal will be subjected is foremost. Corrosion resistance of a particular alloy is related to the amount of chromium alloyed with the steel. Additions of nickel and molybdenum will further improve the corrosion resistance, particularly in environments exposed to chlorine and chloride salts.

Chromium is the most important alloying element added to steel, and an amount in excess of 10% is needed to qualify a steel alloy as stainless steel. The chromium will form a clear hydrous oxide layer over the surface of the steel alloy. This oxide film is stabilized by the chromium and is passive or nonreactive to many corrosive conditions. The film that develops is continuous and nonporous. When scratched, it rapidly heals to ensure the continuous passive barrier.

Passivity is the corrosion-resistant behavior achieved by the thin hydrous oxide film. The passive state of the film is maintained even when an oxidizing condition exists. Under normal conditions stainless steel exhibits a passivity approaching that of the noble metals. However, when the oxide film is breached and prevented from redeveloping by inhibiting oxidizing agents, the stainless steel surface becomes active and will corrode like ordinary iron.

Metallurgical Basics

There are four main groups of stainless steel. These are defined, in part, by the basic structure the crystals form in different alloy states and by certain properties unique to these structures. The first group comprises the *austenitic* types of iron-carbon alloys. The S30000 series alloys are austenitic types, made of the iron-chromium-nickel combinations. A few of the S20000 series alloys also fall into the austenitic group of stainless. The austenitic stainless steels are characterized by being nonmagnetic. Slight magnetism will occur after cold working the materials as the metals begin to work harden. Nickel is the main element that varies within this type. The S20000 series having very low levels of nickel, manganese is substituted and the nitrogen levels are increased. The amount of nickel alloyed in this group ranges from 4% to 22%. As chromium is increased to improve the corrosion resistance of the metal, nickel must also be increased to sustain the austenitic structure of the alloy. Carbon is kept to low levels in the austenitic group, particularly to assist in the reduction of carbide precipitation. The austenitic crystal structure of this group is a face-centered cube with carbon interstitial.

The next group includes the *ferritic* types of stainless steel alloys—the iron-chromium alloys. The more common alloys of this type are the S43000 and S40900 stainless steel alloys. An increase in chromium improves the corrosion resistance of these alloys. They are magnetic in all occurrences. The ferritic alloys are less ductile than the austenitic group and have a crystal structure that is a face-centered cube with a small amount of dissolved carbon.

The *martensitic* types of stainless steels constitute the third group. Like the ferritic group, these are iron-chromium alloys; they are magnetic in all conditions and can be hardened by heat treatment. Alloy S41000 is the most common of the martensitic stainless steels. These alloys are lowest in corrosion resistance of the stainless steel types. Cutlery and stainless steel fasteners of the self-drilling variety are made of this alloy group. The martensitic crystal structure is a body-centered cube with trapped carbon.

The fourth group, rarely considered as architectural alloys, are the *precipitation hardened* stainless steels. These stainless types are made of the iron-chromium-nickel combinations, with corrosion resistance similar to that of the austenitic group. They are magnetic in all occurrences and are very hard. Shearing and cutting sheets of precipitation hardened alloys is difficult with conventional sheet metal equipment. Alloys of precipitation hardened stainless are considerably more expensive than the alloys of other groups.

Choosing the Correct Alloy

A specific alloy should be matched to the intended use of the metal. For instance, a specifier may consider Type S41000 stainless steel, which contains approximately 12% chromium and no nickel. When Type S41000 is exposed to an industrial environment, for instance, as ornamental trim or stainless roofing, a superficial rust film will develop. This rust film is in itself protective to the base metal, but its appearance is considered objectionable. If Type S43000 stainless steel is used, which contains 17% chromium, the rust will take several months to develop. Type S43000 stainless is used frequently for automotive trim, but since most people clean and wax their cars, the rust never appears.

The addition of 7% to 9% nickel to an S43000 alloy stainless creates the architectural alloys S30100, S30200, and S30400, also known as the 18-8 stainless steel alloys. These alloys will not develop rust in most industrial exposures; therefore, they are commonly specified for architectural uses.

Nickel is the second most important alloy added to the steels. Nickel greatly enhances the corrosion resistance of stainless steel and increases the ductility, or shaping ability, of the metal. Scaling resistance is improved during hot forming and reducing. Scale is the rough oxide material that develops on the surface of steels when they are hot rolled and processed. On steels, scale can be a very adherent, rough layer requiring treatments in potent reducing agents, such as hydrofluoric and hydrochloric acids. Scaling resistance will improve color consistency from one batch of stainless to another. Nickel also improves strength when high temperatures are applied.

In coastal environments and other locations where chloride and sulfur deposits might predominate, Type S31600 stainless should be used. Deposits of chloride salts will pit S30100, S30200, and S30400 stainless steel alloys. Type S31600 stainless steel has molybdenum added to the alloy mix. Molybdenum, when alloyed with the chromium steels, increases the corrosion resistance further, especially when the stainless steel is exposed to sulfite, sulfate, acetic acid, and chloride environments. Molybdenum expands the range of passivity of the stainless steel alloys and counteracts pitting of the surface. Because of the addition of molybdenum and its high nickel content, Type S31600 stainless steel is more expensive than S30200 and S30400 alloys.

Other alloying elements added in very small quantities are also used in the formation of stainless steels, but with little corrosion-resistance improvement:

- *Silicon*—to increase resistance to scaling
- *Manganese*—to improve hot-working properties
- *Sulfur, phosphorus,* and *selenium*—to stabilize carbon and carbon precipitation

Basically, for architectural stainless steel applications, Types S30100, S30200, S30400, and S31600 are the alloys commonly used. Type S41000 is a common alloy for many self-drilling fasteners used in both stainless and aluminum construction, as well as in steel panel construction. The S30000 series alloys are not hard enough to drill through structural steel, whereas the hardness provided by the S41000 alloy stainless is. Type S31000 stainless steel is a common self-tapping screw alloy, and fasteners of this alloy are often used to anchor aluminum or galvanized steel wall and roof panels and aluminum curtain wall extrusions.

The S30000 Series Stainless Steels

Alloy S30100

Austenitic Stainless Steel
18-8 Stainless Steel

Alloy Mix—Nominal Composition

Chromium	16% to 18%
Nickel	6% to 8%
Carbon	0.15% maximum
Manganese	2.00% maximum
Silicon	1.00% maximum

Alloy S30100 is the least corrosive resistant of the S30000 series alloys because of its low chromium content. This alloy can be used in mildly corrosive environments or when periodic surface cleaning will be performed. Mildly corrosive environments such as rural or interior applications will not pose a problem.

Yield Strength	30,000 psi (206.8 MPa)
Ultimate Strength	75,000 psi (517.1 MPa)
Hardness—Rockwell B	92

Alloy S30100 is available in sheet and strip. Standard tempers are:

Annealed
Quarter Hard
Half Hard
Three-Quarter Hard
Full Hard

Alloy S30200

Austenitic Stainless Steel
18-8 Stainless Steel

Alloy Mix—Nominal Composition

Chromium   17% to 19%
Nickel   8% to 10%
Carbon   0.15% maximum
Manganese   2.00% maximum
Silicon   1.00% maximum

Alloy S30200 is the common architectural alloy. This alloy possesses excellent corrosion resistance and good workability. Its strength is not as high as that of alloy S30100. Alloy S30200 is almost interchangeable with alloy S30400. Many specifications even designate the stainless steel alloy as 302/304 because of their similarities in composition.

This is the alloy that has shown excellent long-term results on the top of the Chrysler Building in New York. It is used for general-purpose architectural applications, both external and internal to a building.

Yield Strength   30,000 psi (206.8 MPa)
Ultimate Strength   75,000 psi (517.1 MPa)
Hardness—Rockwell B   90

Alloy S30200 is available in sheet, strip, and bar forms. Standard tempers are:

Annealed
Cold rolled

Alloy S30400

Austenitic Stainless Steel
18-8 Stainless Steel

Alloy Mix—Nominal Composition

Chromium   18% to 20%
Nickel   8% to 12%
Carbon   0.08% maximum
Manganese   2.00% maximum
Silicon   1.00% maximum

Alloy S30400 is usually the alloy of choice for most architectural work. This alloy possesses excellent corrosion resistance, good workability, and because of its low carbon content, welding processes will generate only moderate amounts of chromium carbides. Alloy S30400 is used for exterior roofing, wall panels, and other architectural features requiring superior corrosion resistance.

Yield Strength   30,000 psi (206.8 MPa)
Ultimate Strength   75,000 psi (517.1 MPa)
Hardness—Rockwell B   90

This alloy can be provided in sheet, strip, and bar stock and is typically available in the annealed and cold rolled temper.

Alloy S31000

Austenitic Stainless Steel
25-20 Stainless Steel

Alloy Mix—Nominal Composition

Chromium	24% to 26%
Nickel	19% to 22%
Carbon	0.25% maximum
Manganese	2.00% maximum
Silicon	1.50% maximum

Alloy S31000 is not in common architectural use. This alloy is very corrosion resistant and, designed for high temperature use, weldable and very ductile. Often it is used for self-tapping fasteners used to attach aluminum and steel fabrications. This alloy is higher in cost than the other architectural alloys.

Yield Strength	30,000 psi (206.8 MPa)
Ultimate Strength	75,000 psi (517.1 MPa)
Hardness—Rockwell B	95

This slightly harder alloy is available in sheet, strip, plate, and bar forms. Available tempers are the annealed.

Alloy S31600

Austenitic Stainless Steel
18-8 Stainless Steel

Alloy Mix—Nominal Composition

Chromium	16% to 18%
Nickel	10% to 14%
Carbon	0.08% maximum
Molybdenum	2.0% to 3.0%
Manganese	2.00% maximum
Silicon	1.00% maximum

Alloy S31600 is an architectural alloy, frequently used where pitting corrosion is a concern. The added molybdenum aids in resisting chloride attack on its surface. This alloy is commonly used in Europe and the Far East for exterior metal applications exposed to urban or coastal environments. Alloy S31600 is the most corrosion-resistant alloy of the common architectural alloys.

Yield Strength	30,000 psi (206.8 MPa)
Ultimate Strength	75,000 psi (517.1 MPa)
Hardness—Rockwell B	95

This alloy is available in sheet, strip, plate, and bar forms. Available temper is the annealed.

The Low Carbon Alloys

If welding of stainless parts is required, a modified alloy type, designated by the letter *L* following the alloy number, is used, such as Type 304L stainless. The designation *L* is the AISC form description and stands for low-carbon alloy stainless steel. When stainless steel is welded, carbide precipitation occurs around the weld. Carbon precipitation is a condition created when chromium carbides form at the grain boundaries near the weld. Chromium combines with the carbon in the stainless steel when high temperatures are reached, such as welding temperatures.

When the chromium carbides form, the area adjacent to the formation is stripped of the protective chromium and localized integranular corrosion can occur. Designating a low-carbon stainless steel alloy reduces the formation of chromium carbide. In the late 1920s carbide precipitation was a major problem with stainless because 0.15% to 0.25% carbon was common in these steels. Low-carbon stainless steels today contain a maximum 0.08% carbon. There is one caveat regarding use of the low-carbon versions of the 304 and 316 alloys: the low carbon steels have a reduced yield stress.

Alloy	Yield Stress
S30400	30,000 psi (206.8 MPa)
S30403 (304L)	25,000 psi (172.4 MPa)

Alloy S30403

Austenitic Stainless Steel
18-8 Stainless Steel
304L

Alloy Mix—Nominal Composition

Chromium	18% to 20%
Nickel	8% to 12%
Carbon	0.03% maximum
Manganese	2.00% maximum
Silicon	1.00% maximum

Alloy S30403 is the low-carbon version of alloy S30400. Carbon is reduced to allow for welding with a reduction of carbide precipitation. Carbide precipitation develops when the heat of welding causes chromium carbide to form at the intergranular boundaries. Note the lower strength of this alloy in relation to the higher carbon content alloy S30400.

Yield Strength	25,000 psi (172.4 MPa)
Ultimate Strength	70,000 psi (482.6 MPa)
Hardness—Rockwell B	90

Alloy S30403 is available in sheet, strip, plate, and bar forms. Standard temper available is the annealed.

Alloy S31603

Austenitic Stainless Steel
18-8 Stainless Steel
316L

Alloy Mix—Nominal Composition

Chromium   16% to 18%
Nickel   10% to 14%
Carbon   0.03% maximum
Molybdenum   2.0% to 3.0%
Manganese   2.00% maximum
Silicon   1.00% maximum

Alloy S31603 is the low carbon version of alloy S31600. This alloy retains the high corrosion resistance of the S31600 alloy but with a reduced carbon content. The lower carbon content is useful for reducing carbide precipitation developing from welding processes. Alloy S31603 has significantly lower strength than the higher carbon content version.

Yield Strength   25,000 psi (172.4 MPa)
Ultimate Strength   70,000 psi (482.6 MPa)
Hardness—Rockwell B   95

This alloy is available in sheet, strip and bar forms. Available temper is the annealed.

The S20000 Series Stainless Steels

Used to a far lesser degree are the 200 series stainless steel alloys. These alloys were developed during the Korean War when nickel became scarce. They tend to show up as alternative alloys whenever the nickel market starts to fluctuate upward. The 200 series alloys are austenitic stainless steels that contain reduced amounts of nickel and have the added element of manganese. These alloys have good corrosion resistance in most environments and are slightly more reflective than the S30000 series alloys.

Alloy S20100

Austenitic Stainless Steel

Alloy Mix—Nominal Composition

Chromium   16% to 18%
Nickel   3.5% to 5.5%
Carbon   0.15% maximum
Nitrogen   0.25% maximum
Manganese   5.5% to 7.5%
Silicon   1.00% maximum

Alloy S20100 was developed to reduce the nickel requirements. This alloy is corrosion resistant but lacks some of the formability of its nickel-bearing cousins. These alloys are stronger and slightly harder than the S30000 series alloys.

Annealed Temper

Yield Strength	38,000 psi (262.0 MPa)
Ultimate Strength	95,000 psi (655.0 MPa)
Hardness—Rockwell B	90

This alloy is available in sheet, strip, and plate forms. Available tempers are:

Annealed
Quarter Hard
Half Hard
Three-Quarter Hard
Full Hard

Alloy S20200

Austenitic Stainless Steel

Alloy Mix—Nominal Composition

Chromium	17% to 19%
Nickel	4.0% to 6.0%
Carbon	0.15% maximum
Nitrogen	0.25% maximum
Manganese	7.5% to 10.0%
Silicon	1.00% maximum

Alloy S20200 is similar to alloy S20100 with added chromium. Its corrosion resistance is good, but workability is still not to the levels of the S30000 alloys.

Annealed Temper

Yield Strength	38,000 psi (262.0 MPa)
Ultimate Strength	90,000 psi (620.5 MPa)
Hardness—Rockwell B	90

This alloy is available in sheet, strip, and plate forms. Available tempers are:

Annealed
Quarter Hard
Half Hard

The S40000 Series Stainless Steels

Alloy S41000

Martensitic Stainless Steel

Alloy Mix—Nominal Composition

Chromium	11.5% to 13.5%
Nickel	0.50% maximum
Carbon	0.15% maximum
Manganese	1.00% maximum
Silicon	1.00% maximum

Alloy S41000 is a martensitic alloy with good corrosion resistance. Although it will rust when exposed to exterior environments, the rust is superficial and will serve to protect against further corrosion. Most stainless steel cutlery is fabricated from the hardened form of S41000. This alloy is magnetic.

Annealed Temper

Yield Strength	32,000 psi (220.6 MPa)
Tensile Strength	60,000 psi (413.7 MPa)
Hardness—Rockwell B	95

Hardened Form

Yield Strength	140,000 psi (552 MPa)
Tensile Strength	180,000 psi (1,241 MPa)

This alloy is used for manufacturing nuts and bolts, springs, kitchen tools, fasteners, and cutlery. It is available in sheet and forging stocks.

Alloy S43000

Ferritic Stainless Steel

Alloy Mix—Nominal Composition

Chromium	14.0% to 18.0%
Nickel	0.50% maximum
Carbon	0.12% maximum
Manganese	1.00% maximum
Silicon	1.00% maximum

Alloy S43000 is a ferritic alloy with superior corrosion resistance owing to the increase in chromium. It will rust when exposed to exterior environments, but at a slower rate than the S41000 alloy. This alloy is used frequently as automotive trim. Interior ornamentation can also be fabricated from this alloy. Alloy S43000 is magnetic.

Annealed Temper

Yield Strength	35,000 psi (241.3 MPa)
Ultimate Strength	60,000 psi (413.7 MPa)
Hardness—Rockwell B	95

This alloy is available in sheet and strip, as well as bar forms. The available temper is the annealed.

Super Alloys

Alloy developers have recently sought to develop better, more corrosion-resistant alloys of stainless steel. Alloys that resist localized corrosion (such as pitting corrosion), known as "super stainless steels," have, to date, limited uses in architecture, but they are worth discussion because of their potential. In most cases, stainless steel can achieve better corrosion resistance through an increase of alloying elements such as chromium and molybdenum. The problem develops as increasing amounts of these alloying elements also increase the potential occurrence of inclusions, precipitates, and phase changes in the metal. These occurrences adversely affect corrosion resistance.

Steel mill producers seek to minimize the adverse conditions by using modern techniques of manufacture. Careful control of alloying constituents and elimination of impurities are the main results of these modern techniques. New stainless steels with properties previously believed unachievable are being obtained.

Since stainless achieves its corrosion resistance from the protective oxide film, reduction of impurities during manufacture will assist in the development of an uninterrupted layer. Interstitial elements found in the manufacture of steel are carbon and sulfur. These impurities affect the continuity of the oxide layer at certain times and under certain fabrication conditions. Carbon steals chromium from the protective layer of stainless, developing chromium carbide precipitation around welds. When the amount of chromium is decreased, the protective film is greatly reduced.

Sulfur forms sulfide inclusions that combine with alloying elements such as manganese. Manganese sulfide will damage the performance of the protective oxide layer. Manufacturing techniques, such as vacuum-melting processes, produce steels with low sulfur content. Other steel-refining processes, such as argon-oxygen decarburization, reduce the amount of carbon. These modern techniques make possible the reduction of impurities in all types of stainless steels. When all of these techniques are combined, the "super" characteristics of stainless steel develop.

An empirical method used by stainless steel producers is the *Pitting Index* (PI). The Pitting Index is established to provide a comparison of different alloys according to resistance to localized corrosion. The occurrence of pitting is the main cause of premature deterioration of stainless steel. This index, an empirical relationship of the different alloys of stainless based on their alloying constituents, reflects the likelihood of pit development. The higher the number, the more resistant the alloy is to pitting corrosion. The Pitting Index is determined as follows:

$$PI = \text{Pitting Index} = (\% \text{ Cr}) + (3.3)(\% \text{ Mo}) + X(\%\text{N})$$

where $X = 0$ for ferritic alloys and 30 for austenitic alloys.

% Cr = percent of chromium

% Mo = percent of molybdenum

% N = percent of nitrogen

As seen in Table 4-3, alloy S31600, one of the superior architectural alloys, has a PI of 25.3, whereas the super stainless alloy S34565 has a PI of 52.4!

TABLE 4-3. Comparison of Pitting Indexes for Various Stainless Alloys

ALLOY	C MAX	Cr	Ni	Mo	N	OTHER	Pi
S20100	0.15	18	5.5	—	0.25	8.5	25.5
S41000	0.15	13.5	0.5	—	—	2.0	13.5
S43000	0.12	18	0.5	—	—	3.0	18.0
S30200	0.15	18	9	—	—	3.0	18.0
S30400	0.06	19	11	—	—	—	19.0
S31600	0.08	17	12	2.5	—	—	25.3
S31254	0.02	20	18	6.0	0.20	0.6 Cu	45.8
S34565	0.02	24	17	4.5	0.45	—	52.4

The current high price and limited availability of the super alloys should change as the modern techniques are more universally adopted by the mill producers of stainless steel alloys.

Tempers and Hardness

When austenitic stainless steels undergo cold rolling, they work harden. This work hardening is measured and described as cold-rolled tempers. Annealing is usually necessary to bring the level to a particular temper. There are five basic tempers specified for stainless steels. The degree of temper is determined by the amount of cold rolling and annealing. The tempers are categorized by the minimum values for tensile and yield strengths (Table 4-4).

Stainless steel used in most architectural applications require the annealed temper. There is a dead soft temper available in some alloys. This temper has a reduced yield strength and lacks spring-back tendencies. Care should be taken when using this temper because of the softness of the metal.

As for hardness, even the annealed version is extremely hard in relation to most metals. As the temper increases, the hardness increases to the point that special tooling and equipment are required in order to process the metal.

TABLE 4-4. Standard Tempers for S30100 Stainless Steel Sheet and Strip

TEMPER	TENSILE STRENGTH		YIELD STRENGTH		ROCKWELL Hardness b
	ksi	MPa	ksi	MPa	
Annealed	75	517	30	207	92
Quarter Hard	125	862	75	517	102
Half Hard	150	1034	110	758	107
Three Quarter Hard	175	1207	135	931	110
Full Hard	185	1276	140	965	—

FINISHES

Following the final annealing and pickling of the stainless steel sheet, the mill applies the finish ordered. Depending on the end use and on the additional processes specified to follow, this finish can be adjusted.

Mill Finishes

The mill finishes are applied to strip and sheet stock forms of the material. Plate and bars are available from the mill in a hot rolled condition only. Additional finishes are applied to these products by other processors. The initial, and sometimes final, surface finish on the stainless steel is applied at the stainless mill or secondary metal processor. A secondary metal processor is a plant that purchases No. 2 finish material from the steel mill and adds a mechanical or chemical finish through additional processes. This processor may be the end fabricator, a division of the mill, or an independent company specializing in the finishing of metals. Table 4-5 lists the standard mill finishes available.

No. 1 Finish

A No. 1 finish is characterized by a rough, dull appearance. The dull surface is produced by hot rolling the stainless steel slab to the desired thickness, then annealing and descaling by pickling. This finish is available only on thick stainless steel plate, sheet, and bars, because it is a hot-rolled finish and not acquired by reduction on cold rolls. It is not typically an architectural surface, because the color and texture are not consistent and because the surface on large plates is mottled and uneven. This finish often contains streaks running in the direction of the rolling process, which are darker lines in contrast to the rest of the surface.

The No. 1 finish is also known as a hot-rolled and pickled surface. The minimum thickness available is 0.063 inches (1.6 mm). Because the sheet thickness is reduced by hot rolling, variations in thickness of as much as 10%, plus or minus, may occur. Small quantities are not normally stocked or available from the mill. For large orders, the surface finish can be improved and certain quality criteria can be established with the mill before any processing has begun.

No. 2D Finish

A No. 2D surface is a cold-rolled finish, applied to the surface by passing the coil of stainless steel through dull cold rolls after annealing and pickling.

TABLE 4-5. Standard Mill Finishes

FINISH DESIGNATION	DESCRIPTION	AVAILABLE FORMS
No. 1	Dull and mottled	Plate, bars
No. 2D	Dull, matte, consistent	Sheet, strip
No. 2B	Dull, reflective	Sheet, strip
No. 2BA	Highly reflective	Sheet, strip

Figure 4-3. Reflectivity of No. 2D finish stainless steel.

The result is a matte surface with low reflectivity. Care must be taken if this finish is to be the final architectural surface. The mill must control the annealing process to ensure color uniformity. The finish surface of the sheet must be protected with a paper or plastic coating soon after the final pass through the dull rolls. This surface is a nondirectional finish that cannot be reproduced by shop finishing. Used more and more frequently in architectural applications, 2D finishes are used when less reflective surfaces are required. When using this finish for architectural applications, the mill should be warned to avoid printing the alloy designation on the exposed finish surface. Often this surface must be ordered in mill quantities. It is not typically stocked by stainless steel service centers.

No. 2B Finish

A No. 2B surface is a cold-rolled finish with a more reflective surface than a 2D finish. The No. 2B finish is obtained by passing the pickled sheet through a final set of highly polished rolls. This finish is a smoky, reflective surface, not often used for architectural applications, except as flashings or as in the case of the roof surface of the Chrysler Building. The No. 2B finish is commonly used as the mill sheet prior to the application of more polished surfaces by secondary finishing processes. Further passing through cold rolls improves the finish but sacrifices the ductility of the thin sheet.

No. 2BA (Bright Annealed) Finish

Bright Annealed finishes are highly reflective surfaces produced at the mill by final annealing the stainless steel in a controlled-atmosphere furnace. The purpose of the controlled atmosphere is to prevent the development of surface oxidation and scale. The Bright Annealed finish cannot be matched in

Figure 4-4. Reflectivity of No. 2B finish stainless steel.

a fabrication shop. Attention to detail and control of welding should be exercised when using metal with this finish.

The 2BA finish is a common ornamental surface used on reflective trim parts and strips and ornamental metal panels and fascia. It is not a true mirror finish, but has the greatest level of reflectivity of all mill products. The reflective image often has a slight "smoky" appearance or may be very clear and reflective, approaching a mirror-polished surface. Samples should always be required when specifying this surface finish. The lack of apparent polishing lines gives the 2BA finish a nondirectional appearance.

Polished Finishes

The most common architectural stainless steel surfaces are the mechanical surface finishes. Mechanical surface finishes are applied to sheet or strip stock from the mill by a secondary processor. A metal-polishing house or metal fabricator will start with a No. 2B mill finish. If plate or bar steel is to receive the final mechanical finish, then the fabricator would start with a hot-rolled mill surface.

The mechanical finishes are applied to the surface of the metal by successive passes of polishing wheels or belts across the surface. In the case of plate or bar steel, the rough mill surface must first be ground smooth to remove the scale or leveled to eliminate the inconsistent surface.

Table 4-6 lists the various standard polished finishes used on stainless steel.

No. 3 Finish
A No. 3 polished finish is characterized by short parallel lines, running the length of the sheet. The lines, known as grit lines, are mechanically applied

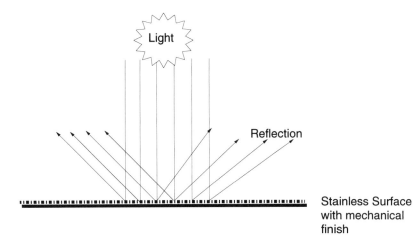

Figure 4-5. Light-scattering effect of satin finish textures.

by abrading the surface with fine-grit abrasive belts or wheels. The No. 3 finish is the coarsest of the standard polished surfaces. The abrasive range used is from 80 to 150 grit, with 100 grit the standard for this finish. Typically, this finish is produced by passing the metal first through an 80-grit aluminum oxide belt, and then through a 100-grit aluminum oxide belt.

Polishing fabricated parts to this finish is difficult. Matching one polished surface provided from a fabricator or polishing house to that of another No. 3 polished surface provided from a different source is difficult to impossible. Depending on the lighting, the finishes may appear as different metals. The characteristic short, skipping grit lines are difficult to match or repair on large flat surfaces. When the short skips are not closely matched for depth and length, the rework will stand out. Weld repairs at corners and edges can be effectively matched when polishing is kept to a minimum by masking and blocking the weld.

The short, parallel grit lines tend to break up the reflected light, which assists in concealing minor scratches and imperfections on the surface of

TABLE 4-6. Standard Polished Finishes for Stainless Steel

FINISH DESIGNATION	DESCRIPTION	AVAILABLE FORMS
No. 3	Reflective, coarse grit lines	Plate, sheet, strip, bar
No. 4	Reflective, finer grit lines	Plate, sheet, strip, bar
No. 5	Nonstandard	
Hairline	Reflective, fine parallel lines	Plate, sheet, strip, bar
No. 6	Reflective, long parallel lines	Plate, sheet, strip, bar
No. 7	Very reflective, mirror with lines	Plate, sheet, stip, bar
No. 8	Mirror finish	Sheet, strip, bar
No. 9 (No. 8 special)	Buffed mirror finish	Sheet, strip

Figure 4-6. Reflectivity of No. 3 finish stainless steel. Note chatter marks in reflection. Chatter is considered a defect.

stainless steel. For large areas constructed from multiple sheets, it is critical to use sheets polished by new belts with matching grits.

No. 4 Finish

A No. 4 finish is similar to a No. 3 finish, except that the parallel grit lines are finer. This mechanical surface is developed by using 120 to 240-grit abrasive belts or wheels. Following the same steps for polishing the No. 3 finish, follow a pass through a 100-grit aluminum oxide belt with a pass through a 180 or 240-grit silicon carbide abrasive belt. The No. 4 finish is a common architectural surface because of the more lustrous appearance obtained by the finer grain. The fine lines can be repaired and reproduced to match welded or fabricated sections, as in the No. 3 finish. The matching of the finish is difficult, but achievable. Care should be taken when mixing different suppliers of this finish; slightly different grits will create differences in the final appearance.

No. 5 Finish

The No. 5 finish was never produced in the United States, and records of the true nature of this finish are scarce. It was said to be a European finish produced by polishing with a nonwoven abrasive belt—a rather broad description. A guess would be that this finish is what is now known as "hairline," "long-grain," "fineline," or any of a number of names relating to the finish produced by fine-line brushing with a Scotch-Brite[3] pad.

[3]Scotch-Brite is an abrasive nonmetallic pad manufactured by 3M Corporation, St. Paul, Minnesota.

Figure 4-7. Reflectivity of No. 4 finish stainless steel.

No. 6 Finish

A No. 6 finish has a soft satin appearance. It is produced by tampico-brushing[4] a No. 4 finish in an abrasive and oil medium. This finish has very fine grit lines, but its reflectivity is lower than that of a No. 4 finish because of the many minute lines on the surface.

A common variation of this surface uses the fine range of the No. 4 finish applied by using Scotch-Brite pads or Scotch-Brite rolls. This finish is characterized by long parallel lines running the length of the sheet, which appear almost as fine grooves in the surface of the metal. Common in the Far East, this finish is known as "long-grain" or "hairline." The reflectivity is greater than a No. 4 finish, and the scratch lines are longer, running the length of the sheet or part.

This variation of the No. 6 finish can be repaired and matched with greater ease than the other directional polished finishes. Minor surface imperfections on the base stainless steel will disappear when viewed from certain angles, but may appear when viewed from others. It is important, when inspecting the sheet stainless steel, to view it from different lighting angles, preferably under the lighting conditions in which the final installation will be viewed.

The benefits of the directional finishes on stainless steels lie in the degree and form of light reflectance from the surface. The linear texture of the stainless surface created by the small ridges acts as a defraction grating. This property causes a perceived iridescence by scattering reflected light off the closely spaced ridges. The scattered light conceals minor surface imperfections on

[4]Tampico is a stiff fiber of the *Agave rigida* plant. This hard but pliable fiber is used in circular power brushes. Desired for the fiber's ability to hold polishing compounds.

the stainless surface. The directional surfaces, such as the No. 3 and No. 4 finishes, reflect light at right angles to the grit line direction regardless of the light source.

No. 7 Finish

A No. 7 finish is a highly reflective surface. The grit lines are still apparent, but the surface starts to take on a mirrorlike aspect. This finish is produced by successive buffing operations on a finely ground surface. It must be polished through successive grits up to and including 320-grit silicon carbide. After the abrasive polish, a buffing operation is performed to develop the mirrorlike surface. Light is reflected from the No. 7 surface in intense hot spots, as in reflective glass or the more highly polished No. 8 finish.

No. 8 Finish

The mirrorlike surface of a No. 8 finish is produced by polishing a finely ground sheet of material with successively finer grit abrasives up through 320-grit silicon carbide. After abrading, the surface is buffed repeatedly with very fine buffing rouges.

Typically, there is fine "angel hair" visible in the reflective surface, particularly when viewed from an angle. The angel hair is very small grit lines that were not removed in the buffing process. Removal of these grit lines is very

Figure 4-8. Reflectivity of No. 8 finish stainless steel. Note very slight grit lines in reflection.

difficult with conventional buffing equipment. To achieve a flat surface, the back side of the polished sheet should be roughly ground and polished too.

Highly Reflective Surfaces

Japanese stainless steel processors introduced a mirror-surface product they call a No. 8 finish. Because of the superior nature of the surface, it is sometimes called a "super No. 8" or "No. 9" finish. This finish shows no grit lines whatsoever and has a reflective surface indistinguishable from a glass mirror. Currently there are no U.S. firms producing this finish. Japanese, Canadian, and Chinese sources are producing this finish.

The mirror-finishing process uses special high-speed buffing equipment with special rouges. This is followed by a thorough inspection of the surface and repeated buffing of areas that show any signs of grit or surface marking.

There are many other custom or proprietary finishes used to provide special reflective properties to a stainless steel surface. Some of these finishes can be intermixed to produce interesting reflective effects. Table 4-7 summarizes the families of custom finishes produced mechanically.

Selective Polishing
Selective coating of mirror-finished stainless steel, then applying a directional finish over the unresisted regions, followed by removal of the resist will

Figure 4-9. Reflectivity of No. 9 finish stainless steel. Note complete absence of grit lines.

TABLE 4-7. The Basic Families of the Custom Finishes Produced by Mechanical Means

CUSTOM FINISH	DESCRIPTION
Selective Polishing	Uses resists to protect certain surfaces during application of polishes or abrasion to adjacent surfaces.
Swirl Finishes	Application of grinding disc at various angles, overlapping to produce surfaces that have a three-dimensional appearance.
"Engine Turn"	Produces small circles applied in grids or overlapping layers. They can also be developed by slipping along the length of the sheet or part.
Distressed Finishes	Characterized by deep, multidirectional scratches across the sheet, usually in random arches.
"Angel Hair"	Characterized by very small, fine scratches applied randomly across the surface.
Abrasive Blast	A matte surface created by abrading the surface of the metal using various media applied at high pressure. Glass bead, clean sand, quartz, and steel shot are just a few of the media forms.

Figure 4-10. Selective polishing of fine satin over No. 8 finish.

produce contrasting reflective surfaces. Selective finishing can produce very decorative surfaces on stainless steel. By application of special resists to the surface of stainless steel, exposed regions of the surface can receive textured polishes by blasting or brushing. The resulting surface is highlighted by the contrast created by the different reflectivity of the textures. Fine detail can be achieved by using photoresist materials. The resist materials used are often rubber films cut by hand and carefully applied to the stainless surface, or plastisols applied by silk screen or by photoresist techniques.

Swirl Finishes

Certain finishes, through the use of grinding wheels and stainless wire brushes, can be created on a stainless surface. For mechanical brushed finishes, definition between mirror and brushed transition lines is critical in producing distinctive designs. Brushed lines can be counterpoised to present contrasting appearances caused by light reflecting at different angles.

Using a grinder in a random fashion will, in the hands of a craftsperson, produce elegant surfaces of alternating swirls. The swirls have random biases in relation to one another and reflect light in a similar fashion, producing an illusion of contour.

Figure 4-11. Reflectivity of custom ground/swirl texture.

These are wholly custom finishes, usually dependent on the abilities of the applicator. It is necessary to begin with flat 2B or 2D stainless steel. A light directional polish is applied to the surface by passing the sheet through a polishing roll. A grinder is applied by hand or by using a CNC-assisted frame, to control the movement and bias of the grinding head. As the grinder passes across the surface, small areas receive the texture in a slightly curved direction. Alternating the grinder disc edge will apply the curved brushing in similar fashion to a painter using a small brush. Larger swirls can be produced as well by multiple grades of disc grits. Small areas and prewelded assemblies can also receive custom swirl ground patterns. It is important to have control samples to work from if any level of consistency is required.

Engine Turn

The "engine turn" custom finish is characterized by small circular lines, either in rings or in overlapping complete circles, applied over a mirror or No. 2 finish. The circles can range in diameter from as little as 12 mm to as much as 300 mm. The circles are applied by CNC-operated machines, which control a bank of spinning motors. The motors have a disc of silicon carbide imbedded in rubber or resin or in a block of the same material. While the disc is spinning, the motors lift the spinning disc from the surface and move it to some preprogrammed location, each time raising and lowering the disc to apply the decorative finish in repeating swirls.

Distressed Finishes

A distressed finish is produced by randomly applying a scratch or series of scratches to the metal surface using a small grinding wheel. The application is sometimes performed by hand, using a wheel to scratch the surface in random patterns.

Angel Hair

An "angel hair" finish is similar to a distressed finish, except that the grit lines are small and very fine. Produced by a wire wheel applied at various angles

Figure 4-12. Reflectivity of "distressed" finish stainless steel.

Figure 4-13. Reflectivity of "angel hair" finish stainless steel.

to the surface, this technique achieves an attractive nondirectional finish on stainless steel.

Abrasive Blast Finishes

Another finish desired for its nondirectional texture is the shot peened or abrasive blast surface finish. Blast or peened finishes produce a soft satin reflection without directional grit lines. The process of applying the texture work hardens the surface and can cause some surface distortion, particularly on thin gauges. Refer to Table 4-8 for a list of various abrasive blast media used on stainless.

The abrasive blast finish is applied by spraying the surface of the metal with very fine, homogeneous particles. The particles can be glass beads, aluminum oxide, clean sand, steel shot, or any number of other materials. Different materials create different tones, ranging from the soft silver sheen produced by glass beads to a dark gray coarse texture produced by fine sand.

Simply applying the finish by sand blasting the surface of a stainless fabrication will usually lead to an imperfect, stained, and distorted metal. Care should be taken to ensure that the stainless surface is free of all contamination, such as fingerprints, oil, weld marks, and grinding lines, prior to blasting.

TABLE 4-8. Blast Media and Finishes

BLAST MEDIA	FINISH PRODUCED
Fine Sand	Dark, coarse surface
Glass Beads	Lighter, smooth, grainy surface
Silicon Carbide	Very coarse surface, dark tone
Aluminum Oxide	Very coarse surface, medium tone
Stainless Steel Shot	Honed surface, tiny curved impacts
Ground Quartz	Coarse, shiney surface, angular impacts

Figure 4-14. Reflectivity of glass bead finish stainless steel.

Blasting the stainless surface will not remove such surface defects, but permanently fix the contamination in the surface.

To eliminate or reduce surface distortion, the stainless sheet should be blasted on both sides. Laminating the stainless to a rigid backer such as honeycomb board will reduce distortions when only one side is blasted. Use a thicker stainless sheet or add shape to the form to stiffen the object if necessary.

After blasting the stainless steel surface, passivating will render the surface clean of damaging iron particles and help it to resist fingerprinting. When the stainless steel is blasted, the surface becomes susceptible to fingerprinting and smudging. The protective chromium oxide layer has been removed by the blasting, and the surface is grainy, resulting from the induced texture. Passivation will occur naturally when the metal is left to oxidize under typical atmospheric conditions. Chemical treatments using nitric acid also can be used to accelerate the passivation process. The natural formation of the corrosion resistant film can be enhanced by treating in nitric acid. Nitric acid will remove free iron particles from the surface of the metal.

Nitric Acid Passivation
Nitric acid: 10 to 15% by volume in water

Better results will occur if the solution is heated to 150°F (65°C) for the austenitic stainless steels, 120°F (50°C) for the ferritic or martensitic alloys. Allow the part to be immersed in the solution for approximately 30 minutes. Thorough rinsing should follow. Exercise extreme care when using acids. Use eye and hand protection. Nitric acid fumes are extremely dangerous.

For in-place fabrications, rust particles can be removed using a conventional rust removing cleaner. An oxalic acid (wood bleach) paste applied over the surface and gently rubbed into the surface may remove some surface particles of corrosion. If this fails, the only solution may be to reblast the surface using an abrasive media such as glass bead. The glass bead will remove a portion of the surface contamination and leave a smoother, more passive surface behind. The glass will not contaminate the stainless surface.

Aluminum oxide can damage the surface of the stainless by implanting small particles of aluminum in the outer layer of the metal. Exterior applications will expose the metal to the possible development of corrosive cells.

Shot peening applies tiny steel ball bearings or steel shot under pressure to the metal surface. The transfer of energy to the stainless surface by the shot will warp thin stainless steel sheets. The resulting surface is usually coarser than that produced by abrasive blasting, and the shaping of the metal from the impact is more severe. Passivation of the surface with a dilute nitric acid solution is critical to dissolve the microscopic iron particles imbedded by the steel shot. These finishes should be provided by companies experienced in the application of custom finishes and the handling of acid solutions. Samples should be obtained and examined prior to the final application. Surface flatness characteristics should be discussed and evaluated before beginning any project using abrasive and shot peen finishes.

Electropolishing

Another process used to enhance the reflective levels and finish quality of stainless steel is electropolishing, or electrolytic polishing, of stainless steel. Electropolishing is an electrochemical process that removes some of the microscopic deformations on the stainless steel surface, leaving a brilliant, reflective tone.

Unlike mechanical finishes that remove metal from the surface by using abrasives, electropolishing levels the surface of a metal by creating an electro-potential between the surface of the stainless and a cathode placed in a chemical bath. The microscopic surface high points of the metal are etched away by concentrating the charge at the microscopic peaks, leaving a more even and thus smoother surface. The chemical bath is a strong acid solution,

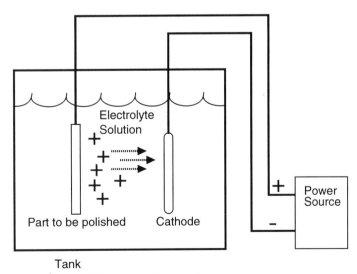

Figure 4-15. Electropolishing technique.

heated to a predetermined temperature. A current passes through the stainless steel part and travels across the electrolyte solution to the cathode. The high points on the surface of the metal part concentrate the electrical potential and are dissolved at a quicker rate than the lower parts. The result is a smoother, more even surface.

Small parts manufactured from stainless are commonly electropolished to produce smooth, reflective surfaces. Large sheets of stainless can also be electropolished. No. 3 and No. 4 satin surface finishes can be electropolished to enhance the reflectivity and remove some of the surface metal damaged and stressed from the mechanical finishing operation. The sheet emerges with a more reflective surface, and the polish lines are less distinct. The resulting surface is very passive and resists fingerprinting.

Acid Etching

The acid etching surface finishing process uses chemical treatments to arrive at a custom finish. Acid etching actually removes a thin layer of stainless by dissolving the metal. Ferric chloride is the corrosive used to dissolve stainless at a controlled rate. The longer the exposure to ferric chloride, the more metal is removed. Acid etching of stainless steel can produce an attractive surface on the metal. Etching can be selective, by coating the surface of the stainless with a resist material capable of withstanding the corrosive behavior of the ferric chloride. Designs can be silk-screened onto the surface or photoresisted to produce fine detail in the acid-resisting layer. The stainless steel sheet with the resist selectively applied is exposed to a spray of corrosive material. The corrosive material removes the portion of the surface not protected by resist. Corrosives sprayed onto the stainless at angles perpendicular to the surface will flow off, removing some of the metal. Incorrect etching will undercut the resist, creating rough edges. Resist materials that are not correctly adhered will also cause undercutting and imperfect edges to the design. Once the stainless steel surface is etched to the depth desired, usually from 1 to 3 mils, the surface is thoroughly rinsed and the resist coating removed.

The etched surface can be brightened or left with the tone produced by the etching process. Brightening can be achieved by chemical polishing. Chemical polishing will dissolve surface metal but not to the degree of the etching process. The etched surface has a dull silver sheen because of the way the etching process eats away portions of the metal at different rates. The surface of the etched metal is microscopically rough. Further coating of the etched surface is not necessary; however, it is wise to electropolish the surface after processing.

One problem that often arises from incorrect procedures is surface finish smutting. If the rinse cycle is not adequately maintained, or if the rinse has some contamination, a dark stain, like a water stain, appears across the etched surface. This stain cannot easily be removed.

Embossing and Coining Textures

Embossing of stainless steel is an effective way of creating a decorative surface texture that acts both to stiffen the surface and conceal surface mars and

Figure 4-16. Etched stainless steel door.

scratches. Embossing allows the use of thin-gauge stainless steel sheet in situations where apparent flatness and durability are desired.

The technique is relatively inexpensive, assuming the choice is a stock pattern and not a custom surface. Embossing or texturing is produced by passing the metal in sheet form, usually coil stock, between two large embossing rolls. These rolls contain the texture to be induced into the metal. One roll has the texture engraved into its surface, and the other roll has the negative of the texture raised from the surface. The surface of the flat sheet is selectively stretched by the raising of the pattern as it passes through the rolls. The stretching and pinching of the metal work hardens and induces rigidity into the thin metal sheet. This rigidity helps to conceal oil canning, and the pattern distorts the light reflection to assist further in concealing any imperfections.

Coining is produced in a similar manner to embossing by passing the stainless sheet through rolls, only one of which has a texture. Many different patterns are available. Diamond shapes, ridges, and smooth bumps are just a few of the standard patterns. One of the more popular architectural coined surfaces is the "Linen" pattern, offered by the German company, Krupp. A very similar pattern is known as "Cambric" and is rolled onto stainless steel sheet by British Steel/Avesta. The "Cambric" pattern was used on the Canary Wharf office tower in London, designed by Ceasar Pelli. Through embossing and coining the stainless surface is made more rigid, thus allowing thinner gauges. The surfaces diffuse the reflected light differently and effectively conceal minor scratches and mars.

Any of the mill or polished finishes can be textured. Textured surfaces can also be repolished to create interesting contrasts. Selectively polishing the raised portion of the surface and leaving the recessed portion with the original polish is a technique known as highlighting.

Embossed textures, like most roll-applied finishes, are directional in appearance. This is because of the single-directional nature of the rolls applying

Figure 4-17. Reflectivity of rolled "shadow" finish stainless steel.

the texture. The textures are applied down the length of the sheet and repeat as the roll comes full circle.

Such directionality must be taken into consideration when using embossed materials. Nondirectional satin finish textures are sometimes applied with the use of coining rolls. The texture produced can be very similar to a light blast texture. Heavy rolls can be shot peened to imprint a negative of the pattern into stainless steel sheet. There are a few proprietary finishes with striking, consistent, nondirectional satin finishes. These finishes are usually available in only relatively heavy gauges because they are applied on a Sendzimir mill, a set of very high pressure reducing rolls. The tremendous pressures are necessary for imprinting the texture onto the stainless steel surface.

COLORS IN STAINLESS STEEL

A natural, chromelike surface has the unfortunate stigma of giving a clinical or sterile appearance. It lacks warmth. The bright white of reflected light from a mirror surface, in some circumstances, lacks the richness projected by polished brass or gold leaf. Various techniques are available to produce astonishing colors with the natural luster of stainless steel.

Organic and inorganic coatings have been applied to stainless, as well as other metals, not so much for protection, but for aesthetic appeal. The Japanese call these products resin-coated stainless steel and use stainless as the base material more often than galvanized steel or aluminum. The costs associated with applying paints to stainless steel are not much greater than for similar treatment of aluminum.

There are other techniques available for adding color to stainless steel, which fall into three common categories. Each method produces remarkable metallic color tones by the phenomenon of light interference. Light interference creates spectral colors by separating a wavelength of light into its various components.

Heat tinting is one of the oldest metal coloring techniques used on steels and brasses. Through heating the metal, an oxide film layer is generated, which acts as an interference film. Heat tinting of stainless steel can produce colors ranging from a bluish hue to a brownish black. The colors are rich and very striking, but they are uneven, difficult to reproduce, and not possible on large parts or sheets. Heat tinting is more an artistic medium for smaller, controllable stainless steel parts.

Immersion in molten salt baths is another coloring process. This technique uses fused salt baths that react with the surface of the stainless and create colorful oxides. Colors are limited to black, whitish yellow, magenta, and gold. These colors are striking, but are limited to small ornamental features. The finish is durable and uniform. The colors are obtained by chemical reactions with the chromium content of stainless steel. Coloring stainless steel enhances its corrosion resistance beyond that of the uncolored version. The added film adds to the passivity of the natural film.

Coloring of stainless steel using the acid bath method is the most commercially viable process. The product is known as "colored stainless steel." This process, which takes the development of interference films to a commercial art form, uses technology developed by International Nickel Co.,

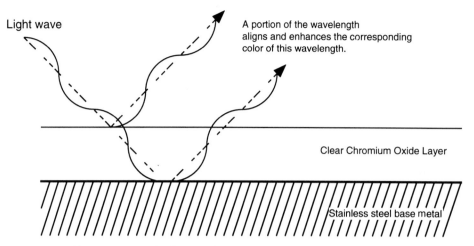

Light wave

A portion of the wavelength aligns and enhances the corresponding color of this wavelength.

Clear Chromium Oxide Layer

Stainless steel base metal

Figure 4-18. Light interference phenomena.

Ltd (INCO). INCO has licensed many firms around the world, in particular European and Japanese companies as well as a few North American companies, to perform the process.

There is no color in architectural use today that equals the iridescence produced by light interference on stainless steel. The acid bath process produces interference colors by closely controlling the generation of the chromium oxide film layer. This oxide film layer is clear, yet remarkable colors are produced by the physics of light interference.

The colors generated are those obtainable by separation of the light spectrum. Light reflects off the dual surfaces, the surface of the metal and the surface of the oxide film. Depending on the thickness of the oxide film or the angle of the light reflecting back to the eye, different colors can be obtained. Portions of the light wave are enhanced in relation to other adjacent wavelengths, or portions are counteracted by the misalignment of these wavelengths.

The INCO process consists of initially degreasing the stainless steel sheet or part, then immersing it in a hot chemical bath. Degreasing is performed in a heated caustic soda bath. The chemical bath is a mixture of chromic and sulfuric acids. Immersion induces the formation of a clear oxide film similar to, but much thicker than, the natural film produced by atmospheric exposure. Depending on the film thickness developed, interference colors are produced. The film thickness ranges from 0.02 microns for the bronze color to 0.36 microns for the green color. The interference oxide film is 50 times greater in thickness than the natural oxide film that develops when stainless is exposed to the atmosphere. Both sheet and fabricated parts, of any alloying form of stainless steel, can be colored through this process.

The INCO process allows for careful regulation of the surface film and adds stability so that the color will not alter. Control and regulation of film growth is achieved by monitoring the electrical potential developed by the thickening surface layer. A potentiometer is positioned to read the thickness of the oxide film as it grows. The object or sheet is pulled from the immersion tank, and the color is noted at the particular electrical potential reading. Subsequent coloring of additional parts is compared to the reading originally obtained.

Immersion time and chemical composition determine what color is achieved. The time of immersion is very sensitive to bath temperature variations. It is critical to ensure that the bath is of equal temperature throughout.

The surface finish on the stainless also affects the color monitoring, since various surface finishes have unique electrical potentials. A satin finish stainless steel will color differently than a mirror finish stainless even when time and temperature are constant. Likewise, fabricated parts can have different colors if the surface finish is not consistent throughout. As the interference film is generated, the colors on the stainless steel surface appear in the following sequence:

Bronze
Blue
Gold
Red
Purple
Green

Intermediate levels are possible, but consistent control is extremely difficult. The intermediate colors occur at the transition from one color to the next. Depending on the finish, microscopic sections may color differently at the intermediate levels, producing, for instance, specks of red in a blue satin finish panel. Maintaining consistency from one panel or part to the next at the intermediate levels can be a nightmare. There are no realistic controls for the intermediate colors.

When a sheet is removed from the coloring bath, the color is usually different than planned. Sheets to be colored green, for instance, appear red or blue. As the chemical process continues for a few moments and the sheets dry, the green color begins to appear.

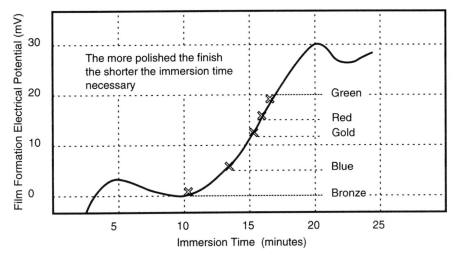

Figure 4-19. Graph of coloring process on stainless steel.

After coloration occurs the film is hardened. Initially, the film is porous and unstable, prone to easy damage. Pores of 100Å to 200Å are prevalent across the surface. A hardening treatment applied to the colored stainless is required to stabilize this thin interference film. This step consists of immersion in dilute phosphoric and chromic acids and application of an electrolysis treatment, using lead as an anode and the colored sheet as a cathode. The electrolysis and immersion lasts for approximately 5 to 10 minutes and fills the pores with a hydrated chromium oxide layer. The film, after hardening, has a chemical makeup of chromium and iron in combination with H_2O. The resulting film has a spinel structure, hydrated and enriched with chromium, which adds durability to an abrasion-susceptible surface. This technique is required for all commercial colored stainless steel applications. Without it, the color of the surface will change with the simplest contact.

Black Stainless Steel

Producing the color black requires a slightly different process than that used for the other interference colors. A black color is achieved by initial immersion in a chromic and sulfuric acid bath solution at 212°F (100°C) for 5 to 10 minutes, followed by the hardening process. Usually Type S30400 stainless steel is used to arrive at the various colors. Black is the only color difficult to achieve with Type S30400 stainless; the color produced is a dark charcoal gray. Type S43000 stainless, however, will produce a rich, dark black color.

Custom Modifications

The coloring process can be performed on any stainless finish, even the heavy-textured embossed patterns. A bright, glasslike finish will result from the treatment of a highly polished stainless like a No. 8 or a No. 2 Bright Annealed. Likewise, a matte finish will result when a satin stainless finish such as a No. 3 or a No. 4 finish is treated.

Selective polishing, etching, and photoresist techniques can all be used to further enhance the decorative surface created. Improved quality is achieved when the flat sheet or fabricated part is electropolished prior to coloring.

Selective polishing mirror stainless steel with satin finish wheels or brushes removes the color at the area of polishing and produces decorative contrasting surfaces. Abrasive blasting also removes the colored stainless steel surface film to produce decorative contrasts.

Color stripping is the selective removal of the film, exposing the natural color below. By subsequent resisting and recoloring of the surface, multicolors can be obtained on the same sheet or fabrication. Color stripping is performed by coating the sheet of colored stainless steel with a masking using an acid resist film. The panel is then immersed in an acid bath of hydrochloric or sulfuric acid. The colors of areas not protected by the acid resist film will be removed. Typically, the resist is then removed by immersion in a caustic acid bath.

Fabricated parts as well as flat sheets, strip, foil, rods, and tubes, can receive interference coloring. The limiting factors are the dimensions of the tank used for processing. Care should be taken to ensure that the stainless steel parts are of the same alloy type and finish. Different alloys will color differently.

NEW TECHNIQUES OF COLORING

Light Interference Phenomenon

The phenomenon of light interference is the interaction of reflected light from different surfaces. Light is unique in that it acts physically as both a particle and a wave. Light waves traveling in parallel paths, reflecting off the surface of a transparent film and the surface of a base material, will, if given the right circumstances, interact in such a way that a constructive reinforcement of the light wave or a destructive cancellation will occur. As the light travels through a transparent film, even over very minute distances, the wavelength is altered. Thus, through regulating the thickness of the transparent oxide film on the base stainless steel surface, the wavelength of light reflecting off the base metal is altered in relation to the wave of light reflecting off the film surface.

For colored stainless steel it is necessary to consider a light source emitting the full spectrum of visible light. A full spectrum will ensure the cancellation or constructive reinforcement of the color desired. The cancellation of all colors will produce the color black.

Fabrication Considerations

Colored stainless steel can be cold worked much as natural stainless. Special care, because of the fragile nature of the film, should be taken. The film is hard, so that when it is formed small microscopic cracks develop along the formed edge. These cracks are not a detriment; however, there is a slight color change at the break line. When severe forming operations are performed, the colored stainless experiences a reduction in gloss and appears slightly frosted, much the same as natural stainless. It cannot be refinished without removing the color.

Welding colored stainless will discolor the region around the weld. Welding prior to coloring is possible, depending on the nature of the fabrication. When this occurs, special care must be taken in finishing the surface prior to coloring. Electropolishing the surface after grinding and polishing the weld joint will usually provide best results in color consistency.

Spot welding and resistance welding are more difficult because of the increase in resistance the thicker film provides. This type of welding will also alter the interference film layer around the point of joining.

Soldering is an excellent way of joining colored stainless steel parts. Because of the low heat involved, soldering does not damage or remove the film. Soft soldering of the Pb - Sn alloys without chloride soldering fluxes will work without damage to the color as long as heat is kept low. Silver soldering generates high temperatures and will discolor the surface. Chloride fluxes remove the film, returning the stainless to its natural silver color.

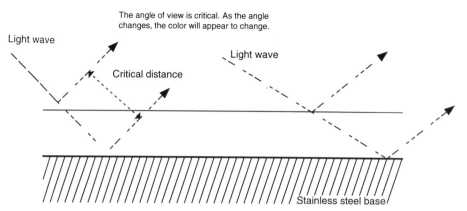

Figure 4-20. Angle of view phenomena on light interference films.

Precautions—Angle of View

The phenomenon of light interference, particularly on reflective stainless steel surfaces, produces iridescent colors. These colors are permanent and stable in most atmospheric conditions. Since the colors acquired are related to film thickness, angle of view will affect the color seen. When objects constructed of colored stainless are viewed from a distance, the color appears most consistent. The same is true for small samples of the finish. But when the small samples are tilted, or when the large fabrication is viewed up close, the color changes because the angle of view has changed. The greens and reds are most susceptible to angle-of-view interpretation. Gold colors can also exhibit this phenomenon. What may look gold when viewed directly may appear red or green when seen at a slight angle. The reason is that the viewer is seeing the object through a different film thickness. A different wavelength results from the angle from which the object is now viewed.

It should be noted that the colors produced on stainless steel by the light interference phenomenon are not anything like anodized finishes on aluminum or paint finishes. The variable color, depending on the viewing angle, is highly metallic owing to the nature of the polished stainless steel surface. Variations will occur from one sheet to the next because of the many factors involved in achieving the color and because of the delicate nature of the clear oxide film.

Another problem with angle of view develops from the method used to fasten the panel or sheet to the structure. Slight variations in surface flatness can make different colors appear on the same sheet. Panels with fixed edges will have a slight bow in the middle resulting from thermal expansion. The edges may appear a slightly different color because of the slight difference in flatness. Removing the sheets or panels and inspecting at close proximity and at a different attachment point may show the panels to be consistent in color. Gold, greens, and reds are sensitive to this phenomenon. The final color is subject to the final placement and to the position from which the object is viewed. There are few finishes that display different appearances from different angles. However, this condition must be understood by both the fabricator of the product and the person who has final acceptance of the product.

Precautions—Staining and Fingerprinting

A surface will stain and discolor when exposed to certain chemical agents. In addition, even after the hardening process, a colored film is easily scratched. The color is achieved only by the action of the film; thus a scratch will appear as the natural uncolored stainless. When the surface is scratched, it cannot be repaired. Damage of the film by abrasive action will require replacement.

Styrofoam, of the rigid variety, can scratch the film and remove the color. The satin directional finishes are more susceptible than the mirror finishes. Even cardboard used to separate colored surfaces will scratch the surfaces. Use a very soft paper interleave, or better still, cover the stainless with a thick film of PVC.

Fingerprinting can be an overwhelming problem as well. The film has a refractive index[5] of 2.2 and will show even minor surface oils. Since the color is obtained from the interference phenomenon of the thin film, if another thin film is added (for example from the perspiration on ones hands), there will be temporary discoloration. Fingerprints and most surface oils can be easily removed, returning the stainless to the original interference color.

When colored stainless steel is delivered to a project site, it is often protected by a thin plastic film applied by the manufacturer. This film contains an adhesive to assist in adhering to the sheet. When the plastic is removed, some of the adhesive remains and can show as a different color. Clean the sheet to remove the adhesive and to prevent dust and dirt from collecting on the stainless steel surface. Leaving colored stainless steel exposed to the weather will eventually even out the color as surface films deposited by nature cover the entire surface.

On newly colored sheets, placing a PVC protective layer over the metal will sometimes change the color. This should be a temporary change and will eventually wear off. The problem occurs in color matching.

Morning dew or other moisture collecting on the surface will alter the film slightly until it evaporates. Colored surfaces, particularly mirror surfaces, are self-cleaning by means of naturally occurring rain. Periodic washing with clean, potable water will help keep the reflective color consistent.

Precautions—Clear Protective Coatings

Clear coating is possible on colored stainless steels. It will, however, make for a less reflective surface. In addition, it is sometimes difficult to get a clear film to adhere adequately to smooth, polished stainless surfaces. Because the color will be altered by the clear coatings, consider test samples before

---

[5]Refractive index is the ratio of the sine of the angle of incidence to the sine of the angle of refraction. This is a measure of the refracting quality of a surface.

coating the full part. Clear coatings have had limited success on the dark blues and black colors. Coatings over red, green, and gold colors have not shown acceptable results.

Precautions—Temperature

Sunlight absorption by colored stainless is much greater than by natural-colored stainless steels. Thermal expansion resulting from heat absorption by colored stainless, particularly the blue and black colors, will be almost five times that of the natural silver color. Temperatures upward of 212°F (100°C) will be reached.

The film is stable even under these temperatures because of its chemical makeup. Failure of the film has occurred selectively across the surface of a sheet because of excessive temperature buildup on one project in the Saudi Arabian desert. It has been suggested that the stainless steel surface became so hot that it "dried out" or dehydrated at locations where insulating materials trapped the heat.

The film is made of hydrated oxides of chromium. Stability under normal conditions will not change the color. Projects in industrial environments have shown resistance to change over many years.

Long-Term Performance

In 1986 a report on the 10-year exposure of colored stainless steel was made available by International Nickel Company Engineered Products Ltd. The report described the surface appearance of colored stainless steel samples after ten-year exposure to the marine environment at the LaQue Center for Corrosion Technology in North Carolina. Analysis of the colored stainless steel samples revealed that corrosion resistance was improved over natural colored control samples and the quality of the color was unaffected. No significant color fade was detected. The uncolored sheared edges of the samples showed indications of rust and surface corrosion, and there were minor amounts of surface pitting.

The automotive industry has tested the coloring process on wheel covers. After 40,000 miles of service, the colors were found to be essentially the same as when installed two years previously.

The process of coloring stainless steel has had wide usage in Japan and Europe. Japan alone is coloring more than 1 million square feet a year. The Reiyukai Temple in Tokyo has 323,700 square feet (30,000 square meters) of black colored stainless steel on the roof, and the ceiling of the great hall is paneled in gold colored formed stainless steel. In Europe, the EuroDisney project, designed by Frank O. Gehry and Associates, has more than 90,000 square feet (8,341 square meters) of gold stainless steel. The color is actually known as "rosy gold" because of its slightly reddish tone.

In the United States there are but a few knowledgeable fabricators working with colored stainless steel materials. Architects and designers are just beginning to see the value and understand the possibility of using stainless with interference colors.

AVAILABLE FORMS AND DIMENSIONS

Sheet

Sheet is the most common form of architectural stainless steel. Sheet is provided in thicknesses up to 0.188 inches (4.78 mm) and in widths of 24 inches (610 mm) or greater. Sheets with widths of less than 610 mm are considered strip material. The finishes available from the mill are similar for both forms.

Maximum available sheet width is dependent on the ability of the polisher. Sheets of stainless steel are available in 72-inch (1829-mm) widths. However, many polishing facilities cannot handle this width.

Maximum available length is also dependent on the ability of the polisher. If the polish is provided in line, then the length is limited by shipping and handling constraints and by fabrication limitations. Mirror finishes are limited by the size of the buffing equipment. Most equipment cannot handle lengths larger than 144 inches (3658 mm). Fabrication equipment is limited to about 144 inches (3658 mm) for most custom architectural metal shops.

Table 4-9 lists the nominal thicknesses available for stainless steel sheet.

Plate

Plate is considered a wrought product in thicknesses greater than 0.188 inches (4.78 mm) and minimum widths of 10 inches (254 mm). Plates are typically provided in an annealed temper and with a hot-rolled surface. The surface is descaled and can receive a satin polish; however, the surface provided on plate from the mill is not smooth or flat but slightly undulating. Often grinding is necessary if a good polished surface is required. The flattest surface available

TABLE 4-9. Approximate Thicknesses for Sheet Stainless Sheet

GAUGE (B AND S STANDARD)	APPROXIMATE THICKNESS MM	APPROXIMATE THICKNESS INCHES
8	4.36	0.172
10	3.57	0.141
11	3.00	0.125
12	2.77	0.109
14	2.00	0.078
16	1.60	0.063
18	1.27	0.500
20	1.00	0.040
22	0.79	0.032
24	0.63	0.025
26	0.50	0.019
28	0.40	0.016
30	0.32	0.013

from the mill is on stainless sheet. The hot-rolled plate material, even after grinding and polishing, is still not as flat as the cold-rolled sheet product.

Stainless Steel Tube and Pipe

Many different sizes of welded tubing are commonly available. Tubing is provided in cross-sectional shapes of circles, squares, rectangles, hexagons, and ovals. These shapes are available with surfaces in any of the standard finishes and many custom finishes. Architectural welded stainless steel tubing is economical because pressure testing is not required. Stainless steel tubing is a common industrial material used for pressure vessels and oil service equipment. Architectural and industrial railings are also uses of stainless steel tubing. Standard sizes range from 0.75 inches to 4.0 inches (19 mm to 102 mm). Stainless steel tubing is available in all the sheet thicknesses and in all the various polishes.

Stainless steel pipe is available in the hot-formed and welded forms, similarly to steel pipe. Pipe is provided with a coarse, rough finish from the mill or stainless steel pipe fabricator. Thicknesses are available as designated by particular schedule numbers. Pipe is considered a structural fabrication and thus is not commonly provided in finishes other than the mill finish.

Stainless Steel Bars

Stainless steel bars of many shapes are available. The bar form of the metal is considered to be those geometric shapes which are small in cross section. Bars are circular, square, rectangular, hexagonal, or oval in cross section.

Hot-finished bar products are provided from the mill in pickled or blast-cleaned condition. Hot-finished products are developed from hot rolling, forging, or extruding. The surface is rough and will require secondary grinding and polishing. Bars can be buffed and polished to induce standard satin or mirror finish textures. Bar stock can be obtained in dimensions as small as 0.125 inches square (3.17 mm square) or with diameters as large as 6 inches (152 mm).

Stainless Steel Extrusions

Stainless steel can be extruded. However, there are certain limitations in dimension, and, depending on the project, cost. Extruding metal requires the metal to flow through a hardened steel die when tremendous pressure is applied to the molten material. Sometimes molten glass is used as a lubricant to insulate the die and allow the stainless to flow through the shape opening. Stainless steel requires special dies and tremendous pressure. The cross-sectional area of the extruded piece is small because of the pressure required to force a large section through. A current limitation is that shapes must fit in a circle of less than 6.5 inches (165 mm) in diameter. Typical shapes are thresholds, angles, and I-beam and channel shapes, as well as T-shapes.

Extrusion methods for stainless are changing constantly. Technological advances will surely outdate this section of the text, so it is recommended to check with qualified fabricators on current techniques.

Stainless Steel Castings

Stainless steel castings are performed in ceramic molds because of the high temperatures of the molten metal. Stainless steel has a melting point in the range of 2552°F (1400°C). Stainless castings are usually sectioned in small pours and then assembled. Statues and sculptures manufactured from cast stainless steel alloys represent the upper limit of architectural uses of this form of the metal. Small ornamental features, particularly when the repetitive quantity is great, can be economically cast in stainless steel. Investment casting and lost wax casting techniques are the typical forms of production. Many small fixing parts and some hardware parts are cast stainless. Some varieties of bolts and screws are cast, then machined to the desired tolerances.

Stainless Steel Structural Members

Stainless steel structural members are available; however, they are not usually considered in architectural applications. Structural shapes are built up members, ranging from thick plates to required sections, or they can be obtained in small sections as extrusions and hot-formed shapes. Extrusions are available in channels, I-shapes, and T-shapes. Special shapes are possible in extrusions, as long as they are kept within a circle diameter of 6.5 inches (165 mm).

On extrusions and structural shapes, the surface is rough and provided in the pickled and annealed condition. Surface polishing and grinding are required to achieve a quality architectural finish on these parts. Because of the thickness and the hot-forming technique used to manufacture these shapes, surface inclusions and pitting are common.

Space frames are available in stainless. Tubes or rods are used for the struts and castings for the nodes. The parts can receive custom finishing to suit the design.

HARDWARE SPECIFICATIONS

Stainless steel items of hardware are available. Fabrications such as hinges, door handles, and pulls are common hardware items manufactured from stainless steel. These fabrications are made from forged stainless steel, machined wrought stainless steel, and stainless steel castings. There are a number of finishes available for stainless steel hardware. These finishes are specified by the Builders Hardware Manufacturers Association (BHMA). Similar to copper and copper alloy hardware, the BHMA specification number replaces the old US numbering system. In addition, there are many chrome-plated and nickel-plated steel hardware parts that mimic or resemble stainless steel.

BHMA Code	Description of Finish	Previous US Number
629	S30000 Series stainless with mirror finish	US32
630	S30000 Series stainless with satin finish	US32D
653	S40000 Series stainless with a mirror finish	none
654	S40000 Series stainless with a satin finish	none

CORROSION BEHAVIOR OF STAINLESS STEEL ARCHITECTURAL ALLOYS

Stainless steel acquires excellent corrosion resistance from the thin chromium oxide film that develops on the surface of the metal. This film develops rapidly when exposed to the atmosphere and stabilizes because of the presence of chromium. The oxide film is transparent and continuous across the exposed surface of the metal. The film is self-healing and nonporous. A scratch on the surface will quickly develop the chromium oxide film, providing continued protection for the metal. The thickness of this natural oxide film is 1.2×10^{-7} inches (3×10^{-6} mm). When this film is passive, stainless steel approaches the corrosion resistance of the noble metals, gold and silver. Passivity is achieved once the surface of the stainless is exposed to oxygen in the atmosphere for two to eight hours.

Passivity is the term given to describe an oxide film barrier that resists corrosion by offering a layer of nonreactive molecules. The chromium oxide layer does not allow metal ions at the surface to migrate into solution. When metal ions go into solution, corrosion cells develop.

Passivity can be lost to the surface when abrasion or chemical etching occurs. Abraded surfaces will regenerate the oxide barrier after exposure to the atmosphere for a few hours. Chemical action can develop from free iron or chlorides on the surface. The iron can come from metal-forming processes and some mechanical abrasion processes. The free iron interferes with the development of the passive film, leaving the surface with localized zones of chemically active metal. Passivity can be achieved by immersion of the stainless steel surface in a nitric acid bath, sometimes with a small percentage of sodium dichromate. This solution dissolves the free iron and leaves the stainless surface unaffected.

Chemical etching of the surface requires additional treatment to achieve passivity. Cleaning and immersion in a passivation bath is necessary to start the oxide layer developing properly.

Stainless steel becomes active when exposed to certain chemical agents that cause the surface or portions of the surface to oxidize. The metal itself, stripped of the protective oxide coating, or when the growth of the film is inhibited, will corrode as if it were plain carbon steel, regardless of the alloy type.

The smooth mirror-polished stainless surfaces perform better than the coarse surface finishes such as the No. 3 or No. 4 polishes. The blast abrasion surfaces, created by applying clean sand or other blast media, tend to pick up and hold contaminants even more effectively. The smoother, mirrorlike surfaces shed moisture and contaminants without trapping minute particles in the grooves or ridges characteristic of the coarser finishes. Thus, the finer the polish, the more corrosion resistant the metal surface is.

Modes of Corrosion

The most common type of corrosion experienced by stainless steels used in architectural applications is surface pitting. Pitting corrosion is characterized by small surface pits accompanied by red rusting and, as seen upon close inspection, a tiny lump of fuzzy material in or around each pit. Pits will first appear on a stainless surface as small rust spots. The spots may be grouped together or distributed randomly across the sheet. Severe pitting can result in complete perforation through a thin metal sheet. In most architectural exposures, perforation of the sheet will take time, since the presence of moisture is necessary to feed the electrochemical reaction.

Pitting is defined as the localized dissolution of a passive metal owing to the development of an electrolytic cell. This cell prevents the protective passive film from redeveloping and causes the base metal to dissolve in a very small localized region. As more of the metal is dissolved and washed away, a small pit develops on the surface. In architectural applications, pitting is almost exclusively caused by exposures to solutions containing chlorides. Coastal environments, particularly those with common occurrences of fog, require special consideration in resisting pitting corrosion in the stainless steels. Currently, one of the worst conditions occurs with the prolific use of deicing salts. In both of these situations, chloride salts are deposited on surfaces. Deicing salts, if allowed to remain even a short time on the surface of stainless steel, will cause rapid and extensive pitting.[6]

Molybdenum, when added to the chromium-nickel-iron alloys, assists in pitting resistance. Alloy S31600, with the addition of 2% to 3% molybdenum, is recommended for conditions where exposure to chloride salts may occur. When stainless has been exposed to deicing salts and corrosion particles are not yet apparent, flush the surface with water, preferably warm and from a hose.

How pitting corrosion is initiated on a smooth stainless surface is the subject of debate; certainly, it is the selective breakdown of the passive layer protecting the metal. Pitting corrosion, once initiated, is self-sustaining and will continue unless the contaminants are removed and the passive oxide layer is restored. This is easier said than done on architectural features. The smoother the surface, the less susceptible it is to catching and trapping

[6]The Weisman Museum of Art at the University of Minnesota, designed by Frank O. Gehry Architects in conjunction with the firm Meyer Sherer Rockcastle, initially had a stainless steel skin with an abrasion blast surface. A small section of the building skin was constructed as an example of the final appearance and located near the East-West Campus bridge. The metal skin on the small structure was exposed to deicing salts which had been applied to the bridge surface and then sent airborne from passing vehicles. The mock section of the building was installed in August 1992 as construction was starting. By the following March, major staining and surface pitting had set in from exposure to the deicing salts.

The alloy used on the mockup was S30400. The metal was changed to S31600 alloy and given a No. 4 satin finish. The finish is smoother and thus less susceptible to particles from the atmosphere resting on the surface.

particles. However, if a mirror finish surface has started to show signs of corrosion, removal of the contaminant will require repolishing of the surface. Corrosion stains can be removed, but the pits will remain. Satin finish stainless can be cleaned using a mild abrasive cleaner, a phosphoric acid cleaner, or an oxalic acid cleaner. Rinse the surface thoroughly before and after and follow by polishing in the direction of the grain, using an abrasive cloth or wheel.

One characteristic of pitting corrosion is the limited area of attack. A pit will not spread across the surface, but will just continue to dissolve the metal in the area where it started. Generally, when pitting occurs on a stainless surface, it occurs at different areas where contamination has established itself.

Pitting corrosion can start when transferred corrosion particles generated from adjacent metals run off onto the stainless surface and damage the passive layer. Transferred rust particles from iron fabrications such as grates or rails will damage and stain the surface of stainless steel. If they are left in place for a moderate period of time, perhaps a few months, the surface may develop minor pitting corrosion. Aluminum and copper corrosion particles will affect the passive surface in the same way.

Iron contamination, not yet established in the surface of stainless, can be cleaned by using a solution of 20% to 40% nitric acid (by volume) and at temperatures ranging from room temperature to 140°F (60°C) for periods of 30 to 60 minutes. Caution must be used when working with acids. Proper protective clothing, eye safety, and ventilation are essential. Other types of corrosion behavior experienced by stainless steels are intergranular corrosion, crevice corrosion, galvanic corrosion, and stress corrosion cracking.

Intergranular corrosion is rare and limited to severe environments. Almost always related to regions around welds, intergranular corrosion occurs where carbide precipitation has weakened the passivity of a surface. The corrosion attack occurs at the grain boundaries of the stainless steel. Reduction of carbon content in the alloy will eliminate the potential for intergranular corrosion by decreasing the occurrence of the chromium carbides. Low-carbon stainless steels such as Types S30403 and S31603 were created for this purpose.

Crevice corrosion occurs at joints or at junctions with nonmetal materials, such as gaskets and seals. The passive film generation is inhibited by the confinement and subsequent contamination within a crevice or seam. The size of the seam must be very small, usually less than 0.08 inches (2mm). Foreign matter, in particular chloride or sulfide salts, may migrate and collect within the crevice and lead to the development of an electrolytic cell. Corrosion occurs within the crevice with little indication on the exposed surface. If moisture is prevented from collecting in the crevice, corrosion usually ceases or proceeds at a lower rate. Elastomeric materials, such as silicone sealants, provide good seals if their elastic recovery does not generate a crevice. Sealing materials that show poor stress relaxation characteristics or gasketing with high porosity will be prone to crevice corrosion.

Crevice corrosion will show itself as rust particles leaching out from a seam. Corrective measures require the disassembly of the parts and the repassivation of the surface. Use sound welded connections in lieu of bolted connections to eliminate potential crevices around the bolt head and washer. Good design, frequent inspection and cleaning of deposits, and good rubber

TABLE 4-10. Results of Galvanic Coupling of Various Metals with S30400 Stainless Steel

DISSIMILAR METAL	OBSERVATIONS
Copper—commercially pure	No galvanic attack. Stains from copper salts.
Aluminum—commercially pure	No galvanic attack. Heavy staining. Shallow pitting.
Yellow Brass 70/30	No galvanic attack. Minor tarnish.
Lead	No galvanic attack. White spots. Patina stains.

or silicone gaskets, instead of absorbent materials, will prevent the development of crevice corrosion.

Galvanic corrosion involves the interaction of dissimilar metals. There are many factors that determine whether galvanic corrosion conditions will develop. For most architectural metal uses, galvanic corrosion of stainless steel is not an issue. Urban environment exposure tests performed by the International Nickel Company on stainless steel coupled with other metals and subjected to more than 30 years of exposure were evaluated visually for any damaging effects.[7] Reference Table 4-10 for results of the exposure tests on galvanic coupled metals.

Samples of stainless steel and other dissimilar metals were overlapped and riveted, then placed on racks and left to experience 30 years of a semi-industrial environment in Pennsylvania. The samples were arranged so that moisture would run off the dissimilar metal and over the stainless steel. Samples of Type S30400 stainless were coupled with commercially pure copper, commercially pure aluminum, brass made of 70% copper and 30% zinc, and commercially pure lead sheet.

In general, S30000 series alloy stainless steels, when coupled with other metals and allowed to develop the protective passive layer, will resist galvanic corrosion effects. It is recommended, however, that certain precautions be exercised when joining stainless steel with dissimilar metals. These precautions concern the vehicle of galvanic corrosion and apply to all dissimilar metal couplings. Galvanic corrosion requires three basic conditions to be present: (1) electron flow from one metal to the other by means of an electrolyte solution or direct coupling; (2) oxygen or an oxidation agent; and (3) moisture to create the solution. If one of these factors is eliminated, galvanic corrosion is also eliminated.

There are certain advantages to stainless steel and the passive nature of the oxide film when used architecturally with other metals. Because stainless steel is so corrosion resistant and can be easily soldered, it makes good built-in or concealed gutters on aluminum roofing or prefinished galvanized steel roofing. Since gutters made from aluminum or prefinished galvanized steel must depend on sealant at the joints, ends, and tubes, the ability to solder

[7]Baker, et al. *The Long-Term Atmospheric Corrosion Behavior of Various Grades of Stainless Steel.* Philadelphia: American Society for Testing and Materials, 1988, pp. 52–67.

stainless and the galvanic resistance of passive stainless makes it an attractive choice for this type of application.

Stainless is very strong and resilient. Clips made of S30000 series austenitic stainless steel are sometimes used on copper roofing to provide extra strength. There is a danger here if the stainless-to-copper coupling is exposed to chloride salts; however, the clip usually has limited exposure. The extra strength of the stainless clip seamed into the copper rib provides excellent uplift resistance. For aluminum or galvanized metal roofing, the use of stainless steel clips and fasteners provides excellent strength with little concern for galvanic coupling of the metals in most architectural exposures. Chloride exposures, including coastal environments, are the primary areas of concern.

Stainless steel fasteners are often used in conjunction with aluminum and galvanized steel panels. The fasteners can be of the S41000 alloy if self-drilling is desired, or of the softer, but more corrosion resistant, S30400 alloy. The S30400 alloy requires predrilling of the holes before attachment since pilot points cast onto the S30400 screws are soft and cannot drill into the harder steel supports. Concern has often been expressed about the use of the magnetic S41000 alloy for fasteners, even when they are used to engage aluminum and steel materials. As previously mentioned, the S41000 stainless may develop a superficial rust layer that serves to protect against further corrosion. Often the fasteners are even painted to match a particular siding or roofing color. Inspectors have been known to drop a magnet into a bucket of fasteners to determine alloy type. If they are not of the specified austenitic alloy, the fasteners may be removed and replaced, sometimes at the expense of the siding in place.

New techniques using a nitriding process hardens austenitic fasteners. Nitriding places nitrogen into the surface of a metal. Nitrogen will harden the surface of S30000 series stainless sufficiently to allow fasteners to drill steel and other stainless steels.[8]

Stress corrosion cracking is another process that can have an effect on stainless steels. Although rare in architectural uses, stress corrosion cracking will occur if the fabricated stainless steel part undergoes fatigue or other stress build up and then is exposed to chloride contamination. The result is a complete failure of the metal, signified by a crack at the point of stress buildup.

Compatibility with Other Materials

Stainless steel is compatible with most every common building material. Austenitic stainless steel is used as exposed ornamental trim or cladding and as concealed throughwall flashing to deflect moisture from inside masonry cavity walls. Stainless steel is resistant to masonry alkalinity.

Chloride salts or hydrochloric acids used in some construction and maintenance operations are the common potential sources of problems. Masonry and stone construction materials are often washed down with builder's acid. Builder's acid, or muriatic acid, is dilute hydrochloric acid. If this acid splashes

---

[8]Takashi Muraoka, "Nitriding hardens austenitic screws." Nickel, December 1994.

onto nearby stainless steel surfaces, reddish rust spots will quickly appear. In enclosed spaces, if the fumes from this acid are allowed to drift over the surface of polished stainless, corrosion particles will develop rapidly. The rust spots are superficial, at least initially, but should be removed before severe pitting corrosion occurs. Removal of the rust particles can be difficult on mirror polished surfaces. Mild abrasives are usually necessary to remove the rust, and these will damage the finish. Satin polishes can be cleaned more easily since mild abrasive cleaners, applied in the direction of the grain, can be used to match the existing finish.

Some soldering fluxes used to pretreat a stainless steel surface for sealing joints are made of chlorides, typically zinc chlorides. These fluxes are used to remove the oxide growth and to allow the solder to adhere to the metal. After soldering, the excess flux should be thoroughly rinsed from the surface. Better yet, use a nonchloride flux on stainless. There are a number of good products on the market for preparing a surface for soldering.

In northern climates, exposure to deicing salts is damaging to a stainless steel surface. Salts of this type, sodium chloride, potassium chloride, and calcium chloride, will create localized pits when allowed to remain on a stainless steel surface. The salts must be removed, and any red rust spots that form must be cleaned from the surface. Mild abrasive cleaners are required, followed by a thorough rinsing. Persistent stains will require the use of phosphoric acid, oxalic acid, or sulfuric acid, followed by rinsing with water. Once rusting has begun, cleaning the surface of contamination is very difficult, particularly on large items.

Some sealants used on stainless steel require priming for proper adherence. Review the particular sealant application with the sealant manufacturer. Caution against the use of chloride primers, as they will cause the stainless steel surface to corrode. Be sure long-term exposure applications are examined. If the primer removes the protective oxide layer, applying a sealant may keep oxygen from redeveloping the film. If this occurs, the stainless steel at the sealant joint may be nonpassive and subject to corrosive attack. Make certain to use good elastic sealants. Sealants that lose elasticity may expose a joint to crevice corrosion.

Austenitic stainless steels are resistant to corrosion when exposed to natural fresh waters. If stainless steel surfaces are installed in a passive state, they should remain that way for the life of the building.

Along the seacoast, where chloride salts will deposit on the surfaces of the stainless steel, it may be necessary to clean the metal periodically. Chloride salt deposits can form localized corrosion cells, which will create small pits. The first indications of a pit on a stainless surface are small brown rust spots.

CARE AND MAINTENANCE

The Problem with Reflective Surfaces—Oil Canning

As designers found more decorative uses for stainless steel, certain characteristic problems with reflective finishes and visual flatness became apparent.

Initially, architectural uses of stainless steel were limited to trims and folded sections along storefronts, or dull finishes used as architectural roofing. But the desire for large flat sections, using the most reflective finishes, created new problems not previously encountered with earlier uses.

The typical complaint in using large flat sections of stainless steel is observed "oil canning." *Oil canning* is the metaphorical term given to describe the tendency of flat sheets to undulate in and out of the average plane of the surface. The resulting lack of optical flatness is more apparent on reflective surfaces such as No. 8 and No. 4 finishes.

The phenomenon described as oil canning is a basic characteristic of thin sheets, regardless of the material. Thin sheets act as diaphragms and are unstable across large flat surfaces because internal stresses across such sheets are not consistent. Sheets of metal may look smooth and flat when lying in a stack prior to any fabrication or shearing, but fabrication operations can induce stresses in the surface that create the oil-canning effect. If dies used to form the sheets are not true and free of edge defects, the final product will reflect an enhanced version of the minor edge defect on the forming tool.

Moreover, minor production variations at the mill can induce differential stresses across the flat surface that are apparent only after the sheet metal goes through temperature variations. These temperature variations may be the difference between the chill of a cold winter night and the absorption of afternoon sun on the surface. Rarely will temperature variations permanently damage a flat surface. Usually, as the temperature changes, the panel will return to flatness, or at least to the level of flatness as initially installed. The temperature change and the resulting thermal expansion or contraction of the sheet metal surface usually accentuates any waviness already present in the surface.

Installation practices may also induce oil canning in the finish product. If the flat panel is incorrectly confined along the edges and not allowed to expand and contract during temperature cycles, then waviness and, if severe, localized buckling may occur. Long flat panels should always be fastened by using clips or cleats along the edges. This fastening technique should allow for unrestricted movement, at all edges, caused by expansion and contraction generated by thermal changes. On large flat surfaces, even with clipping the edges and allowing thermal expansion to occur unrestricted, some surface deflection will occur.

Sheets of stainless steel, improperly stacked or shipped in flat containers will be subject to differential stresses in the metal. When the sheets are formed, the waviness can be set into the sheets and will be more pronounced when subjected to temperature variations. At different times of the day, when temperature and light reflecting off the surface have changed, often the oil canning or lack of optical flatness will vary. What appeared acceptable in the morning may look dreadful in the afternoon.

Welding and some forming operations will develop unequal stresses on flat surfaces. Folded edges may work harden slightly while the center of the sheet remains annealed. Roll forming on machines with relatively short distances between shaping stations may stretch the edges, causing the flat center of the pan to want to "puff" up or down out of plane.

Even handling of a flat sheet to and from the shear and to the press brake, or during field installation of the product, can rack and warp the flat surface or induce stresses that subsequent operations set into the surface.

On laminated panels oil canning may occur if differential expansion occurs between the face stainless sheet and the backing material. Delamination of the stainless face sheet will cause oil canning as the delaminated portion of the panel attempts to expand and contract out of the plane of the more restricted, laminated portions of the surface.

The directional polished finishes, the No. 8 mirror finishes and the No. 2 Bright Annealed finishes, show minor surface deviations from true flatness. The less reflective surfaces—the No. 2D, the rolled textured surfaces, and, to a small degree, the No. 2B surfaces—tend to show better optical flatness. They are not actually flatter, they just reflect light differently.

Reflective stainless surfaces greatly enhance even minor distortions in a metal surface. Distortions may not be readable when the sheets are examined in the flat. Standing the sheets on edge and viewing the metal with zoned or concentrated lighting will make the minor distortion readily apparent. Often these distortions appear as major dents, yet on close examination, they seem to disappear. Distortions of this type cannot be repaired and will require replacement. Such distortions are sometimes caused by stacking the sheets onto uneven surfaces, for example, a wooden skid with knots or exposed nails. As more sheets are stacked onto the skid, the lower sheets are imparted with a reflective distortion. The skids should have a layer of masonite material or another smooth, rigid surface barrier without nails.

Heat differentials can also cause reflective distortions of this type. Soldering on the back side or using catalyzed adhesives to attach stiffeners to thin sheets will appear on the face side as a dent, sometimes looking as if a ball peen hammer was used on the reverse side of the sheet. Actually, it is a distortion set into the reflective surface. This visual distortion is not measurable. It cannot be felt by running a hand over the surface. The flaw is in the metal and induces the distortion seen by the eye. Because of the way they reflect light, these ghost distortions are more common in directional satin finishes. Often, viewing from different angles will make the "dents" invisible. Before rejecting the sheet fabrication, view the assembly in its finished position; you might find the "ghosts" have left. The frustrating part of these distortions is that they cannot easily be caught in even the best quality control program. They are not visible until the sheet has been stripped of its protective film, which is usually after the expense of fabrication has been incurred.

Flatness Specifications

The American Institute of Steel (AISC) specified tolerances for flatness are usually not adequate for most architectural uses of flat stainless steel surfaces. The AISC tolerances for flatness state a deviation from the flat plane, but give no limit on how long or how short this plane is. NAAMM (National Association of Architectural Metal Manufacturers) specifications describe a method of measuring flatness in thin sheet surfaces. The angle of the slope of the deviation from the plane of true level is taken. The NAAMM specifications state that the angle, as a percentage of 180 degrees, must be within the following limits:

- For surfaces having a finish of high reflectivity (No. 4, No. 7, No. 8) 1.00% maximum
- For surfaces having a finish of medium reflectivity (bead) 1.25% maximum
- For surfaces having a finish of low reflectivity (No. 2D) 1.50% maximum
- For embossed textured surfaces 2.00% maximum

Using these criteria will reduce some of the occurrences of oil canning, but will not eliminate it. A 1% deviation for a reflective surface will allow some major variations in surface flatness, and optical flatness will not be acceptable. It is advised to use 0.5% as the maximum allowable slope created by an undulating surface, even the lower reflective surfaces. This limit is possible and will reduce the appearance of undulating surfaces.

Some specifications request for "visual flatness" to be the criterion for acceptable flatness. This criterion is impractical, if not impossible. "Visual flatness" is a subjective term. Depending on the observer and the conditions when the viewing is performed, visual flatness is an impossible criterion to achieve.

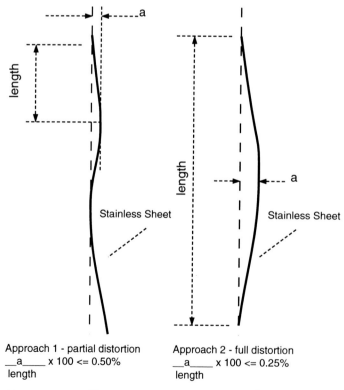

Figure 4-21. Flatness criteria—measuring technique.

Since stainless is used on prestigious projects, expectations are usually higher for this metal than for others. Visual flatness under these circumstances may be unobtainable.

Designing for Flatness

When flatness is a major concern, the designer should take into consideration the following guidelines:

1. Specify stretcher level quality from the mill producer.
2. Use a heavier gauge. Increase the thickness of the metal, particularly when a reflective finish is being used. The cost may be higher because of the increase in metal, but there is no substitute for thickness when flatness is a high priority of the design. Keep in mind that plate thickness material greater than 0.188 inches (4.77 mm) is not as flat as sheet thickness material of 0.188 inches (4.77 mm), but even less so because the thick plates are hot rolled. Thick, hot-rolled plates have an uneven, wavy surface which, when polished, provides a distorted, reflective surface.
3. Use a nonreflective or textured finish to conceal the apparent waviness of a flat surface.
4. Inducing a slight curve in a large flat region will eliminate oil canning across the surface.
5. Use an attachment system that allows a metal panel to expand and contract without the buildup of stresses around fasteners.
6. Avoid the use of long or wide flat surfaces. Provide stiffener ribs or fluting when possible, or laminate to a flat, rigid material. Lamination of stainless is a very difficult process and should be performed only by those experienced and familiar with current technology.

When the metal surface is not laminated to a rigid backing material, then the lateral distance between stiffening members is recommended as follows:

Finishes	Width-to-Thickness Ratio
Highly reflective surfaces	150 maximum
Medium-to-low reflectivity	200 maximum
Textured surfaces	200 and higher

For instance, assume that the gauge of a No. 4 finish stainless steel sheet is 16 gauge, or 0.063 inches (1.60 mm) nominal. The furthest space between stiffeners or panel sections, according to the preceding ratio for the reflective No. 4 finish, is 150×0.063, or 9.45 inches (240 mm). For a No. 2D finish of the same gauge, the stiffeners can be placed 12.6 inches (320 mm) apart. In practice this guide is useful to reduce apparent oil canning of the surface. Be cautious in the attachment of stiffeners using epoxy. The "ghost" deflections mentioned earlier may appear when the epoxy heats up from the catalyzed reaction.

The following are tolerances of flatness allowed by the American Society of Testing Materials. These specifications are generous and allow for a very undulating surface.

ASTM A480 Flatness Tolerances
Cold Rolled Stainless Steel Sheet
Austenitic S20000 and S30000 series
¹/₄ and ¹/₂ hard tempers

Thickness	Sheet Width	Flatness Tolerances inches (mm)	
		¹/₄ hard temper	¹/₂ hard temper
0 to 0.016 inches (0 to 0.41 mm)	24 inches to 36 inches (610 mm to 914 mm)	0.5 inches (12.7 mm)	0.75 inches (19.05 mm)
0.016 to 0.030 inches (0.41 to 0.76 mm)		0.625 inches (15.88 mm)	0.875 inches (22.22 mm)
0.030 inches and above (0.76 mm and above)		0.75 inches (19.05 mm)	0.875 inches (22.22 mm)

The flatness tolerance is defined as the maximum deviation from a horizontal flat surface. For this specification, the center of a stainless steel sheet can bow out as much as 0.875 inches (22.22 mm) and still meet flatness criteria set out by A480.

If stretcher leveled standard of flatness
(not including hard tempers)

Thickness	Width	Length	Flatness
All thicknesses	0 to 48 inches (0 to 1219 mm)	0 to 96 inches (0 to 2438 mm)	0.125 inches (3.18 mm)
	0 to 48 inches (0 to 1219 mm)	Lengths > 96 inches (Lengths > 2438 mm)	0.25 inches (6.35 mm)
	Widths > 48 inches (Widths > 1219 mm)	All lengths	0.25 inches (6.35 mm)

The ASTM A480 flatness tolerance levels are far too liberal to achieve any real quality flat architectural surfaces. Most supply houses will accept, at worst case, half these limits. The specifier should work with a local fabricator or representative of a mill to determine what is achievable. These parameters should not be used as an excuse for less than quality architectural metal work.

In general, architectural stainless steel applications require more stringent criteria for flatness quality than those set forth by commercial standards. At the very minimum, uniform surface appearance—that is, an absence of surface streaks or lines and no discoloration or scratches—should be accepted. Inspection for chatter marks created during the polishing operation should be performed. Any chatter marks in the surface finish should be rejected (Figure 4-6). Coil stop and start marks should be marked on the sheet to ensure fabrication does not set this deviation into the finished product. The stainless should be packaged and handled in such a way to eliminate skid marks resulting from packaging and dents and edge damage caused by handling. Minor distortions should be prevented by careful packaging and skidding. Often, on directional textures, the minor distortions will not appear when the

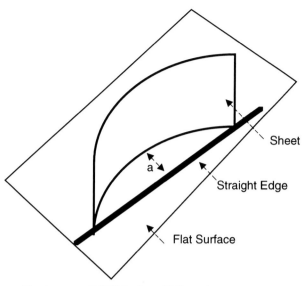

a Maximum = 0.375 inches (9.5 mm)

Figure 4-22. Measuring technique for conoe set in sheet metal material.

sheets are installed with the grain direction vertical. Try to inspect the sheets in the same direction and similar lighting in which they are to be installed.

Additional inspection criteria for sheet and plate varieties of wrought stainless steel relate to the initial receipt of the skidded material. Certain physical characteristics should be examined for minimum quality levels. When the skids of material are received, pull off the top three sheets and lay them individually on a flat inspection table with the prime sides up. Examine the corners and the sides of the flat sheets for any deviations from the flat plane of the table. The maximum deviation from the flat table should be 0.065 inches (1.65 mm). Visual inspection of the prime side finish can also be performed at this time. Then flip the sheet over, protecting the prime side from scratching, and perform the same examination. A maximum deviation of 0.09 inches (2.29 mm) should be considered acceptable. A sheet with any edge wave or undulation and any distortions, just in from the edge, is considered low quality and not up to architectural standards.

Stand the sheet up with the long edge down on a flat floor. Place a straightedge along the sheet and measure the deviation from this edge. Any deviation from the straightedge is considered coil set and should be a maximum of 0.375 inches (9.5 mm). Run a square along the sheet and perpendicular to the floor. Measure any "conoe set" in the sheet. Maximum conoe set should be 0.25 inches (6 mm) either in or out of plane.

If, on inspection of the top three sheets from a skid of metal, there are any deviations falling outside the range considered, the entire skid should be sent back and replaced. Accepting conditions outside these ranges will afford less than the desired architectural quality.

Stainless flat sheet should be tension leveled with a protective film over the prime surface. Film should be used to protect this surface throughout all cut-to-length, fabrication, and forming operations. A second film may have

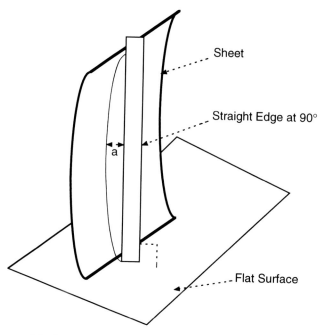

a Maximum = 0.25 inches (6 mm)

Figure 4-23. Measuring technique for coil set in sheet metal material.

to be applied during installation. This second film, if exterior, should be UV resistant and should remain in place until all subsequent work is completed. Exercise caution when using adhesive-applied vinyl films. Sunlight tends to bake these films onto the metal if they are not UV resistant. The film shreds and comes off in small segments when the adhesive dries.

Distortions from Attachments

For large flat panels, stiffeners are often applied to the back side of the stainless steel sheet. Capacitor discharge stud welding on the nonexposed side of the sheet will translate to the exposed side when thin gauges are used or when too much heat is developed at the point of discharge. Samples should be tested and the current supply monitored to ensure consistency. Using catalyzed adhesives to apply the stiffener can also impart distortions to the reflective finish surface. This problem is often difficult to see and becomes apparent only under certain lighting conditions.

Care During Fabrication

Surface contamination of stainless steel can occur during fabrication or surface finish application. During fabrication, small particles of the die or shear may imbed in the stainless, only to rust or form galvanic corrosion cells. Keeping a protective plastic film over the stainless during the fabrication processes will reduce or eliminate such contamination and assist in keeping the metal scratch and mar free.

Stainless steel sheets should not be dragged over other stainless or metal surfaces before, during, or after fabrication. Forming dies should be clean and free of all nicks and dents. Slightly damaged dies used to form stainless will translate the damage to the reflective metal. Obvious indications of this situation are repeated imperfections that occur on different parts but at the same place on each part. Neoprene protection during forming assists in reducing imperfections, and urethane dies will produce clean surfaces, free of mars and dents.

Surface finishing with wire brushes or steel shot will contaminate a metal surface with fragments of steel. Steel or iron particles that become imbedded in the surface of stainless steel will rust, and rusting will continue as long as the iron particles are present. Corrosion cells may develop around these particles if they inhibit the growth of the passive oxide film. Steel particles can be removed from a surface by immersion in a bath of heated nitric acid. Always use stainless steel wire brush or nonmetallic abrasive pads.

Immersion in a bath of approximately 20% to 40% nitric acid solution heated to a range of 130°F to 150°F (54°C to 65°C), for 30 to 60 minutes, then thoroughly rinsing in warm water, will passivate the surface of the stainless and remove most foreign materials, particularly free iron. On abrasive blast surfaces, a quick dip in a dilute nitric acid bath, followed by a thorough rinse in clean water, will help reduce the tendency of the surface to fingerprint. Type S30300 stainless and the martensitic or ferritic stainless steels alloys should not be passivated in this manner. This treatment may produce slight surface etching on these alloys. For these configurations, use 20% (by volume) nitric acid with 4% to 6% (by weight) sodium dichromate. This is applied at temperatures ranging from 70°F to 110°F (21°C to 43°C).

For in-place fabrications, rust particles can be removed using a conventional rust removing cleaner. An oxalic acid (wood bleach) paste applied over the surface and gently rubbed into the surface may remove some surface particles of corrosion. If this fails, the only solution may be to reblast the surface using an abrasive media such as glass bead. The glass bead will remove a portion of the surface contamination and leave a smoother, more passive surface behind. The glass will not contaminate the stainless surface.

Note: Caution should be exercised when working with acids. Proper disposal and handling should be left to those who know and understand the dangers of acids, as well as the required disposal procedures.

Restoring a Damaged Finish

After stainless steel parts are welded, the finish needs restoration and blending to renew the uniform appearance. For all finish work on stainless steel it is very important to use dedicated equipment. Equipment used for aluminum and steel should not be used for stainless steel. Rusting on restored satin finish surfaces is the result of imbedded free-iron particles introduced into the surface. Coated discs and sanding belts do not contain particles that can cause rusting. Such particles are transferred from disks and belts used previously on steel parts. An adequate quality assurance program must be established to eliminate this transferrance. Color coating the discs, or complete separation, is advised as a starting point.

For satin finishes, it is important to verify the finish to be matched. Polished sheets supplied by different venders rarely match. Moreover, material supplied from opposite ends of the polish runs have different levels of grain depth because of wear on the belts. To restore satin finishes damaged by welding, use zirconia aluminum discs and belts at slow speeds, about 600 rpm. Use progressively finer grits, starting at 80 and finishing in the 180-grit range. Use 80-grit material to reduce the weld area, instead of the coarser 35 grit. A 35-grit material may work more quickly in removal of the excess weld; however, the rest of the day may be spent in removing the very coarse 35-grit lines.

To restore a satin finish, use the following steps:

1. Determine the finish to be matched. Sample a small section to determine the grit levels required.
2. Choose an abrasive. Zirconia alumina abrasive is recommended for the initial grits.
3. Remove the excess weld. Use an 80-grit, 7-inch (178-mm) by $^7/_8$-inch (22-mm) fiber disc and air-cooled backup pad. Control heat buildup as the weld is removed. Use fine grit to remove the weld. In fact, the more inexperienced the operator, the finer the grit should be.

   Do not concentrate on the width of the weld. Focusing on just the weld will produce a hole. Begin at the perimeter of the sheet and work in strokes of 8 to 12 inches (203 mm to 305 mm), adjusting the angle with each pass.
4. Remove the fiber disc scratches. Use a finer-grit disc, 120 grit is recommended, on an 8-inch (203-mm) blank paper disc with an 8-inch (203-mm) stick and sand backup pad. Also use a foam backup pad to cushion the disc. Operate at 4,500 rpm.

   Begin at one edge and apply strokes at a 90 degree angle from the 80-grit lines. Remove the 80-grit lines and leave the 120-grit lines. Then apply the abrasives again at 90 degrees to these lines.
5. Remove the 120-grit lines. Use the same tool and backup pads, but with 150-grit abrasives. Cross the grain and repeat step 4.
6. For the No. 4 satin finish, use a silicon carbide, medium-grit clean and finish wheel, 6 inches (152 mm) in diameter by 1 inch. Use a straight shaft air-powered drill at 400 to 600 rpm. Feather in and out of the repair area.

For the highly reflective No. 7 and No. 8 finishes, follow similar steps, but continue to approximately 320-grit silicon carbide abrasives. Follow this with extensive buffing using "cut and color" compounds to eliminate the minute scratches. It is nearly impossible to remove all the "angel hair" scratches in the reflective surface; however, most domestic No. 8 polished surfaces can be matched. When repairing welded regions, the distortion around a weld on a polished surface cannot be eliminated, regardless of the level of grinding and polishing. The distortion is in the surface metallurgy of this stainless. The grains take on different directions, and the reflection follows these changes.

Care of the Highly Polished Finish

There are problems in working with a highly polished finish. Repairing any minor damages or scratches is difficult and requires repeated buffing and polishing with jeweler's rouge to remove all polish lines. Any heating of the surface, for instance, by attaching clips with solder on the back side of the polished surface or by attaching clips with catalyzed epoxy, can produce distortion on the polished surface. The distortion may not be readily apparent until the material is viewed in a certain light.

Care of the Abrasive Blast Surface

The surface created by abrasive blast is coarser than the polished satin and mirror finishes. This increased roughness will collect contaminants and hold them despite the natural cleaning benefit of rain. Thus, when using an abraded surface out of doors, expect higher maintenance requirements to keep the surface clean and free of particles that may create corrosion cells. Abraded surfaces are of particular concern in coastal regions where salt deposits develop. Road salts, applied nearby and sent into the air by passing vehicular traffic, will also cause rapid deterioration if not quickly removed from the surface.

Another surface-staining condition may arise when abrasive blast surfaces are used as panels on the exterior of a building and sealant is applied at the joints. The sealant surface has a tendency to collect dirt particles from the atmosphere. During light rains, these particles are washed onto the adjoining stainless steel surface. The abraded surface holds the staining particles around the sealant joints. This type of stain can be very difficult to remove, requiring strong detergents and mild abrasive brushing with a nylon bristle scrub brush.

Cleaning

General maintenance of stainless steel surfaces should be carried out periodically, according to a schedule similar to that used for glass surfaces. For architectural stainless steel, use a solvent that has low toxicity and is water soluble. Wipe or brush the solvent onto the surface with long-fibered nylon brushes. Brush in the direction of the grit lines if possible.

Next, apply a mild liquid detergent—5% ammonia solution is adequate—again using the nylon brushes. Rinse the detergent from the surface with water, squeegee excess water, and allow the surface to air dry.

Refer to Table 4-11 for suggested cleaning techniques.

FABRICATION CRITERIA

Cutting

Austenitic stainless steels are difficult to machine. Stainless steel is a poor heat conductor, and cutting or drilling of this material creates local temperature increases. If tooling is dull and tends to push the stainless steel rather

Figure 4-24. Corrosion cell development possibly from embedded iron resulting from sand blasting or exposure to de-icing salts.

than cut, the stainless will undergo strain hardening at the point of deformation. Subsequent cutting will encounter an extremely hard surface layer. For cutting and milling, use sharp tooling with a large positive rake and adequate lubricant. The machine used to perform the cutting and piercing operation should be very rigid and operate at the correct cutting speed.

Cutting operations for stainless steel are the same as for steel. Shearing, blanking, and perforating require a 30% to 50% capacity increase because of the toughness of the stainless steels. One problem developed with shearing of stainless steel is stress relief across the sheet. When stainless steel is decoiled into flat sheets without tension leveling, stresses are often induced in the flat sheets. Shearing the flat sheet relieves some of the stress, and the sheet curves inward toward the center. What was once a relatively square sheet now has an inward-curved side. This condition usually occurs on thin gauges.

TABLE 4-11. Suggested Approach to Cleaning Stainless Steel

PROBLEM	SOLUTION
Fingerprints, light surface oil, dust, and dirt	Remove by wet cloth with a mild neutral detergent, such as glass cleaner. Rinse and dry.
Surface streaks (left after drying; particularly troublesome on satin finish surfaces)	Use a nonstreaking glass cleaner or stainless steel cleaner. Often the stainless steel cleaner will leave a slight film of lemon oil. This will eliminate the streaks.
Adhesives left from protective covering	Initially use a very low tack PVC or other protective film. If the adhesive is thick, use mineral spirits or alcohol. Lacquer thinner will also work. Follow with a thorough rinse, then glass cleaner to cut down the streaks. Hot water and detergent will take off most of the adhesive as well. Follow with the nonstreaking glass cleaner.
Adherent protective film, hardened glues and sealants	MEK (if allowable by federal regulations on commercial solvents). Try a copper knife to scrape the particles. Copper usually will not scratch stainless steel.
Heavy stains such as rust and surface-pitting remnants	Remove as much as possible with commercial detergent and nylon brushes. Follow with a rust removing cleaner and a mild abrasive powder, such as calcium carbonate. Practice on a small area since this will scratch the surface.
	Use an oxalic acid paste. Rub the paste into the surface in the direction of the grain.
	If the parts are small, immerse in a heated bath of 10% to 15% nitric acid. Use extreme caution when using acids. Rinse and neutralize.
	If the stains are deep and stubborn, refinish the surface using mechanical means. For satin finishes, use an abrasive pad. You will probably need to perform this over the entire surface. For nondirectional surfaces, such as a sandblasted surface, reblast with clean dry glass beads. Do not use sand.

Brake Forming

Press brakes are common sheet metal fabrication equipment. Complex shapes, even small quantities, can be adequately formed on press brakes. Forming stainless steel with brakes requires properly designed dies and adequate pressure. Hand brakes do not have the capacity to form Quarter Hard stainless into very elaborate shapes. Hydraulic or mechanical press brakes are the required equipment. Standard press brakes can accommodate sheet and plate material in lengths up to 12 feet (3.66 meters) and thicknesses up to 0.188 inches (4.76 mm).

Designing formed stainless steel depends on an understanding of minimum bend ratios. Minimum bend ratios are the minimum radii developed at a bend formed by press braking a stainless steel plate or sheet. The temper of the metal sheet plays an important part in the minimum bend ratio possible. The following list shows the minimum radii for the austenitic stainless steels:

Temper	Minimum Radius at Bend
Annealed	0 to 1.5 times the thickness
Quarter Hard	1 to 2 times the thickness
Half Hard	2.5 to 4 times the thickness
Full Hard	4 to 6 times the thickness

Another consideration when press brake forming stainless steel is the increased spring-back characteristic. The angle formed in the metal should be applied at 2 to 3 degrees greater than the desired angle.

Stainless steel fabrications require special care and handling during press brake operations. Because of the high reflectivity, any imperfections on the die, no matter how minor, will be accentuated on the stainless steel fabrication. Proper die covers should always be used to separate the stainless steel surface from the die surface.

V-Cutting

Stainless steel can be V-cut to produce the appearance of sharp corners on heavy thicknesses. V-cutting removes controlled amounts of metal from the back side of the sheet. A portion of the material remains and develops the corner. Because of the better corrosion resistance of stainless, there need be little concern for the lack of material at the edge.

Roll Forming

Roll-forming operations on stainless products are common for shapes that are repeatable over large quantities. Roll-forming operations on stainless are identical to those performed on other metals. Roll forming is a continuous forming operation in which long lengths of stainless steel are developed by passing a strip or coil of metal through matching rolls set in stands, sometimes as many as 20 stands, with each stand performing a sequential forming operation.

Other forming operations typically used on metals can be used on stainless steels. With stainless steel used for architectural applications, the single greatest concern is with the quality of the forming equipment. Equipment

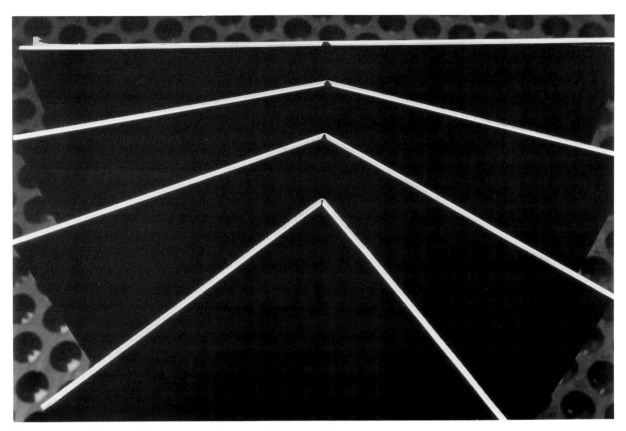

Figure 4-25. V-cut technique in stainless steel sheet and plates.

used for black iron or rough steel materials are not usually up to the minimum level of quality needed to form stainless steel fabrications.

Adhesive Application

To eliminate the tendency of light gauge stainless steel panels to "oil can," designers and specifiers often require lamination to a more rigid substrate. For stainless steels, this can be difficult. There are some excellent adhesives on the market today that will stick to just about everything, even when installed improperly. For good, consistent results, look to those firms involved with the lamination of stainless steel panels—this is not a process that can be learned overnight.

Basic problems to be overcome:

1. The more reflective the surface, the more difficult it will be to achieve consistent flatness. Initial flatness may occur, but eventually differential movement from thermal influences will produce distortion on the surface.

2. Selectively adhering clips or stiffener supports, even on heavy gauges, will produce distortion on the surface when it is viewed from certain angles. Most adhesives require a chemical reaction to activate. A chemical reaction develops some heat, which can translate to

distortions on the surface. The distortions may not even be measurable, but will be apparent.

3. Using honeycomb fillers on thin gauge can telegraph through to the reflective surface. It is often better to use continuous boards with full contact to eliminate this effect. Lamination to plywood, gypsum board, or cement board requires a thick elastic adhesive for good results. Exterior uses of these core materials can be prone to failure because of the differential movements of the materials involved.

FASTENING AND JOINING TECHNIQUES

Seaming and soldering of stainless steel is a common means of joining thin sheets. With stainless steel sheet, just about all the common mechanical sheet metal seams can be used.[9]

Solder Joints

Stainless steel is a poor conductor of heat. For this reason, when soldering stainless steel, the heat is generally concentrated at the joint or location where the iron is applied. The soldering iron must be kept longer at the joint, and the heat will remain longer at the joint. It is recommended that a 95-5 solder (95% tin, 5% lead) to be used to seal joints in stainless when strength is a requirement. The higher tin content matches the color of stainless better. The joint to be soldered must be clean and free of all excess oils. A commercial zinc or ammonia chloride flux is often used for pretreatment of the stainless steel surface at the location of solder. Care should be taken to remove all excess flux and rinse the joint of all residue. Chloride fluxes are corrosive and, if not neutralized, will damage the metal. Another concern is crevice corrosion at the solder joint. If excess chloride flux is trapped in the solder joint, crevice corrosion can take effect. Phosphoric acid base fluxes are the preferred type for pretreating the solder area. Phosphoric acid fluxes are activated only when temperatures required for the melting of the lead-tin solder are reached.

Welding

When stainless steel is joined by welding, the temperature of the metal surrounding the point of weld reaches the melting point of the metal. For the austenitic grades this temperature is about 2,600°F (1427°C). At this point the metal adjoining the weld undergoes a metallurgical change. How these changes affect the desired properties and the finish surface of the metal depends on many factors; one of high importance is the skill of the welder. Other factors to be considered are the alloy makeup, the weld method used, and the thickness of the metal.

In regard to corrosion, carbon precipitation at the weld is of main concern. At elevated temperatures, temperatures surpassed by the welding operation, chromium and carbon develop chromium carbides. These carbides

[9]*Architectural Sheet Metal Manual*, 5th ed. Chantilly, VA: SMACNA, 1993.

precipitate out at the grain boundaries, removing the needed chromium required for corrosion protection. Reducing the carbon present by qualifying the alloy type as a low-carbon alloy, such as 304L or 316L, is recommended when welding is to be performed on stainless steel parts subjected to corrosive environments.

One precaution when using the 304L and the 316L finishes is the reduction in yield stress from 35 ksi to 30 ksi (241 MPa to 207 MPa). The reduction in carbon produces a marked reduction in strength. Another concern to be addressed when welding stainless steel sheet product is warpage of the surface. Stainless does not conduct heat well, but isolates the heat in a region in close proximity to the weld. This zone is known as the heat-affected zone. Distortion is severe nearest this zone as stresses develop from heat buildup. To reduce heat buildup, copper chill bars are placed along or directly behind the weld zone. Copper is used because of its superior heat conductivity, reducing the temperature at the weld zone. To further reduce temperature, dry ice can also be used. Dry ice will develop into carbon dioxide gas as it thaws. The work space should be ventilated. Techniques such as skip welding or stitch welding should be used to reduce the buildup of stress zones on large flat regions. This technique requires welding the flat section in small, short, welds spaced a meter apart. Once the flat section is held in position, weld the area between these first fixing points. Skip around the part so as not to build up stress zones.

With mirror polished surfaces, another item to address is the distortion that occurs in the polished weld zone. Because of the metallurgic change that occurs at the grain level, repolishing even thick surfaces will not eliminate the distortion. Satin finishes can be blended to conceal the weld, but mirror polishes will always have the telltale distortion at the region around the weld, no matter how much polishing and grinding is done.

Stainless steel can be welded with conventional welding equipment. The electrical conductivity of stainless steel is less than that of most other metals. With this higher electrical resistance, less current is required to spot weld (resistance weld) stainless steels.

Shielded metal arc welding is recommended for stainless steels with thicknesses of 0.040 inches and less (18 gauge and thinner). Use AWS E 3xx - 16, $5/_{64}$-inch or smaller diameter. Mig welding is also an excellent means of welding stainless steels. Use an electrode with a matching metal to the base metal to be welded.

5

Lead and Zinc

INTRODUCTION

Lead and zinc have been in use as architectural materials for many centuries. The use of lead as an architectural metal dates back to the Roman times. Zinc, on the other hand, is comparatively new, having been in use only since the sixteenth and seventeenth centuries. Both lead and zinc are dark metals. Initially they both have a slight blue cast. Gray in color, zinc has a slight green or dark blue tone when aged. Lead darkens, almost to black.

Why discuss the two metals together? Perhaps because the industrial group of mining companies is called the Lead and Zinc International Association. Or possibly the metals are considered together because they often occur in nature together.

These two metals find their widest use in architecture as coatings over other metals. Zinc is coated as galvanized steel and lead is coated in its pure form on to copper or as a lead–tin mix for terne coatings on steel and stainless steel. This chapter deals with the pure forms of the metals. Both lead and zinc are typically cast and rolled thin into dull gray sheets in the more pure unalloyed forms.

LEAD—THE SILENT METAL

HISTORY

Lead has been used by humankind for at least 5,000 years. Lead was a byproduct in the early mining of silver. The lead oxide or sulfide ores, in combination with silver, would be "melted away" from the silver ores.

Historically, lead is one of the oldest architectural metals. Used as linings and roofs on ancient Roman structures, lead sheets have shown unsurpassed endurance in architectural applications. There are lead roofs today, such as the one on Hagia Sophia in Istanbul which has experienced more than 1,400 years of environmental impact. Lead was also used to transport water during the Roman era. Sheets of lead would be rolled into oval shapes, and the top seam was pounded together to form a pipe. The elemental symbol for lead is Pb, derived from the Latin name for lead, *plumbum*. The word "plumber" is also derived from this old term and from the relation of lead to water transportation.

Lead ores were easily refined because of their low melting point—621°F (327°C). Native Americans mined and used lead for ceremonial and eating utensils, refining the metal down from ore and casting it in rock or sand molds.

In the early 1500s, the English commonly used lead in roofing and roof-draining systems, decorating the lead fabrications with ornate designs. Lead could be easily rolled thin and shaped into pipes. The seams in the pipes were heated and joined. Lead is still used in England today to fabricate intricate roof forms and shapes.

In early America, the colonies imported much English lead for flat roofs, built-in gutters, and other rain-diverting devices. Lead pipes became a common means of transporting water into the homes of early America. During the Revolutionary War, lead fabrications were removed from homes and collected to be melted down and converted into gunshot for the rebellion. Another interesting characteristic of lead is its surface tension. When in the molten state and passed through a gauged sieve, lead forms perfect little spheres. The spheres cool as they pass through the sieve and drop in the air or water. These little spheres made for perfect, reproducible gunshot.

By the early 1800s the toxic nature of lead became apparent to the medical societies, which attempted to warn the public of the hazards of drinking and eating from lead utensils. Awareness was slow, but the use of lead in water pipes, as well as the use of leaded pewter in eating utensils, began to decline.

One of the first architectural uses of lead in America was in a Virginia plantation house. The roof was constructed of imported lead sheets in approximately 1726. The roof lasted until 1838, when it was replaced by tin-plate roofing sheets. Lead was also extensively used as protective flashing and for guttering in much of New England through the 1800s.

Lead, in the commercially pure form, is used today mainly as roofing, waterproofing membranes, sound-deadening sheets, or shielding for radioactive applications. The joints are soldered or "burned" to create complete seals. Lead has also had wide use as a coating or alloy coating for other metals, such as lead-coated copper and terne-coated steel.

PRODUCTION AND PROCESSING

Although lead deposits are found throughout the world, this metal is actually quite rare, making up less than 0.002% of the earth's crust. The lead deposits, however, occur in concentrations that simplify mining and refining operations. Moreover, because the metal is so inert and corrosion resistant, it does not change substantially. This makes for ease of recycling and reprocessing.

The United States is the leading producer of both primary and secondary lead. Lead mines in Missouri produce as much as 80% of the domestic lead supply. The major lead ore is Galena, or lead sulfide. This metallic rock is found in major concentrations in the southeast portion of Missouri.

Lead is easily refined from its sulfide and oxide ores. With such a low melting temperature and its inert nature, lead smelting operations are relatively easy. Lead ore is initially crushed and put through a floatation process to concentrate the ore in a frothy mass on top of a tank of water. The froth is removed and dried to develop a more concentrated lead sulfide. Next, the lead goes through a sintering process to remove most of the sulfide as sulfur dioxide. This poisonous gas is recovered, converted to sulfuric acid, and resold as a by-product. The impure concentrate of lead is then placed in a smelter and further refined. In this step other valuable metals are removed and recovered, such as copper and zinc. Final refinement melts the lead in a large kettle to remove additional impurities such as antimony, tin, zinc, copper and silver. The resultant lead is 99.9% pure.

ENVIRONMENTAL CONCERNS

Lead is considered to be a hazardous and toxic material. Lead toxicity, however, requires consumption. A person has to absorb lead into the blood system by oral ingestion or by inhaling large quantities of lead oxides or other lead salts. Lead is an accumulative poison. It stays in the blood system for a period of time, accumulating to levels that have a toxic effect on the body. The effects of lead poisoning are reversible, and complete recovery is possible. However, large doses can cause seizures, coma, and death.

Lead should not be used in or near food preparation areas. It should not be used where children can come in contact with the metal or the metal salts. Children are particularly susceptible because of the way lead toxicity inhibits brain development. Lead paints are the main hazard. If lead is used on walls of homes or on children's toys, the paint and paint dust can contaminate areas occupied by children.

Another potential hazard is lead oxide runoff from lead-coated roofing and paneling. There has, to date, been no formal testing of the amount of runoff generated from lead and lead-coated roofing products. However, tests performed by running water over small samples showed significant amounts of lead contamination in the collected water. The levels of lead coming from the metal surface were initially very high, but after exposure for a short duration, the runoff levels were reduced significantly.

When working with lead, wear protective clothing. Do not sand or cause the lead surface to disintegrate into dust that might be inhaled. Do not smoke cigarettes or eat food while working with lead or lead-coated materials. Doing either may allow some of the metal to be consumed. Thoroughly wash hands and clothing after working with lead, and before breaking for lunch or the occasional cigarette.

When soldering lead, work in well-ventilated areas. The Occupational Safety and Health Administration of the United States (OSHA) requires air sampling for some applications where soldering takes place. Check all current regulations for handling and disposal of lead and lead products.

In recent years, the use of brass containing lead has been discontinued for pump devices in wells. As previously mentioned, lead does not go into solution when mixed with other metals. Lead will wear away and dissolve in soft waters. Soft waters containing CO_2 will develop lead carbonate, which is slightly soluble.

ARCHITECTURAL ALLOYS

Lead quality standards are established in ASTM B29-79 Standard Specification for Pig Lead. This standard establishes the limits of impurities within the commercially pure alloy. For most architectural sheet lead applications, use hard lead alloys containing the following elemental constituents:

Antimony   4% to 6%
Lead   94% to 96%

These alloys fall under the following classifications:

Lead alloy series: L52900–L53000
ASTM B 749 Sheet Lead

The antimony-bearing leads are known as Hard Lead or Regulus Lead in Europe. The added antimony increases the strength and hardness of pure lead. Pure lead has a tensile strength approximately of 3,600 psi (25MPa), and the Hard Lead has a tensile strength of 4,100 psi (27MPa).

In Europe, the British standard BS 1178, "Milled Lead Sheet and Strip for Building Purposes," is used. In this specification, the thicknesses vary slightly from that produced by the United States standards.

Sheet lead is cold rolled or continuous cast to a particular thickness. Lead is specified in pounds per square foot. One pound of lead equates to one square foot of lead with a thickness of 0.0156 inches (0.40 mm). For most architectural applications, 4-pound lead is used. This equates to a sheet thickness of 0.0624 inches (1.6 mm).

Refer to Table 5-1 for the thickness-to-weight relationship of lead sheet.

AVAILABLE FORMS

Sheet Lead

Lead sheet is cold rolled or continuous cast, then rolled to the required thickness. Sheets are available in maximum widths of 96 inches (2,438 mm).

Maximum sheet thickness is considered to be 0.188 inches (4.78 mm). Minimum thicknesses commonly available are 0.005 inches (0.127 mm). Previously, lead was cast into sheets. Many of the old structures had these large cast sheets for roofing skins. The rolled lead lacks the same texture as these previously sand-cast sheets, having a smoother, less grainy surface.

For typical architectural waterproofing applications as metal building skins or as water membrane containment sheets, the 6-pound lead sheet is used. Thinner, 4-pound sheets are used for sound-deadening sheets located inside walls.

TABLE 5-1. Standard Thicknesses

POUNDS PER SQUARE FOOT	APPROXIMATE THICKNESS (INCH)	APPROXIMATE THICKNESS (MM)
2.0	0.031	0.79
2.5	0.039	1.00
3.0	0.047	1.19
3.5	0.055	1.39
4.0	0.063	1.60
5.0	0.078	2.00
6.0	0.094	2.38
8.0	0.125	3.18
10.0	0.156	4.00
12.0	0.188	4.76
14.0	0.219	5.56
16.0	0.250	6.35

Plate

Lead is available in plate thicknesses to 2 inches (50.8 mm). All thicknesses greater than 0.188 inches (4.78 mm) are considered to be plate.

Extrusions

Lead can be easily extruded into many shapes and forms. Lead pipes, angles, rods, and bars are all typically extruded. The maximum diameter circle that can be extruded is 12 inches (304.8 mm). Extrusion lengths are limited. Long-length extrusions will collapse if not fully supported, both during the process and during handling.

Castings

The low melting point and the fluidity of lead makes it one of the easiest metals to cast. Large objects as well as small, thin-walled objects, can be cast with lead. Lead melts at about 621°F (327°C) and can be poured into a casting mold at 700°F (371°C). Sections of smaller castings can be "burned" together, with the edges smoothed, to appear as one large casting.

Lack of strength and problems with creep make lead a difficult exterior metal when large sections are considered. Expansion and contraction can move the casting until ruptures develop.

Many statues made from lead use a combination of castings and thick sheets molded and hammered to the art form's shape. The edges are welded together and burnished to blend the seams. The interior of the statue is usually a painted steel or iron cage.

FINISHES AND COLORING

Lead does not polish well. The surface can be brightened by removing the oxide layer using fine sandpaper followed by buffing. The appearance is a blue-gray metallic surface. This appearance, however, does not last long. Lead will quickly develop an oxide layer of tarnish, regardless of interior or exterior exposure.

Lead can be cleaned by using a trisodium phosphate cleaner to remove the oxides. A wax application can be used to seal the surface for a period of time.

Lead can be darkened with different chemical treatments, such as mild acids. But on large projects this technique is difficult, if not useless. Lead is so nonreactive to chemical treatments that getting the conversion colors to take hold requires gentle heating. It is best to leave lead as is and allow natural weathering to occur.

Lead weathers to a whitish-gray patina if the environment is rural and carbon dioxide is prevalent. The white color is lead carbonate. Quite often, on vertical surfaces, lead develops streaks or patches of white oxide. These patches may not be continuous over the entire surface but may appear as if the surface has been partially washed.

In urban regions where pollutants are more prevalent, lead develops a darker, almost black patina. This patina is made of both the carbonate and lead sulfate. The darker sulfate tends to be more even across the sheet.

CORROSION CONSIDERATIONS

Lead is very inert. Atmospheric corrosion hardly effects this dense, dark material. The rate of corrosion resulting from sulfur dioxide pollutants is very slow. Lead resists attack from most acids, thus its use as a lining or electrode in anodizing tanks. Hydrochloric acid will attack lead, as will nitric acid.

Galvanic Corrosion

Lead, and the runoff of lead oxides, will corrode aluminum. Lead will stain the surface of aluminum and deposit a whitish patina on stainless steel. The stain can be cleaned from a stainless steel surface; however, the effects on an aluminum surface may be deeper. Corrosion cells can develop, which will pit the surface.

The oxide can also stain glass. Runoff from lead surfaces onto glass can produce an adherent stain. The stain can be cleaned by using denatured alcohol and mineral spirits.

Effects of Other Materials on Lead

Lead is attacked by free lime. Free lime is found in fresh concrete and will rapidly corrode lead. To guard against this effect, coat the lead surface with a black asphaltic emulsion.

Lead can be stained by the corrosion products of steel. Rust stains will imbed into the oxide layer, making for a difficult cleaning project. Stop the corrosion from occurring, and wipe down the sheets with trisodium phosphate. If this does not work, try a mild nitric acid solution. Use caution when working with acids. Acids will etch the surface slightly. Start with a small area and assess the result.

Discoloration of Lead Surfaces

Lead will develop a white film in patches, particularly on vertical planes. It can also develop rustlike streaking with a reddish color. This is red lead oxide, which develops when lead is exposed to alkalis such as lime or airborne fertilizer. Red lead oxide appears as streaks of rust stains. The carbonate form does not develop when this occurs. Cleaning of the surface is difficult and patchy. Once cleaned, the carbonate needs to develop or the red oxide will recur.

CARE AND MAINTENANCE

Environmental issues notwithstanding, lead construction poses some unique concerns. On sloped and vertical surfaces, the weight of lead and the expansion of the metal causes creep effects. As lead expands down-slope, the added effect of the denser lead sheeting restricts the return action as the metal contracts. This tendency is known as "metal creep." During the early years of the United States, many governmental buildings had lead roofs installed. The temperature extremes of the Virginia and Washington D.C. climate caused the lead to expand and contract in excess of what lead was experiencing in the more moderate European climate. The roofs were found to have large rips where the stress generated from thermal movements exceeded the yield strength of the metal.

Lead is better suited for flat, fully supported applications. Steep or unsupported applications will buckle and tear as the metal expands and contracts. Lead has one of the highest coefficients of expansion, higher than aluminum and just below zinc. Couple this characteristic with the lack of columnar strength, and there will be buckling. The first signs are wrinkles and buckles occurring in the center of the sheet. Tears will develop along or near seams.

Decorative formed and stamped lead sheet allows for shapes with deeper cross sections than simple flat sheets. The deeper cross section will help to resist buckling stresses generated from thermal movements.

FABRICATION AND INSTALLATION CONSIDERATIONS

Lead is the simplest metal to cold work. It will not work harden and can be hand folded and malleted into the most intricate shapes. Lead is heavy; working with lead is like working with giant noodles. The material lacks any meaningful stiffness. There must be people at every corner and edge to lift it. It is usually easier to drag the metal. Lead cannot be worked like a

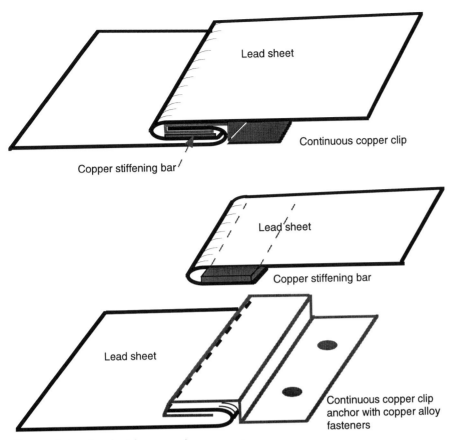

Figure 5-1. Lead with copper bar supports.

conventional metal. It does not shear well; a brake is useless. It is best to cut and fold by hand.

When working with large panels, it is more effective to use a copper bar as a stiffener. Fold the lead over the bar, and use this as the support. The panels should clip onto a continuous rigid support, and they should interlock into one another. The panels should be clipped or hung from the top or elevated side of the lead sheet. This will allow gravity to work with the expansion and contraction stresses.

FASTENING AND JOINING

Use copper nails and clips, wherever possible, for fastening. Brass and stainless steel screws will also work well.

Lead can be easily soldered by tinning the joint. However, a soldered joint is weak. The preferred method is to "burn" the joint together. *Burning* is an old colloquialism used to describe the melting and fusing of lead sheets together. The membrane sheets used as waterproofing are burned together at the seams into large sections. Lead roofing and ornamentation is typically formed at the site. Batten and standing seams can be hand turned and folded to create seams similar to those used in copper work.

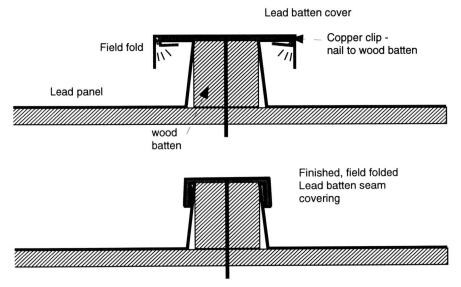

Lead batten cover

Field fold

Copper clip -
nail to wood batten

Lead panel

wood
batten

Finished, field folded
Lead batten seam
covering

Figure 5-2. Lead forming operation.

ZINC

INTRODUCTION

Zinc is an architectural metal from out of the past. Its use as an architectural metal through the centuries is instantly apparent in the weathered gray rooftops of Paris. This gray-blue metal was one of the first architectural metals used for roofing and water diversion during the beginnings of the industrial revolution.

Zinc is one of the transition elements, along with copper, silver, gold, mercury, and cadmium. Zinc is rarely found naturally in a pure state, unlike many of the other transition elements. However, the oxide, sulfate, and carbonate mineral forms of zinc are abundant in certain regions.

Zinc has a ductility equal to that of copper. It is a dense metal with a specific gravity of 7.14, which is very close to that of steel. (Copper has a specific gravity of 8.96.) Zinc will not spark when other metals strike against it, making it a choice metal for hazardous material containment. Zinc, like the

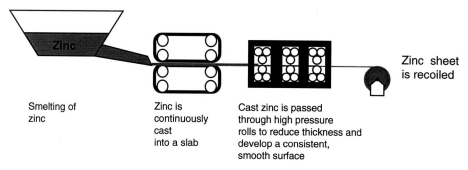

Zinc

Smelting of
zinc

Zinc is
continuously
cast
into a slab

Cast zinc is passed
through high pressure
rolls to reduce thickness and
develop a consistent,
smooth surface

Zinc sheet
is recoiled

Figure 5-3. Zinc roll casting technique.

other transition elements, has excellent corrosion-resistance characteristics and can be coupled with other metals to enhance their performance through the phenomenon of galvanic protection.

Modern alloys of zinc, some containing small amounts of copper and titanium, are common architectural metals used for building materials. Other forms of zinc are chrome plated and used for decorative trim on automobiles and appliances. Zinc can be cast into hardware, such as door handles, or into large ornamental grilles and art features.

When properly detailed and installed, zinc alloys perform well in most atmospheric exposures. The metal surface will slowly age to a uniform gray blue color as it absorbs carbon dioxide from the atmosphere. The gray-blue patina enhances the corrosion resistivity of the base metal. Scratches into the metal surface will quickly "self-heal" as a new oxide film develops to seal and protect the base metal.

Zinc has had extensive use in the European countries and should, because of the concerns about lead, make major headway in the United States market for natural gray color tones in architectural metal. In the United States, zinc is more commonly associated with the ubiquitous galvanized coatings on steel. As a coating, zinc in many different alloy forms has had enormous architectural use. This chapter, however, is devoted solely to the commercially pure metal and the various forms available for architectural use.

HISTORY

Zinc has been used in copper alloys since the Roman era. The first indications were recorded in 200 B.C., when zinc was added to copper to develop brass. The initial production of zinc began in Gosler at about 1550, with commercial production of the metal occurring at about 1650. Zinc, in sheet form, was developed by William Champion of England. The first pure zinc sheets were brittle and easily cracked during rudimentary forming operations.

It was soon discovered that these brittle cast zinc sheets could be hot rolled at temperatures in the range of 212°F to 302°F (100°C to 150°C) to improve formability. The hot rolling improved the grain structure by stretching the grain along the length of the sheet. Stretching the grain structure greatly reduced the brittleness of the zinc sheet metal and increased the tensile strength.

One of the first successful architectural uses of zinc occurred in 1814 when the royal castle in Berlin was roofed with zinc sheets. The zinc performed so well that all future royal buildings were required to be roofed with this metal.

Formed into tiles and fabricated into elaborate ornamentation, this ductile metal became a common architectural material throughout Europe. The second World War interrupted European zinc production, which did not really recover until the 1960s. At about this time commercial production was improved by new production methods such as electrolytic deposition and continuous casting.

Large deposits of zinc ore, known as hemimorphite, were discovered in Flanders during the eighteenth century. After the French took control of the region, a large zinc fabrication facility was established. The first zinc sheets were hot rolled from the facility in 1811.

There are large deposits of zinc in the United States, but it was not until the late 1830s that the metal was exploited domestically. Prior to this time, the United States was a major importer of zinc. The St. Charles Hotel in New Orleans was one of a number of projects successfully roofed in imported zinc sheets in the early 1800s.

The United States developed domestic zinc sheet production in 1838. The supply of ore had always been in the States, but exploitation of the metal lagged behind that of copper. Large quantities of high-grade ore were found in New Jersey and Missouri to bolster the supply of the metal, and soon rolling mills were set up to produce sheets primarily for the canning industry, which at this time was more a jar industry. Zinc lids for mason jars, made by the Ball Brothers Glass Manufacturing Company (now Alltrista Zinc Products Company), are common artifacts from this previous century.

PRODUCTION AND PROCESSING

High-grade zinc ore, sphalerite (zinc sulfate), is mined in Germany and other regions of central Europe. Sphalerite is also heavily mined in Missouri, around the Joplin area. The ore zincite is mined in New Jersey.

Zinc refining is similar to that used to refine copper. The zinc ore is crushed into fine particles, then concentrated by floatation. The floatation process concentrates the mineral form of zinc by soaking it in an agitated pool of water while bubbling air through the mix. The minerals attach to the

Figure 5-4. Directional nature of rolled zinc surface—preweathered.

air and float to the surface, where they are skimmed. Waste particles sink in the agitated soup.

Once concentrated, the zinc sulfide mineral form is roasted. Sulfur is removed and turned into sulfuric acid. The sulfuric acid is used to dissolve the zinc and create a powerful electrolyte. The zinc is next electrolytically deposited out of the electrolyte and onto an anode. Finally, the nearly pure deposit is removed from the anode to be cast.

Today most zinc sheet is continuous cast onto a rotating cylinder. While still hot, the metal is rolled through pressure rolls and reduced to a specified thickness. When commercially pure zinc is initially cast, it has very low tensile strength, on the order of 9,000 psi (62 MPa). But, through subsequent rolling of the sheet to reduce thickness and develop grain directionality, the tensile strength improves to the 24,000 psi (165 MPa) range.

Modern processes utilize an alloy of zinc with small additions of copper and titanium to enable better sheet and strip characteristics in the metal. The grain structure is stabilized and strength is improved with the addition of titanium. Minute amounts of copper reduce the brittleness of the metal sheet.

Sheet metal roofing and flashing material made from commercially pure zinc, often referred to colloquially as "rolled zinc," provide excellent enclosures when properly used. The term *rolled zinc* developed in the early 1800s, when production techniques developed in Belgium used rolls to cast the thin, unalloyed zinc sheets.

ENVIRONMENTAL CONCERNS

In the United States, zinc has experienced heavy competition from other gray-colored metals, such as lead-coated copper and terne-coated stainless. With growing concern for the environmental hazards posed by lead, its producers should be able to capitalize on the nontoxic nature of this metal. Runoff from zinc poses no hazard to the surrounding environment. Zinc and its natural oxides are nontoxic to human and plant life and can be handled without health risks.

"Zinc fever" is a major hazard, however, when the vapors of zinc oxide are produced and inhaled during welding or brazing operations. This illness is characterized by headache, chills, fever, and nausea and can be quite debilitating. The symptoms may be delayed 4 to 12 hours after exposure. Symptoms disappear upon intake of fresh air and rest. The illness usually lasts from 24 to 48 hours. Perform welding and soldering operations only in well-ventilated areas, with the use of respirator equipment. Cutting with plasma torches will also produce zinc oxide fumes. Avoid performing this operation in confined spaces. Use a respirator.

Zinc is used in many ointments and lotions for protecting and treating topical ailments of the human body. Another name for the zinc oxide ore is calamine. The bug bite and poison ivy lotion with the same name contains zinc as an itch inhibitor. Zinc sulfate is used as an aspirant ingredient in many antiperspirants.

Zinc is recyclable. Of zinc ingot, 5% to 10% is derived from recycled metal. This percentage should increase with time as more of the metal is recovered and recycled. Current scrap value of the metal is low. The metal's dull gray color makes it difficult to identify and distinguish from other scrap metals. As the value of scrap increases, the identification and separation of zinc from other metals will improve and so will recycling. Zinc refinement and casting processes use less energy than refining and casting operations for aluminum and copper. The low melting point of the metal will keep the recycling costs down in comparison with most other metals. Zinc uses lower amounts of energy during production and processing and produces less pollution than copper production.

ARCHITECTURAL ALLOYS

Alloy 700

Common Name : Rolled Zinc Sheet

Alloy Composition—Nominal by Weight
Zinc   99.80%
Copper   0.7 to 0.9%
Titanium   0.08 to 0.12%
Traces of cadmium, lead and iron are also present.

Wrought Alloy—Available Forms
Strip   ASTM B69
Sheet   ASTM B69, Federal Spec. QQ-Z-100A
Plate

Natural Color   Bright, slightly bluish silver
Oxidized Color   Dull, blue-gray
Weathered Color   Dull, blue-gray

The 700 alloys of zinc contain copper and titanium to reduce brittleness and to help modify grain structure for better strength. These sheet alloys are grain sensitive; that is, they perform differently depending on the direction of the grain as the metal is cold worked. For instance, the coefficient of thermal expansion is greater with the grain than across the grain. Tensile strength is also a function of grain direction. Many cold-rolled metals have this characteristic, zinc more than most.

Coefficient of Thermal Expansion
Along the grain direction 13.8×10^{-6}in/in/°F 9.26×10^{-4}mm/mm/°C
Perpendicular to the grain 10.8×10^{-6}in/in/°F 7.26×10^{-4}mm/mm/°C
  direction

Tensile Strength
Along the grain direction   24,000 to 28,000 psi (165.5 MPa to 193.1 MPa)
Perpendicular to the grain direction 30,000 to 35,000 psi (206.8 MPa to 241.3 MPa)

There is a rolled zinc product made in Europe that has properties and makeup similar to those of alloy 700. This product is specified under the

European standards for Electrolytic High Grade Zinc. This rolled zinc product meets the following specifications:

DIN 1706 Electrolytic High Grade Zinc
DIN 17 770 Mechanical Properties
BS 6561A

This alloy has a degree of purity of 99.995% zinc. The remaining amounts are titanium and copper, included to develop better creep resistance. Many successful projects across Europe are covered with this product.[1]

There are a number of other sheet and strip alloys of zinc available with various properties designed for specific applications or forming operations. From the standpoint of architectural sheet applications, the 700 series alloys and the European alloys are preferred. The following list describes other commercially pure alloys and their characteristics:

100 Series Alloys
Good formability; high coefficient of thermal expansion
Used in batteries, badges, and automotive stampings
Contains zinc, lead, cadmium, and iron

300 Series Alloys
Poor formability; high coefficient of thermal expansion
Used for etching plates in the printing industry
Contains zinc, aluminum, and magnesium

Special Alloys

There are superplastic alloy forms of sheet zinc. These alloys, when heated to the fully annealed level, can be formed with vacuum or compression processes. Tremendous elongation limits, as great as 2,500%, have been demonstrated.

Die Cast Alloys

Zinc alloys can be readily cast using sand and permanent mold methods. The low melting temperature and the good fluidity of the alloys enables very intricate forms and shapes to be cast. In addition, the alloys can be easily finished and honed to produce very decorative surfaces.

Each of the common cast alloys of zinc contain aluminum and small quantities of copper, along with some other alloying metals. The alloys all have adequate strength. Ultimate tensile strengths in the 40,000 psi (248 MPa) to 64,000 psi (441 MPa) range are achievable. Their strength is directly proportional to the amount of aluminum in the alloy mix. The aluminum makes the casting appear lighter in color and reduces the overall weight.

Refer to Table 5-2 for a list of suggested cast alloys.

---

[1]One of the European products is known as Titanzink, Zinc-Titane, or Titanium Zinc and is produced by Rheinzink of Germany.

TABLE 5-2. Cast Zinc Alloys

ASTM	UNS NUMBER	GRADE	ALUMINUM %	COPPER %	COMMENT
B86	Z33520	Alloy 3	3.5 to 4.3	0.25 max	Good castability; common cast alloy
B86	Z35531	Alloy 5	3.5 to 4.3	0.75 to 1.25	European alloy; good castability
B669	Z25630	ZA-8	8.0 to 8.8	0.80 to 1.30	Decorative applications; excellent finishing
B669	Z35630	ZA-12	10.5 to 11.5	0.50 to 1.25	General purpose; good castability
B669	Z35840	ZA-27	25.0 to 28.0	2.00 to 2.50	Light weight; good castability

AVAILABLE FORMS

Sheets

Sheet zinc is covered under ASTM B69, "Rolled Zinc," and further qualified under Federal Specification QQ-Z-100A, "Zinc Alloy Sheet and Strip."

Thickness

Zinc sheet metal is available in thicknesses specified by order. The sheet is rolled cold to the specific thickness desired within the zinc mills' tolerances.

	English	**SI Units**
Minimum	0.012 inches	0.30 mm
Maximum	0.049 inches	1.25 mm
Standard Thicknesses	0.024 inches	0.60 mm
	0.026 inches	0.65 mm
	0.028 inches	0.70 mm
	0.032 inches	0.80 mm
	0.040 inches	1.00 mm
	0.050 inches	1.20 mm

Tolerance range is ± 0.035 mm.

Width
Standard sheet widths are as follows:

English	**SI Units**
20.0 inches	508 mm
24.0 inches	610 mm
28.0 inches	711 mm
39.4 inches	1000 mm

Variable widths within those dimensions shown are available in strips and slit coils.

Length

Zinc is available in coil form or can be obtained in decoiled sheet. Limitations are imposed by the shipping and packaging of the skidded sheet. Ten meters is about the upper practical limit for shipping sheet zinc. For lengths beyond this, trucking and packaging become a difficulty.

Standard Temper Characteristics of Zinc

Cold-Rolled Zinc Sheet or Strip
Tensile Strength   21,750 psi   150 MPa
Yield Strength   14,000 psi   100 MPa
Hardness-Rockwell B   40

Extrusions

Zinc is available in extruded shapes. The cross-sectional dimensions are limited and should be designed with thicker walls. Thin-walled extrusions tend to collapse. Usually, small cross sections are solid shapes to avoid this problem. The extrusion is pushed through a steel die, similarly to an aluminum extrusion, and stretched to straighten the shape. Lengths of extrusions are limited to approximately 4 meters. The ends of the extrusion will deform and are usually cut off and recycled.

There are four main extrusion alloys available. The zinc-aluminum alloys contain 11% to 14.5% aluminum and are light in color. The zinc-copper-titanium alloys contain approximately 1% copper and 1% titanium.

Alloys   **Trade Names[2]**
Zinc-Aluminum   Korloy 2570
                              Korloy 2573

Zinc-Copper-Titanium   Korloy 3130
                              Korloy 3330

FINISHES AND WEATHERING CHARACTERISTICS

Standard Mill Finishes

Zinc sheet metal is obtained from the mill with a surface finish either of a reflective silver color or preweathered to a dull grayish-blue color. The reflective sheen will gradually tarnish after atmospheric exposures of three to six months. This tarnish is a natural, protective patina, which should appear eventually as an even gray-blue color, similar to that of the preweathered sheet.

The reflective silver color is produced as the zinc is passed through polished reducing rolls. The finish has a directional nature, produced by small scratch lines in the surface and by the stretching of the grains in the direction of rolling.

The preweathered finish has a directional nature as well. The in-line oxidation process enhances the appearance of the grain structure. Initially, this appears as slightly lighter or darker streaks or lines in the sheet. These lines weather out upon further exposure. Both mill surfaces have a grainy appearance, similar to that of unpolished pewter.

---

[2]Zinc extrusions are manufactured by Commenco of Canada. Korloy is their trade name.

Zinc can be polished and buffed to develop highly reflective surfaces, but is not typically used in this form unless it is to be a base for plating other metals. The polished and buffed surfaces are slightly "smoky" in their reflectivity. They lack the reflective chrome color of mirror-polished stainless steel. For the sheen to remain, clear coatings are necessary to resist further tarnishing of the surface. Do not expect these finishes to last in exterior exposures.

Zinc fabrications will continue to age naturally, even preweathered sheets. As zinc weathers, it develops a thin oxide layer that slows further corrosion of the surface, similarly to other natural metals such as aluminum. Zinc oxide, unlike copper oxide, will not stain adjacent surfaces. The oxide is clear and insoluble in water. Once it forms on the zinc surface, it develops a very adherent bond to the base metal.

Zinc, exposed to the atmosphere, undergoes a complex oxidation process. Initially, the shiny surface develops a tarnish of zinc oxide. This occurs rapidly upon exposure. Water from rainstorms and dew collecting on the surface of the metal changes the oxide to a hydroxide. Carbon dioxide in the atmosphere combines with the hydroxide to develop zinc carbonate. Zinc carbonate is an adherent, insoluble film, which becomes the final corrosion-resistant surface.

Custom Finishes

Because zinc is soft, with a Rockwell Hardness in the range of 54 to 66, it can be hammered and embossed with ease. Other finishing techniques, such as satin or mirror polishing, can be performed; however, when used on the exterior of a structure, the surface will weather over, negating the polishing effects of these techniques.

Zinc shapes and surfaces can be engraved with relative ease. Zinc has a nongalling feature; it tends to self-lubricate, reducing tool wear, and allows sharp, distinct cuts into the metal surface. If decorative engravings are desired, specify a harder alloy such as the 800 series alloys. If the zinc fabrication is to undergo forming prior to engraving, one of the softer alloys can be used.

Coloring Techniques

Different techniques have been used to develop color on a zinc surface. Dilute chromic acid treatments will develop zinc and chromium oxides on the surface. The colors are different hues of blue and gold. Complex oxides can be developed on the surface of zinc to produce pale yellow-gold colors.

Surface treatments using oxidation techniques, causing reaction with the metal to form custom colorations or surface etchings, are possible. These coloring techniques require the surface to be sealed if the color is expected to remain. Darkened surfaces, achieved by pickling treatments, can be used outdoors without coating. Such finishes may develop some streaking effects, depending on the environment of the exposure.

Corrosion Characteristics

The surface oxide, when allowed to develop correctly, will show excellent corrosion resistance even when exposed to urban and coastal environments. The surface corrosion rate is very slow. Pitting of the surface is usually negligible, if venting and natural washing are included in the design. If condensation is allowed to concentrate on the underside of the zinc sheet, particularly when coupled with detrimental material such as acidic wood or bitumen, or dissimilar metals such as copper, expect rapid and severe corrosion.

Some metals, such as copper or copper alloys, generate corrosion compounds that will attack zinc. Galvanized steel (zinc-coated steel), aluminum, lead, and stainless steel will not have adverse effects on zinc. These materials are close to zinc on the electromotive scale (reference Table I-2), and the concern for galvanic corrosion is limited. When coupled with aluminum, lead, or stainless steel, zinc acts as the sacrificial metal and will decay, as compared with these other metals, at a very slow rate as long as the protective oxide film is allowed to develop. This is why zinc is used as a protective coating for so many other metals. Copper, on the other hand, will corrode zinc rapidly. Do not allow water to wash from copper surfaces onto zinc surfaces. The zinc surface will first discolor in streaks and then corrode.

A zinc carbonate film is dense and will protect the base metal from further corrosion. The corrosion rate depends on the environmental exposure, urban versus rural, and the bias of the installation. Flat exterior surfaces will corrode more rapidly than sloped surfaces, because sloped surfaces receive the benefit of natural washing.

Sulfur Dioxide

Sulfur dioxide, particularly in environments of high humidity, inhibits the development of the carbonate film. Corrosion rates will be much higher when the metal is exposed to environments containing high sulfur levels. High-sulfur environments are generally found in industrial urban regions and downwind of coal-powered electric plants. When zinc sheet is installed on vertical surfaces or those sloped in excess of 20 degrees, it will resist the deposition of sulfur dioxide particles owing to the washing effect of rains. The corrosion rate of properly installed sloped surfaces of zinc will be slow. The life span of the metal can be expected to extend more than 100 years in many applications.

The underside of a zinc sheet is subjected to different criteria. The carbonate film will not develop to the same level as it does on exterior exposed surfaces. If the underside is allowed to become and remain wet from condensation, localized pitting will occur. Thus, venting the unexposed side of zinc sheet is crucial for installations where condensation will occur. Vertical surfaces and interior applications should not experience the collection of moisture.

Concrete and Mortar

Zinc will not corrode when in contact with concrete or mortar. These materials are more alkaline in nature and do not pose a problem for zinc. The use of zinc as a through-wall flashing is not recommended, however.

Through-wall flashings will trap and hold water leaching through the wall. Over time, the water may corrode the zinc.

Wood Preservatives

Wood preservatives will affect the long-term corrosion resistance of zinc. If they are allowed to collect in areas of moisture, corrosion cells will develop and corrode the zinc. Preservatives used to protect wood from the effects of decay and against fire will corrode many metals, including zinc.

Bitumen

Roofing bitumen will corrode zinc. Bituminous corrosion, which affects all metals, occurs when waste products develop as the bitumen decays because of ultraviolet exposure. Bitumen waste products are very acidic. Concentrated acids, which can develop under certain conditions, will attack all metals.

Painting

Zinc receives most paints well, providing good adhesion and excellent protection against edge corrosion along sheared surfaces. In addition, zinc oxide rapidly develops at scratches through the paint surface, inhibiting further corrosion. Clean the zinc surface of all oils and other foreign materials prior to application of paint.

CARE AND MAINTENANCE

In using zinc as an architectural metal, problems occur when fabrication limitations and installation details are not correctly followed. Zinc will not form well at temperatures below 50°F (10°C). When forming processes are performed on cold metal, it becomes brittle and will crack at simple folds. Working the metal on exterior applications during the winter months is not advised. Simple double-lock seaming when the metal is below these temperatures will crack the metal. Once the metal is formed, or when it is preformed in the controlled environment of a shop, zinc will function adequately regardless of further change in temperature. It is only the forming operation on cold metal that should be avoided. Localized heating of the metal is possible prior to forming; however, this is rather difficult and costly for most applications.

Exterior applications of zinc require ventilation of the underside or nonexposed surface. Nonvented zinc installations can fail if trapped condensation or other moisture is allowed to collect and remain on the underside of the sheets. The trapped moisture may take on an acidic pH, which can dissolve the metal relatively quickly.

The venting criteria require close attention to detail if the application of zinc is a roof or wall panel system. Sheet zinc must not be used in applications where moisture will be trapped or allowed to condense on its undersides. Roofs and parapet flashings are most susceptible, because moisture can travel through building materials and collect on the cool metal. The use of felts help control the movement of moisture, but the felts are not sealed and the clips used to anchor the sheet to the roof substrate penetrate the felts.

On exterior vertical surfaces, make certain that transition details do not create ledges where moisture will be trapped and held. Sloped roofing applications and flat slope flashing applications of zinc must be examined for the potential for condensation to develop on the reverse side of the sheets.

Methods of Venting

To reduce the potential for condensation development on the underside of roofs, special detailing considerations must be addressed. Moisture will condense on the cooler side of an object. If moist, warm air from inside a building is allowed to reach the cool surface of a metal, it will condense. Installing a vapor barrier below the roof skin will inhibit the movement of the vapor to the cool metal. However, this method alone is inadequate, because vapor has a way of getting through most membranes. Cracks, seams, and fastener penetrations all breach the barrier. The most effective method is to vent the space below the roof support system. The building insulation should be installed below this space, with a vapor barrier facing the inside or warm side of the roof. This step should locate the dew point somewhere within the insulation if the correct amount of insulation is used. The dew point is the level at which the air temperature is cool enough to cause moisture to condense. If the vapor barrier is below this line, the moisture from within the building will be prevented from reaching the dew point. Thus, the inside of the zinc sheet will be the same temperature as the inside of the vented space and condensation will not develop.

There were venting systems for zinc roofs that used small corrugated galvanized support liners directly under the zinc. This raised the zinc sheet above the roof vapor barrier, developing a space between the inside of the zinc sheet and the roof support. This space was vented at the roof ridge and eave, using a perforated vent strip. Theoretically, this would allow condensation to occur, but the space created by the corrugated strip would act as a ventilation space. The problem with this method is that it adds cost and difficulty to an otherwise straightforward roof installation. In addition, it puts steel sheets, albeit galvanized steel, in an area subject to moisture. Long-term performance of the zinc may be improved; however, the life span of the roof is predicated on the life span of this thin corrugated galvanized sheet.

Zinc should be used on structures with a positive slope of not less than 3 in 12. At a slope of less than 3 in 12, additional venting must assist the drying of the underside of the sheet. The slope will drain moisture that may pass through the roof construction and collect on the cold surface of the zinc. Felts, at minimum, are necessary, lapped in the direction of water flow. A rosin-saturated building paper should also be used to act as a slip sheet between the zinc and the felts. Bituminous moisture barriers with smooth, slick outer sides will also work well as vapor barriers. Although bitumen will corrode zinc, in this form it is protected from the effects of ultraviolet radiation when it is applied below the metal skin. In addition, the slick, plastic nonadhering side separates the bitumen from the metal. With this material, use a slip sheet of rosin-saturated building paper.

For parapet caps and other flashings installed flat, exercise caution. Build in a detail which allows some degree of slope. Be certain that the wood used

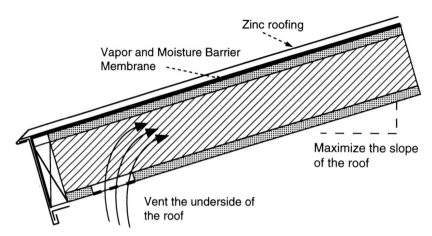

Figure 5-5. Zinc venting requirements—ridge detail.

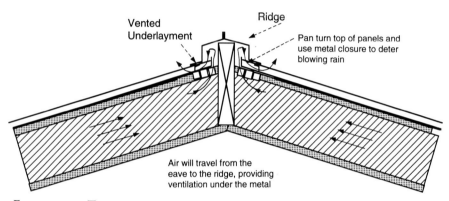

Figure 5-6. Zinc venting requirements—eave detail.

to provide a nailer on the flashing is separated with felts. Some woods will leach acids that can pit and cause premature decay of the metal.

Creep Considerations

Large, heavy fabrications, such as statues, cast from zinc are subject to a condition known as creep. Cycles of expansion and contraction resulting from seasonal temperature changes will create stresses that can crack or tear the zinc shape. A heavy mass of metal in the cast form does not possess the strength to overcome such stresses. This condition does not occur in wrought materials, such as sheet zinc, when they are alloyed. Sheet forms of the metal gain higher yield strength from the cold-rolling operation and the additions of titanium and copper.

FASTENING AND JOINING TECHNIQUES

Soldering

Some specific characteristics of zinc alloys used architecturally are a low melting point, which makes for easy resistance welding; ease of soldering;

their nonsparking nature, relative light weight, good corrosion resistance, and their bias for expansion and contraction with the direction of the rolled grain.

Zinc solders in a similar manner to galvanized steel. Use 50-50 lead-tin solders and neutralize all excess flux. The solders should be antimony free, and even heat should be applied to the joint. Keep in mind that zinc has a very low melting point, 792°F (422°C). Use a noncorroding flux or a zinc chloride flux. Zinc chloride fluxes can be made by dissolving strips of zinc in muriatic acid. The zinc will dissolve and combine with the chloride in the acid. As the zinc dissolves in the acid, hydrogen is released in small bubbles. When, after small amounts of zinc are added, the bubbling ceases, the flux is ready. When soldering is complete, rinse all excess flux from the joint with fresh, clean water. Overdo the rinsing to ensure that all traces of acid are removed from the surface. Use sodium bicarbonate, as well, to neutralize the acid.

Welding

Zinc will spot weld adequately. Welding or brazing the metal is not possible because of the high temperatures involved. Zinc, with its low melting point, will liquefy and flow under the temperatures of MIG and TIG welding.

Other Fastening Techniques

Galvanized nails are recommended for concealed attachments; stainless steel pop rivets or screws will also perform well. Never use copper or brass fasteners. Use clips whenever possible to attach zinc sheets. Clips will spread out the stress at the attachment point and can be designed to allow expansion and contraction. Because zinc has a relatively low tensile strength coupled with a high coefficient of expansion, point loads should be minimized and expansion should be designed into the connection. Use galvanized steel clips, zinc clips, or stainless steel clips to fix the fabricated part back to the structure.

FORMING AND FABRICATION

Uncoated zinc roofing and flashing offers a long-life, maintenance-free material in all environments. The ductile zinc sheet can be formed and shaped with ease using conventional sheet metal equipment. The standing seam or batten rib can be hand folded with simple tools and mallets or machined seamed with modern roll-locking equipment. Zinc is one of the easiest metals to work with.

One benefit of working with zinc is its nongalling characteristic. It does not adhere to itself or other metals as forming operations proceed. Zinc is nonmagnetic and will not spark when other metals are passed over its surface. This characteristic, along with the ductility of the metal, make spinning and stamping exercises relatively simple.

Blanking and punching operations are easily performed on zinc sheet metal. Zinc is somewhat self-lubricating, but water-soluble lubricants can be added to assist. Tool wear is minimal when punching zinc sheet because of the metal's softness. Allow clearances of 5% to 10%. Its nongalling characteristic makes zinc an ideal metal for perforating. It will not hang up on tooling like copper or other metals.

Deep-drawing operations involve very little tool wear as well. No interstitial annealing is necessary to form intricate, deep forms when using this metal.

Zinc has two characteristics that can lead to problems during fabrication and installation. In the wrought cold-worked form, the material has a directional nature. Fabrication activities such as brake forming and roll forming will have different results when performed on metals with different grain direction biases. Zinc cannot be cold worked when temperatures are moderately cool. Brittleness begins to affect the metal when temperatures fall below 50°F (10°C). When cold temperatures really set in, the metal will require heating to be worked.

6

Monel and Titanium

Monel and titanium are two silver-gray metals with incredible resistance to corrosive action. Each metal has a unique background. Titanium is the new metal in architecture, and monel is one of the older ornamental "white" metals. Each metal comes from different periodic table families with different characteristics of atomic makeup. But each metal has attractive similarities for architectural use.

Both metals resist chloride attack, making them uniquely attractive for coastal environments or for urban areas subject to road salt. Both monel and titanium have low thermal expansion coefficients. Architectural use of monel and titanium has been limited to the sheet metal form. Roof panels, wall panels, and thin skins for ornamental column covers are typical uses for these metals. Other forms are available, but cost and workability limit the attractiveness of all but thin sheets.

This chapter presents these and other unique characteristics of monel and titanium along with some limitations on their use.

MONEL

INTRODUCTION

Monel is a nickel-copper alloy developed in the early part of the twentieth century. This reflective silvery-white metal is considered the predecessor to stainless steel. Monel fell from architectural favor during World War II, when sources of nickel were redirected to the war effort. The advent of stainless steel, a cheaper alloy with similar characteristics and appearance, has replaced monel as the more common "white metal." Stainless steel is typically about 50% the cost of monel sheet of equivalent thickness.

The elements nickel and copper make up the major part of this metal. Both elements are mutually soluble; that is, when molten they will dissolve into one another at a particular proportion. There are naturally occurring ores that contain proportions of nickel to copper in a 70% to 30% ratio. For this reason, monel, which contains approximately 63% nickel and 28% to 34% copper, is considered a "natural" metal. Monel shows good strength and toughness, as well as superior corrosion resistance. In fact, monel has better corrosion resistance than nickel and copper have alone. Monel work hardens as it is cold rolled. Annealing is necessary to modify grain size and to eliminate brittleness.

HISTORY

In 1905, Robert C. Stanley, working for the International Nickel Company, smelted nickel and copper ores together. Previously, they were independently refined. The initial cast bar from this smelting was stamped "Monell Metal" in honor of Ambrose Monell, the president of the International Nickel Company at the time. The extra "l" was dropped from the metal's name when the company sought a trademark for the alloy.

Monel was commonly used for roof and ornamentation in the early part of the twentieth century. In the 1920s monel was also used for decorative features, similarly to the copper alloy known as "Nickel Silver." In 1936, the Metropolitan Museum of Art in New York used the metal as a roof and guttering material. Today, inspections of the surface show the metal to be in excellent condition, free of damaging corrosive effects. The St. Louis Science Center, constructed in 1989, used monel as a roof covering over the central dome. With conventional sheet metal details used to construct this intricate covering, the metal roof surface should perform exceptionally. Its appearance fools many a sheet metal expert into thinking that it is mirror-polished stainless steel.

PRODUCTION AND PROCESSING

Monel is produced by reduction of nickel-copper ore to an alloy matte. The ore contains copper in a ratio of approximately 1 to 2. Further smelting of the ore increases the concentration of nickel to approximately 65%. The process of production is similar to that of copper. The metal is taken through hot rolling, then to cold reduction. Annealing is necessary to reduce brittleness in the cold-reduction process.

ENVIRONMENTAL CONCERNS

Monel has no known toxicity; neither nickel nor copper is considered toxic to humans. Yet because of its copper content, monel is somewhat eco-toxic. Aquatic vegetation, in particular, will not adhere to monel until a layer of slime has covered the metal surface. (See Chapter 2, Copper.) Most of the nickel used in the world is mined in Sudbury, Ontario, Canada. The ore is mined with copper. The mining process produces substantial amounts of overburden waste, which is placed back into the stripped-out regions.

ARCHITECTURAL ALLOYS

The conventional monel alloy used in architecture is alloy 400, a wrought alloy. This is an exceptionally corrosion-resistant alloy. Its composition is listed in Table 6-1. There are other varieties, such as Inconel and alloy 250, but they are not architectural forms.

TEMPERS

The temper used for the sheet and strip forms of monel is generally the annealed temper. This provides a surface with a Rockwell Hardness (B-scale) of 60. Three of the common tempers are indicated below. Other tempers are also available.

Temper	Tensile (ksi)	Tensile (MPa)	Yield (ksi)	Yield (MPa)
Annealed	70.0	482.6	27.2	187.5
Half-Hard	75.8	522.6	56.2	387.5
Hard	117.8	812.2	110.0	758.4

AVAILABLE FORMS

Sheet

Monel alloy 400 is available in both sheet and strip forms.

Sheet Specifications
ASTM B 127
QQ-N-281 (class A)

TABLE 6-1. Composition of Monel (Alloy 400)

ALLOYING CONSTITUTENT	PERCENTAGE	
Nickel	63%	minimum
Copper	28% to 34%	range
Iron	2.5%	maximum
Magnesium	2.0%	maximum
Silicon	0.5%	maximum
Carbon	0.3%	maximum
Sulfur	0.024%	maximum

Standard available sheet widths are:

inches	mm
30	762
36	914
48	1,219

Sheet lengths are maximum 144 inches (3658 mm). Strip in coil lengths is available. Sheet thickness range from a minimum of 0.018 inches (0.46 mm) to a maximum of 0.25 inches (6.35 mm). The typical thickness used for architectural applications is 0.021 inches (0.533 mm) for roofing and general sheet metal.

Standard thickness of monel used in architectural sheet metal applications:

inches	mm
0.018	0.457
0.021	0.533
0.031	0.787

Plate

Specification: ASTM B127
Monel plate is available in widths to 96 inches (1219 mm) and in thicknesses ranging from 0.188 inches (4.78 mm) to 2.00 inches (50.8 mm). The surface is provided rough in the hot-rolled, annealed, and descaled condition.

Rod and Bar

Specification: ASTM B164 and ASTM B564
Monel is available in round, square, and rectangular bars. These have various width dimensions, ranging from 0.25 inches (6.35 mm) to 4 inches (102 mm).

Tubing

Specification: ASTM B165
Monel is available in seamless cold-drawn tubing. The wall thickness is relatively great for most architectural requirements. Diameter vary from 1.5 inches (38 mm) to 3.25 inches (83 mm).

Castings

There are cast versions of monel available with consistent elemental makeup. Usually, silicon is added to improve castablity. Because of the higher cost, castings are not typically performed using this alloy.

FINISHES AND WEATHERING CHARACTERISTICS

Monel is considered a white metal. In the polished state, it resembles stainless steel. Monel sheet is available from the mill producer with a cold-rolled finish. The cold-rolled finish has a silver-white color and moderate reflectivity. The color is whiter than the silver-gray tone of stainless steel.

The cold-rolled finish can be further polished to satin and mirror finishes, similarly to stainless steel. When polished, the surface takes on a very bright luster, but lacks a little of the deep metallic chrome appearance of mirror-polished stainless steel. Other decorative finishes can be applied to monel, such as the glass-bead blast texture and the embossed finishes. (See Stainless Steel Finishes, Chapter 4.)

Exposed to the atmosphere, monel weathers to a silver-gray then to a greenish-brown. A patina develops, which is very thin and smooth. In coastal regions, the patina is a gray-green, just coloring the surface slightly. In rural regions, the patina is a very thin, light brown tarnish. Weathering takes years of exposure to environmental pollutants. The process is very gradual and depends on the corrosive environment to which the metal is exposed. The patina does not stain adjacent materials like other copper alloys.

Indoor exposures of the alloy will develop a light tarnish, which is easily removed by wiping with clean, dry cloth and metal polishing compound.

CORROSION AND COMPATIBILITY CONCERNS

As previously mentioned, monel is more corrosion resistant than either copper or nickel. For that matter, monel is more corrosive resistant than stainless steel. Salt water, chlorine, and other corrosive environments will have only minor corrosive impact. A smooth, thin gray-green patina will develop.

Monel is compatible with most every building material. Installed over felts and wood, this material will form a superior roof or fascia system. Follow the same precautions in using this metal with wood preservatives as in using other metals. These compounds, when moist, tend to attack most metals.

Acids. Monel is resistant to attack by most dilute acids and to both hydrochloric and hydrofluoric acid. Monel is used to handle and contain sulfuric acid. Nitric acid will, however, attack alloy 400 because of the lack of chromium in the alloy mix.

CARE AND MAINTENANCE

As a roofing material, monel provides an excellent corrosion-resistant surface. The sheet metal is hot rolled and then annealed and pickled to produce a soft temper. The metal can be hardened by cold working, sometimes to the point of brittleness. For severe shapes, this limits the extent of fabrication a single section can undergo. Spinning deep shapes and deep draw operations are not recommended for monel. Monel can be soldered similarly to other copper and nickel alloy metals; however, the appearance of the solder joint contrasts with the green-gray tone of the metal as it weathers. Monel has a moderate coefficient of thermal expansion, which reduces fatigue on exposure to thermal stresses created by roofing applications. This is a major benefit to the metal when used architecturally. Fatigue stresses resulting from thermal movement will not develop, nor will stresses at soldered joints cause the solder to split and crack. If installed correctly, the metal should perform

TABLE 6-2. Coefficient of Thermal
Expansion—Comparisons of Metals

METAL	COEFFICIENT OF EXPANSION μ in./in.°C
Monel	14.0
Copper	16.5
Aluminum	23.6
Stainless	16.5
Titanium	8.4

and appear consistent with its original installation. Table 6-2 compares monel to other architectural metals with regard to thermal movement.

Exterior surfaces of monel should be allowed to self-clean through the washing effects of rain. Concealed or protected surfaces will, over time, accumulate a layer of oxides and corrosion products. These corrosion products are hygroscopic. They will absorb moisture from the air and appear discolored. This is not a physical problem, but an aesthetic one. Periodic cleaning of these surfaces maybe necessary.

FASTENING AND JOINING

Monel can be welded, brazed, or soldered using most conventional techniques. Use filler metals of similar makeup, such as Monel Filler Metal 60.

Monel can be soldered similarly to copper and copper alloys. A chloride flux should be used to prepare the joint. Use a low-lead solder; this will help match the appearance of the brighter adjoining metal.

Use stainless steel fasteners of the S30000 series alloys or copper nails to attach monel. Do not use fasteners of S40000 series stainless steel, galvanized steel, or aluminum to attach or join monel.

TITANIUM—"WELCOME TO THE TWENTY-FIRST CENTURY"

INTRODUCTION

Titanium could well come into fashion as a serious architectural metal over the next few years. Cost is the only factor standing in the way. Aluminum experienced similar difficulties in entering the architectural market at the beginning of this century. A reduction in the costs of production should make this metal a viable architectural material. Currently, Japan is the only country that has made an effort toward using this beautiful soft-gray metal in architectural applications. Titanium has been in use in Japanese architecture since 1973. It was first used on the roof of a small shrine, and later on Tokyo Electric's domed roof structure known as the "Denryokukan." Since then, architectural promise and interest in the metal have been steadily increasing.

TABLE 6-3. Coefficient of Thermal
Expansion Comparisons

MATERIAL	μ in./in.°C
Aluminum	23.6
Zinc	24.9
Copper	16.5
Titanium	8.4
Glass	12.4
Brick	9.0

Titanium is an element weighing half as much as copper, and 60% of an equivalent volume of stainless steel. The specific gravity of this metal is 4.51, as compared with steel at 7.87, and aluminum at 2.70. Titanium is one of the 14 transition metals in the periodic chart of elements. Other transition metals are chromium, vanadium, and tungsten, each of which adds strength and hardness when alloyed to other materials, most notably steel.

Titanium has $1/300$ the electrical conductivity of silver, making it an embarrassingly poor conductor of electricity for a metal. A definite advantage of titanium over other metals is its extremely low rate of thermal expansion, which approaches that of glass.

Refer to Table 6-3 for various comparisons of thermal expansion.

HISTORY

Titanium was first isolated in 1791 by the English chemist and mineralogist William Gregor. Gregor named the metal "menaccin" after the ore menaccanite, which made up the black sands of Menachin, Cornwall, England. Enthusiasm for his discovery never took hold, perhaps because of the odd name. The metal was rediscovered by the German chemist Martin Klaproth in 1795. Klaproth discovered the metal in the mineral rutile and named this "new" material after the powerful mythological beings known as the Titans. Klaproth gave the metal this name because of the extreme difficulty in removing oxygen and other impurities in order to reduce the metal to a purer form. It is interesting that current knowledge of the characteristics of the metal give the name new meaning.

Early on, the metal was considered to be of little use except as pigments in paints. Minute impurities make titanium brittle and hard to work. Titanium did not yet reveal its true characteristics, as alluded to by its formidable name. More efficient refinement methods had to be developed before the metal would find a commercial purpose other than its use in pigments.

Titanium became commercially available in 1951, when enhanced reduction processes brought out the true nature of the metal, a metal with the ability to develop four times the hardness of steel. Prior to the start of the Second World War, Dr. William Kroll, while working for a German company in Luxemburg, developed and perfected the process for mass production of titanium. He fled Europe to avoid working for Hitler and came to the United States in 1940. The United States initially seized his patent rights under the authority of the alien property custodian when he entered the country. He later was awarded the patent and the associated royalties but not until after a long court battle. Dr. Kroll is also credited with perfecting the process of refining zirconium while working for the U.S. Bureau of Mines.

Since the early 1950s, titanium has had considerable use as an aerospace material, desired for its properties of very high strength coupled with lightness of weight. Considered a strategic material by the United States, architectural uses for titanium in this country were nonexistent. With the rethinking of military needs across the world and the subsequent price reduction resulting from new production techniques, titanium may become a major architectural material.

PRODUCTION AND PROCESSING

Titanium is the fourth most abundant metal on the earth, after aluminum, iron, and magnesium. The metal is mined in the Americas, with Brazil holding a 65% share of the world's supply of rutile, the primary ore. India, Australia, and the United States all have moderate supplies of the ore. The current United States source for a rich titanium ore is off the coast of Australia, where it is mined from the ocean floor. Submarines from the old Soviet Republic are also a major source of titanium; each submarine contains an enormous amount of the metal. The demise of the arms race and the old Soviet military machine has increased the options for using this once-exotic and strategic metal in architecture.

The primary ore rutile, TiO_2, is refined and converted to what is called a "sponge." A sponge is created by mixing TiO_2 with coke and charging the mixture with hot chlorine gas to produce $TiCl_4$. The hot gas causes a continuous fractural distillation of the ore, removing oxygen in the form of carbon monoxide and carbon dioxide. The material is called a sponge because its appearance is that of a large porous block with a spongelike surface. Once the sponge of titanium chloride is formed, it undergoes further purification. Magnesium and sodium are applied in an inert atmosphere to develop a more pure metallic sponge of titanium. These metals are added to form magnesium chloride and sodium chloride, which are removed by flushing the metal to form ions in solution. The waste materials are recycled for the next processing of the ore. This process of ore refinement is known as the Kroll process, named for the man who perfected the process, Dr. William Kroll.

The sponge material is converted into an electrode by welding these blocks of the refined metal together. The electrode is melted in an arc furnace and formed into a slab for further reduction in thickness. Alloying metals, if desired, are added during the process of melting.

Titanium is considered an expensive material, and the greatest expense lies in the production process. The ore itself accounts for about 4%, but the reduction to a slab thickness of approximately 1 inch (25.4 mm) makes up about half the current cost. As titanium production increases, the cost of the reduction process will decrease.

As previously mentioned, many of the architectural projects using the metal have occurred in Japan. Among the more recent projects utilizing the metal as an architectural cladding are the Kawasaki City Municipal Museum, the Kobe Municipal Suma Aquarium, and the Kinzoku Building in Osaka, which is covered with 13,500 square feet (1,350 square meters) of the metal. In the United States, titanium was used as a material in sculptures for the new Denver International Airport.

ENVIRONMENTAL CONCERNS

Titanium and titanium oxide are nontoxic. Prosthesis parts are manufactured from titanium alloys for use within the human body. These alloys do not produce salts or other by-products that may react to a person's immune system. Titanium and its adherent oxide is insoluble in water. The production costs associated with the metal will ensure scrap marketability; scrap recycling is much less costly than ore reduction.

The process of ore refinement seeks to recycle much of the chemical waste created. Titanium mining processes, like most mining processes, have a destructive effect on the environment. Companies involved with mining are seeking to reduce the damaging effects.

Titanium dust is an explosion hazard. Dust created by grinding and polishing operations performed under water or oil may explode upon contact with air. Creating dust in polishing processes should be avoided.

ARCHITECTURAL ALLOYS

Commercially pure titanium is the recommended wrought alloy form of the metal. The commercially pure, annealed form of titanium has a yield strength of 35 ksi (241 MPa), which is equivalent to type 304 stainless steel. The grain is small and tight, giving this form of titanium a Rockwell Hardness (B-scale) of 80.

Alloy Ti-35A (ASTM Grade 1)

The wrought form of alloy Ti-35A comes in sheet and strip forms. Commercially pure titanium in sheet form is specified under ASTM B265-G1 and ASTM 358. The strip form of the alloy is specified under ASTM B 381.

Thickness
Titanium sheet is available in thicknesses ranging from 0.015 inches (0.38 mm) minimum to 0.250 inches (6.35 mm) maximum.

Sheet and Coil Width
Titanium is available in widths up to 49 inches (1,245 mm) maximum. Wrought titanium has an alpha phase grain with a hexagonal structure. The grain structure is stable to 1,650°F (899°C). No benefits are obtained from heat treating the commercially pure alloy.

Yield Strength	35,000 psi	241 MPa
Ultimate Strength	48,000 psi	331 MPa

Hardness	80 Rockwell B-Scale
Specific Gravity	4.51
Melting Point	1675°C
Thermal Conductivity	0.43 cal/cm²·cm·s·°C
Coefficient of Thermal Expansion	8.41 μ in./in.·°C
Electrical Resistivity	56 μ ohms. cm

There are three other grades of commercially pure titanium, grades 2, 3, and 4. Each of these grades is annealed but develops subsequently higher yield strengths and corresponding hardness levels. Grade 4 has a yield strength of 85 ksi (586 MPa). These alloys are not recommended for architectural use because of the increased difficulty in cold working.

Titanium is also available in other forms:

Plate	ASTM B348
Tubing	ASTM F67
Bar	ASTM B 265 G1

Castings

Titanium can be cast; however, architectural uses are limited. Limitations include the cost of molds and the casting process. Cast titanium has the same composition as the wrought forms of the metal; there are no unique casting alloys.

Titanium is cast in precision graphite molds and other generally proprietary investment casting processes. The molds can be very intricate and expensive. Typical uses of titanium castings center around the aerospace industry. Perhaps as more architectural and ornamental requirements develop, different casting processes and techniques will become available.

Extrusions

Extrusion forms of titanium are possible, but impractical for most architectural purposes. Extruded forms are specified under ASTM 348.

FINISHES

There are two mill surfaces available, a low matte surface and a medium matte surface. Both have a smooth light-gray tone. The reflectivity is induced in the process of annealing. Annealing and pickling processes produce the lower reflective matte tone, and vacuum annealing produces a higher reflective tone. Both processes develop a tight grain structure, which imparts the matte appearance. This tight grain structure produces a surface finish on Titanium unlike that of any other metal. The low-matte surface has an appearance similar to that produced by a very fine glass-bead abrasive blast on stainless steel. The natural light gray color and the matte finish resemble a surface generated by a composite material such as graphite.

Titanium does not polish well. Abrasive blasting processes do not produce desirable surface appearances. Titanium dust generated by these operations may be explosive and dangerous. Titanium can be roller embossed. The surface hardness resists deep embossing, but light patterning is possible.

COLORING

An interesting characteristic of commercially pure titanium is its ability to develop color via the light interference phenomenon. Through immersion of the metal in an electrolyte bath, application of an electric charge, and the passage of time, color can develop on the surface of titanium. This is a very strong electronegative metal. With time and constant voltage, different tones of the same color can be produced. Increase the voltage slightly, and different colors can be obtained. As the voltage is applied, the surface of the titanium reacts with the electrolyte and forms a clear oxide layer. The oxide layer acts as an interference film (see "Colors in Stainless Steel," Chapter 4). As light waves reach the surface of the titanium, some of the light reflects off the surface of the oxide film. Other light waves pass through the film and reflect off the surface of the titanium. The color one sees is the interaction of these light waves. Certain wavelengths interact to enhance a particular color band. The resulting color is a very metallic, almost iridescent, matte tone.

Refer to Table 6-4 for coloring possibilities of titanium.

By immersing the titanium into the electrolyte, then slowly removing the sheet, while at the same time adjusting the voltage, a rainbow effect can be achieved. This technique has been used for years on jewelry. It can also be used on tube and sheet forms of the metal if a large enough electrolyte tank and voltage supply are available.

Because the color is an extension of the natural oxide film, it can be expected to remain for a long period of time, even when exposed to atmospheric pollutants. The film has the same hardness and durability of the base metal and can be formed and shaped without damage. Scratches through the

TABLE 6-4. Voltage, Film Thickness and Corresponding Color of Titanium

VOLTAGE	FILM THICKNESS (Å)	COLOR
0	15	Silver
2	25	Silver
4 to 6	105–132	Gold (barely detectable)
8 to 10	160–180	Pale gold
12	270	Rich gold
14	242	Dark gold
16	254	Dark gold with purple tint
18 to 20	260–272	Purple with gold tint
22	349	Blue/purple
24	364	Dark blue
28	469	Medium blue
30 to 32	610–715	Pale blue

interference film will show the base metal. When forming or handling colored sheets of titanium, be certain to protect the surface with a plastic barrier film. Remove the film soon after installation.

The electrolyte used in the coloration bath is hydrofluoric acid. This acid is very corrosive and can produce dangerous fumes. Use extreme caution when working with acids. Follow all recommended safety procedures and OSHA regulations.

WEATHERING CHARACTERISTICS

Upon exposure to air, titanium combines with oxygen and forms a very thin adherent film. This oxide film is highly resistant to further change. Therefore, the color of titanium when installed on the exterior of a structure should remain consistent for many years.

Titanium has the highest corrosion-resistance characteristics of any conventional architectural metal. The metal is extremely durable and has a low coefficient of thermal expansion, qualities that combine to give superior fatigue resistance. In conventional sheet metal applications, the metal skin should perform without the buildup of stresses that fatigue conventional metals.

CORROSION AND COMPATIBILITY CONCERNS

Titanium is one of the most corrosion-resistant metals known. Even in highly corrosive environments, such as coastal and industrial areas, titanium will not be affected. For what other metal can marine environments be considered of little consequence? Chlorides do not affect titanium. Sulfuric acid, nitric acid, and hydrochloric acid have little effect on the metal when they are presented in the dilute state. Hydrofluoric acid, on the other hand, immediately corrodes titanium.

Titanium has a very high affinity to oxygen. A clear, incredibly adherent oxide film develops on the metal's surface upon exposure to air. This oxide film is less than 10 μm in thickness. This tenacious oxide film inhibits further corrosive attack and the flow of electrical current. The oxide exhibits barrier properties similar to, but more resistant than, the chromium oxide formed on stainless steel or the aluminum oxide developed on aluminum. This film has the chemical makeup of the mineralized form of the metal ore, rutile. Metals that seek their mineralized form as protective films become very stable and can be expected to last for many years.

CARE AND MAINTENANCE

Titanium as an architectural metal does have certain characteristics that require attention. From a design standpoint, galvanic corrosion of adjoining metals must be considered. Titanium will corrode other metals. Titanium lies at the far end of the electromotive scale and will corrode aluminum and steel when coupled. Separation of titanium from other metals is important. Rainwater runoff from titanium surfaces will not absorb the oxide, so the staining of adjacent materials is not a concern. Nor should the runoff damage other metals located in the direction of the water flow, because salts are not formed and redeposited. Protect other metals from contact with titanium by isolating them with nonconductive thick paint, silicone, or rubber gasketing.

Titanium sheet metal surfaces showing uneven brightness and variations in color from one lot to the next are common. Because of its durability and the cost, thicknesses of titanium sheet metal are in the lower range, 0.015 inches to 0.024 inches (0.38 mm to 0.60 mm). This thickness reduction generates uneven stresses across the sheet, leading to pronounced "oil canning." Modifying the grain structure by the annealing and pickling process can reduce this tendency. The appearance of the metal is grainy, but fine. Combined with the lower reflectivity, the fine grain size provides a surface appearance with reduced distortion.

Titanium is not affected by other building construction materials. Its extraordinary corrosion-resistant nature allows Titanium to be used with concrete and masonry products, as well as all wood products. Titanium surfaces can be stained by rust corrosion products and can be easily scratched by abrasive action.

Fabricating with titanium requires an understanding of its hardness and strength. Commercially pure titanium can be sheared and pierced similarly to stainless steel. Shear blades and cutting tools must be kept sharp. Forming the metal will require overbending to account for the spring-back of the metal, which is similar to that of Half Hard temper stainless steel. This tendency gives titanium the ability to deflect without permanent deformation. Titanium has a tendency to gall when processed with tool steels. To reduce its tendency to seize up with other metals, use copper-alloy-clad tools. Copper-clad tools will also help resist scratching on the surface of titanium, because copper is a softer metal.

The matte surface of wrought titanium is susceptible to fingerprinting and surface stains. Most nonabrasive cleaners can be used without concern for generating corrosion. However, stains are difficult to remove because of the retentive nature of the porous oxide film.

FASTENING AND JOINING TECHNIQUES

Seaming is a common means of joining thin titanium sheets. With titanium sheets, just about all the common mechanical sheet metal seams can be used.[1]

Welding

Titanium can be formed and welded by conventional sheet-metal methods. TIG and MIG welding require an argon atmosphere, but other than that, resistance welding is similar to that of other metals. One benefit of titanium, when welding, is that carbon precipitation problems do not occur as with stainless steel.

[1]*Architectural Sheet Metal Manual*, 5th ed. Chantilly, VA: SMACNA, 1993.

Soldering

Because of its tight oxide film, soldering of titanium is not possible. Use a good silicone sealant instead of solder.

Fasteners and Seams

Use stainless steel fasteners of the S30000 series alloys to fix titanium fabrications to a structure or backing material. Steel, galvanized steel, and aluminum fasteners will deteriorate rapidly.

Sears Tower. Chicago, Illinois. Stainless steel cladding
for beams and columns in the renovated lobby.
The finishes used are No. 6 (satin finish), No. 9
(highly polished mirror finish),
Swirl (multi-directional ground and polished finish).
Architect: DeStephano and Partners

Sears Tower. Detail.

J.C. Nichols Plaza. Kansas City, Missouri. Dome is covered in flat seam panels manufactured from mirror finish stainless steel with a gold color. The gold color is produced by light interference with the thin, colorless chromium oxide layer.
Architect: Linscott, Haylett, Wimmer and Wheat

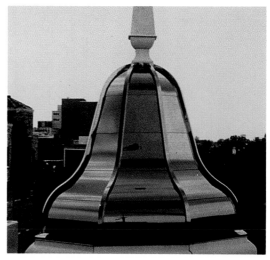

J.C. Nichols Plaza. Detail.

Automated teller machine. San Diego, California. Stainless steel decorative columns and entablature. The stainless steel is a No. 8 mirror finish with a red and blue color induced by the light interference phenomena.
Architect: Krommen Havek McQuowen

Johnson County Detention Center.
Olathe, Kansas.
Satin finish stainless steel band. The black color is produced by light interference with the lettering created by abrasive blasting the surface.
Architect: Shaughnessy, Fickel and Scott Architects

Stainless steel 'palm trees'.
The green color of the 'leaves'
is created by light interference.
The trunk of the 'tree' was created
by milling out cross-hatching lines
into the stainless steel surface.

Barcelona Fish Sculpture. Barcelona, Spain.
Woven gold stainless steel.
Architect: Frank O. Gehry Associates

RLDS Temple Complex.
Independence, Missouri.
Aerial view of the stainless steel roof.
Each stainless steel panel is tapered to
develop a unique part of the roof.
Architect: Hellmuth Obata
and Kassabaum, Inc.

RLDS Temple Complex. Elevation view.

dB Soft Building, Japan.
"Angel hair" finish stainless steel with a black color created by light interference.
Architect: Architect – 5

Fresno City Hall. Fresno, California.
Roof system manufactured from No. 2D stainless steel. Batten system on roof utilizes a bold,
half round stainless steel shape.
Architect: Arthur Erickson Architects with Lew and Patanaude Architects

Fresno City Hall. Detail.

Cathedral of the Immaculate Conception.
Kansas City, Missouri.
Dome of the Cathedral is manufactured from
copper, then coated with 23-karat gold leaf.
Architect: T.R. Tinsley

Cathedral of the Immaculate Conception. Detail.

Bank Building. Coffeeville, Kansas.
This turn-of-the-century bank facade was created by the Mesker Brothers Company of St. Louis.
The thin metal exterior walls and cornice were made from galvanized steel.

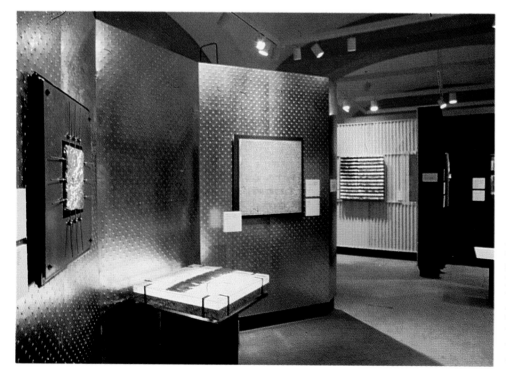

Mondo Materialis
at the National
Building Museum.
Washington, D.C.
Embossed galvanized steel
was used to create
the texture on the
walls of the exhibit.

Liberty Science Center. Newark, New Jersey.
Terne coated stainless steel was used to create the thin metal covering on the spherical enclosure.
Architect: E. Verner Johnson Architects

Sheetmetal Workers 100th Anniversary
Exhibit at the National Building Museum.
Washington, D.C.
Terne coated steel was used to create
the thin flat seam panels on the exhibit
enclosure.
Architect: Frank O. Gehry Associates

University of Tampa.
Tampa, Florida
Ornamental domes
of terne
coated stainless steel.
Architect: Robbins Bell
and Kreher.

University of Tampa. Detail.

Nigata City Aquarium. Nigata, Japan.
Vertical stripe of titanium. The rainbow-like colors
created on this titanium stripe are produced by an
interference film generated by an electrochemical
treatment to the surface.

Suma Aualife Park. Kobe, Japan.
Batten seam roof using thin
titanium sheeting.
Architect: Daiken Architects

Private residence
along the coast.
Titanium standing
seam roofing used.
Finish is the annealed
cold rolled finish.

Canadian Pavilion. Seville, Spain.
Preweathered zinc wall panel system.
Architect: Bing Thom Architects

Multipurpose Hall
for the Carmel Convent,
Graz-Eggenburg, Germany.
Natural weathered
zinc sheet used to form
the intricate roof and
wall cladding.
Architect:
Prof. Dipl.-Ing. G. Domenig

Curved Zinc Roofing in Europe.
Panels are rolled formed with double lock standing seams.
Source: Dorenbach A.G.

Stamped shingles manufactured from zinc.
The ornamentation was created in a wooden die
then transferred to the zinc using high pressure.
Source: Hochbauleitung Staatsbad.

Felix Restaurant. Peninsula Hotel, Hong Kong.
Custom patinated zinc cladding and hammered zinc skirting on "bucket shaped"
wine room and caviar bar. The seat back of the bench is woven zinc strips.
Architect: Philippe Starck

J.C. Nichols Plaza. Kansas City, Missouri.
Lead sculpture under repair.

Hagia Sophia. Istanbul, Turkey
Lead was used as a roofing material for this fascinating
6th Century Byzantine structure.

St. Louis Science Center. St. Louis, Missouri.
Monel standing seam roof cladding.
Architects: E. Verner Johnson Architects

Crown Center Restaurant. Kansas City, Missouri.
5-mm thick aluminum panels with an acrylic,
field applied paint coating over a clear etched surface.
Architect: James Gordon and Associates

Harold Washington Library. Chicago, Illinois.
Aluminum leaf forms and steel cornice railing
are coated with a green polyester powder coating.
Architect: Hammond, Beeby and Babka

Pacific Enterprise.
Los Angeles, California.
3–mm thick aluminum
light fixture is coated
with a pearlescent
polyester powder coating.
Architect: Gensler
and Associates

7

Metallic Coatings on Metals

INTRODUCTION

Copper, stainless steel, and aluminum all consist of homogeneous metal alloys, consistent in their makeup throughout their cross sections. Cut them, shear them, or scratch the surfaces, and the exposed metal matches the surface metal in composition. The cost of copper and stainless steel, or the natural color of these metals, however, may restrict their use. Strength and corrosion resistance can impose additional limitations to the choice of metals when considered in certain applications. Thus, techniques of coating metals with other metals were developed to offer additional alternatives to the designer and to provide acceptable qualities of one metal exhibiting the preferred qualities of another.

Coating one metal with another to enhance the properties of the base metal was in common use in the early 1800s. Wrought sheet iron coated with tin was a common roofing material. Used extensively in Europe, particularly in France, tin coating over iron introduced metal as a roofing alternative competitively priced with wood and tile. Thomas Jefferson, intrigued by the French use of the metal, replaced his worn out wood shingle roof with a tin-coated iron roof in the early 1820s. The pure tin coating provided good corrosion resistance for the iron sheets.

There are numerous techniques for applying one metal to another. One of the oldest and most common commercial techniques involves the application of a metal coating to a metal sheet by hot dipping. This technique consists of briefly immersing a metal sheet into a molten bath of the coating metal. The resulting surface coating is often variable, with thicker, denser regions intermixed with thin, barely coated regions. The coating is typically applied to all sides and forms a very tough, adherent metallurgical bond at the interface of the two metals.

Electroplating is another common technique used to coat one metal onto the surface of another. This technique involves developing an electrical potential between a sheet of metal immersed in a solution and ions of the other metal, which are contained in the solution. The coating metal comes out of solution and plates onto the surface of the immersed metal. The coating is very even and constant across the immersed sheet of material. A metallurgical bond does not occur between the two surfaces, however.

Metal spraying—converting one metal into a plasma and applying it to the surface of another metal by spraying—is a little-used architectural technique. This process requires the surface of the metal to be coated to undergo abrasion prior to application of the molten spray. Metal spraying is not recommended for thin sheets. Thin sheets will show a distortion created by internal stresses tending to curve the sheet. Internal stresses develop because the thermal spray is applied to one side. Spraying both sides will help flatten the shape.

There are other techniques that are not frequently used in architectural applications. One is cementation, which consists of tumbling a fabricated part in a mixture of metal powder while in the presence of a molten flux. The powdered metal will diffuse onto the surface of the fabricated metal. Ion implantation, another such technique, bombards the surface of one metal with ions of another metal while in a vacuum. A thin surface film of the ionized metal is deposited. This technique is also known as vapor deposition. Certain small ornamental features and hardware parts can be finished with the use of this technique.

Electroless nickel plating is used on hardware parts. This process develops an adherent coating of nickel on the base metal by immersing the part in a solution containing nickel salts. The nickel salt solution undergoes a chemical reaction, with the nickel coming out of solution and plating the immersed object.

All the metal-coating techniques produce thin films of one metal tightly adhered to another. The general mechanical properties of the base metal are retained while a different metallic appearance on the surface is achieved. As to corrosion resistance, this thin outer film of metal will act as protection for the base metal.

There are basically two categories of coated metals, those that act as sacrificial metals and those that act as noble metals with respect to the base metal. Sacrificial metals, in most architectural applications, are zinc, cadmium, aluminum, and tin. Noble metal coatings are copper, lead, chromium, and nickel.

In the use of sacrificial coatings porosity is not a concern, since the coating acts to protect the base metal. The surface metal acts as the cathode, providing electrochemical protection to the base metal. Thus, protection of the base metal is afforded at small scratches and along sheared edges.

Noble metal coatings develop galvanic activity at the base of pores along the interface of the metals. Galvanic corrosion along the metal interface will eventually undermine the coated metal, since the base metal is acting sacrificially. Noble metals are often used to provide decorative surfaces, such as chrome- and copper-plated hardware. Exterior uses are not recommended for plated metals unless frequent maintenance is scheduled.

ARCHITECTURAL METALLIC COATINGS

ALUMINUM COATINGS

Alclad

Commercially pure aluminum coatings are commonly applied to less-corrosive aluminum alloys. The pure aluminum coating acts as a sacrificial metal surface to the underlying base metal. Besides the pure coatings, other alloys can be used to clad the surface to provide corrosion protection and retain strength.

This technique is designated as Alclad (see Chapter 1, Aluminum). Alclad coatings are applied by passing aluminum sheet through a molten spray of pure aluminum. The coated sheet is then passed between pressure rolls to assist in the bonding of the two materials. This pure aluminum coating acts as a cathode in protecting the less-corrosive base metal while retaining the structural characteristics of the base metal.

Alclad-coated aluminum surfaces are very reflective and retain a high level of luster for a long period of time. In rural and light industrial areas, the luster can be expected to last for more than 12 years. In coastal environments and heavy industrial areas, the luster will dull in approximately 4 to 6 years.

Forming and fabricating metals with these thin aluminum coatings requires care in handling. The thin, high-purity coatings are very soft and will scratch easily. Forming operations can mar the surface if thick protective films are not used to separate the surface of the Alclad from the tooling.

Aluminized Steel

Aluminum coatings can be applied to other metals as well. A protective aluminum layer on other metals can be achieved through hot-dipping methods and molten spraying. Aluminum is frequently alloyed with other metals prior to being bonded to the surface of a base metal, usually steel.

Aluminized steel is the name given to one common form of coated steel used commercially as a roofing and flashing material. The process of hot dipping steel in molten aluminum baths was developed in the 1940s to provide inexpensive corrosion protection for steel. Available in thin gauges, this reflective coated steel sheet product is generally left unpainted as it is rolled or press-brake formed into the various roofing and flashing shapes.

Aluminized steel is available in two forms, known as Type 1 and Type 2. Both forms are developed by passing a clean ribbon of thin steel continuously through a bath of hot molten aluminum. As the ribbon of steel passes

TABLE 7-1. Coated Architectural Metals

DESCRIPTION	BASE METAL	COATING METAL	COLOR (NEW) COLOR (WEATHERED)	EXPECTED LIFE SPAN
Alclad	Aluminum	Aluminum 99% pure	Shiny white. Very reflective. Slightly mottled grey.	30 years+ in industrial exposures.
Aluminized Steel	Steel	Aluminum	Shiny, slightly spangled. Weathers to dull gray.	8 to 10 years in industrial exposures.
Galvalume	Steel	Aluminum-zinc alloy	Shiny gray with spangle. Ages to a medium gray.	10 to 12 years in industrial exposures.
Cadmium	Steel, nickel	Cadmium	Silver color—typically used for fasteners.	10 years+, depends on exposure.
Chromium	Brass, nickel, steel	Chromium	Bright silver color. Plated parts.	5 years+, depending on maintenance. Cracks form.
Terne	Steel	80% lead, 20% tin	Shiny mottled gray. Weathers to dark gray.	3 years if unpainted, 20 years and longer when painted.
Terne-Coated Stainless	Stainless steel 30400 and 31600	80% lead, 20% tin	Shiny gray—some spangle. Weathers to dark gray.	50 years+. Can develop red lead oxide in rural exposures.
Lead-Coated Copper	Copper	Lead	Blue-gray color. Weathers to dark gray.	75 years+. May develop white oxide streak.
Tin Plate	Steel—interior	Tin	Shiny gray color to dark tone. Interior.	75 years, on interior exposures. Expect rust on exterior.
Tin-Coated Stainless	Stainless steel	Tin	Shiny gray color. Weathers to dark gray.	75 years+.
Pewter-Coated Copper	Copper	Tin	Shiny grey color. Weathers to a green tint.	75 years+.
Zinc (Galvanized Steel)	Steel	Zinc	Spangled gray tone. Weathers to darker gray.	11 years+ in rural exposures. 5 years in industrial applications.
Zirconium	Steel—interior	Zirconium nitride plating	Brassy. Interior use only.	20 years+ on interior applications.

through the molten bath, a thin aluminum coating is applied to all sides. The aluminum coating metallurgically bonds to the underlying steel by forming an alloy layer between the steel and the aluminum surface coating.

Type 1 aluminized steel is used for industrial applications requiring additional corrosion protection. It has a much heavier coating; approximately 0.5 pounds/square foot (2.4 kilograms/square meter) is applied to both sides of the steel sheet. The Type 1 coating is made of an aluminum-silicon alloy.

Figure 7-1. Long-length aluminized steel roof panels.

Type 2 aluminized steel is the common architectural metal product. It has commercially pure aluminum applied to both sides of steel sheet to a nominal thickness of 0.0015 inches (0.038 mm) per side.

Aluminum coating over steel retains the strength and formability of the steel base metal, while offering excellent corrosion resistance in industrial environments, particularly in sulfur dioxide and chloride exposures. Where aluminized steel falls short is along a sheared edge or in deep scratches through the coating. Aluminum, unlike zinc coatings, does not afford good edge protection to steel sheet.

The finish appearance of aluminized steel is bright and reflective. Over time, the aluminum surface will develop a light gray oxide. Some rust may start to appear at exposed edges and around fasteners. To maintain the surface, gently clean it of rust and apply an aluminum paint over the edge. Avoid abrading the aluminum coating.

Paint does not adhere well to aluminized steel, particularly newly applied metal. The product is rarely provided prefinished. Postfinishing is not

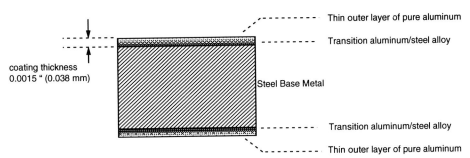

Figure 7-2. Cross section of aluminum applied to steel.

recommended, because the aluminum surface would require etching or priming to allow paint to adhere well.

Available Forms
Aluminized steel is available in thin sheets. Other forms, such as plate and structural shapes, can be obtained by hot dipping in molten baths. On thicker sheets, the thin aluminum coating will flake along formed edges. Consider hot dipping the formed shape after the bending operations are performed.

Hot dipping of large fabricated shapes into molten aluminum baths is not nearly as commonplace as hot dipping into zinc. Available sources are limited.

Soldering and Brazing
Aluminized steel can be soldered and brazed, but only after the protective aluminum oxide film is removed. For soldering, use chloride containing inorganic salt fluxes. These will react at temperatures around 720°F (382°C) and deposit a thin layer of zinc on the surface. The solder type is 100% zinc, or 95% zinc and 5% aluminum. Follow with a coating of aluminum paint to match the color of the surrounding metal.

Brazing produces a stronger joint than soldering. The brazing flux must remove the oxide film and withstand temperatures of 1,100°F (593°C). Brazing rods are 12% silicon-aluminum alloy. Be certain the flux has been neutralized and rinsed from the surface of the joint. Flux left on the joint will cause rapid corrosion of the base steel.

Precautions with Using Aluminized Steel
When using aluminized steel, care in storage is necessary to avoid water stains. Store this metal in clean, dry areas; elevate it and allow condensation to drain from sloped sheets. Do not use strong acids or alkalines to clean the surface. Instead, try mild detergents, without using abrasive brushes or steel wool, which might harm the aluminum coating.

Aluminized steel should not be installed with deep scratches or surface gouges. Once the base metal is exposed, rust will occur in short order. Forming operations should include surface protection to avoid contaminating or marring the delicate aluminum coating.

Aluminum-Zinc Coatings

The aluminum-zinc coating commonly known by the trademark name Galvalume® is another excellent coating product. Galvalume is a trade name for a proprietary coated sheet steel product developed by Bethlehem Steel Corporation. Galvalume was developed specifically for the metal roofing, sheet metal flashing market as an alternative to the widely used galvanized steel sheet. Galvalume is a patented process that involves passing high tensile strength steel through a molten bath of aluminum-zinc alloy metal. Other trade names for the product are Zincalume (Australia), Aluzink (Sweden), and Galvalange S.r.a.l. (Brazil).

Aluminum alloyed with zinc develops an excellent corrosion-resistant coating. By volume, the coating is approximately 80% aluminum, 19% zinc and 1% silicon. By weight, the coating is 55% aluminum, 43.5% zinc, and 1.5% silicon. Like galvanized steel, Galvalume is a coated coil product that will readily receive paint when properly prepared. The coating application is specified under ASTM A 792, "Sheet Steel, Aluminum-Zinc Alloy Coated by the Hot Dipped Process, General Requirements."

There are three coating classes available:

AZ 50 (AZ150)   0.50 oz./sf (150 g/m²) total of both sides
AZ 55 (AZ165)   0.55 oz./sf (165 g/m²) total of both sides
AZ 60 (AZ180)   0.60 oz./sf (180 g/m²) total of both sides

AZ 50 (AZ150) is the specified coating for prepainted Galvalume sheet, and AZ 55 (AZ165) is the specified coating for metal intended to be left bare or in its mill state. AZ60 (AZ180) is a thick coating not typically used for architectural applications.

The surface spangle, similar to that of galvanized steel, is available in regular or extra smooth. Extra smooth provides a smooth surface for paint applications.

The Galvalume process offers advantages over galvanized sheet and aluminized steel coatings. The aluminum-zinc coating has better flexibility and allows more severe forming without surface crazing, as compared to pure aluminum coatings.

In regard to corrosion, Galvalume offers the double benefit acquired from the aluminum and zinc. The aluminum portion of the coating provides an excellent corrosion-resistant film over the steel. The zinc portion of the alloy mix protects the base metal at sheared edges and scratches, like the zinc in a galvanized steel coating. The zinc acts as sacrificial metal, protecting the base metal around small breaches in the coating.

Painting Galvalume sheet steel is similar to painting galvanized sheet. The metal surface is preheated, cleaned with an alkaline cleaner, given a chromate pretreatment, then a minimum of two coats of finish paint is applied on both sides.

Aluminum-zinc coated steels, subjected to accelerated weathering tests, have demonstrated superior performance when compared with G90 galvanized steels.[1] The first signs of visible rust appeared at levels equivalent to 14 years of severe marine exposure. The accelerated weathering tests indicated the aluminum-zinc coated sheets could be exposed to rural and some urban environments for as much as 25 years before rust would appear. In actual weathering tests performed at different locations around the United States, Galvalume was compared with other metals after 11 years of exposure. The observed results are indicated in Table 7-2.

Galvalume has been in service since the late 1970s. Visual inspections of 80 different structures in various atmospheric conditions has provided verification of the results demonstrated through the accelerated tests.[2]

---

[1] H. E. Townsend. "Twenty-Five Year Corrosion Tests of 55% Al-Zn Alloy Coated Steel Sheet." *Materials Performance.* Vol. 32 (April 1993), pp. 68–71.

[2] Data for this claim are presented in further detail via Spec-Data®. Spec-Data® is produced by the Construction Specifications Institute. The information relating to Galvalume can be found in Division 7, Flashing and Sheet Metal; Category: Standing Seam Metal Roofing; Subject: Galvalume.

TABLE 7-2. Visual Observations of Various Coated Steel Products after 11-Year Exposure

METAL	RURAL EXPOSURE	INDUSTRIAL EXPOSURE	MARINE EXPOSURE	SEVERE MARINE EXPOSURE
Galvanized	Good condition	Major corrosion	Good condition	Rust in 4 years Major corrosion in 11 years
Aluminum-Coated Steel	Good condition Some edge rust	Stain on surface	Edge rust	Edge rust
Aluminum-Zinc-Coated Steel	Good condition	Stain on surface	Good condition	Good condition—slight edge rust

Precautions with Using Aluminum—Zinc Coatings

Like galvanized steel coatings, aluminum-zinc coatings should be used in applications that allow the surface to drain. Keep copper products out of contact with aluminum-zinc coatings. Moisture draining off copper surfaces, including condensation lines, will rapidly corrode aluminum-zinc coatings. Do not allow water to stand on stored sheets or panels. Corrosion products will develop and weaken the corrosion resistance of the installed sheet. Stains resulting from moisture are difficult to remove without damage to the coating. If the stains are visibly unacceptable, then replace the sheets or consider painting the surface.

CADMIUM COATINGS

Cadmium coatings are applied by electrodeposition. Thin, protective films of cadmium were once commonly applied to steel fasteners to provide corrosion protection. Cadmium performs similarly to zinc in corrosion resistance by providing sacrificial protection to the underlying base metal. Cadmium is not typically used for architectural wrought or cast products because of its higher cost and the toxicity of the metallic salt. Decorative, selective plating with cadmium is about the extent of architectural applications.

Cadmium-coated parts must not be used in the proximity of food preparation equipment because of its toxicity. The use of cadmium is diminishing rapidly because of this hazard and because the plating operation employed to apply the coating uses a cyanide solution. The fumes generated by welding or soldering cadmium-coated parts are very hazardous. There is currently a "land ban" on cadmium disposal in the United States in landfills.

The finish obtained from cadmium is brighter and does not dull as tin coatings might. Cadmium coatings are limited to small, selective areas and fasteners requiring superior corrosion resistance without a buildup of film. Consider zinc chromate plating as a replacement for cadmium if corrosion resistance is the reason for its use.

CHROMIUM COATINGS

Chromium is applied to other metals through electroplating techniques. Chromium coatings can offer excellent corrosion resistance when correctly applied. These coatings are relatively economical and simple to apply. The surface produced is hard and easily polished to a bright mirror-reflective surface, or the surface can be given a reflective satin finish by mechanically scratching with abrasives.

There are various techniques available for applying chromium plating. The techniques employ an undercoating of nickel, which develops a hard surface that can be given a bright polish. A smooth, polished nickel surface benefits the quality of the subsequent chromium plating. The chromium layer applied to this nickel surface is usually in the range of 0.25 to 0.50 microns. This microscopically thin coating is very hard and provides a decent barrier to corrosive attack as long as it is not breached. The problem is that all chromium plating layers eventually will crack. Cracks occur when the underlying metal undergoes stress. Stress can be from impact or bending, or from temperature variations. The hard chromium layer is not elastic enough to absorb the stress and cracks develop.

Interestingly, chromium plating with a microscopic web of cracks actually performs best. There are proprietary plating techniques developed to produce microscopic cracks across the entire plated surface. The microscopic cracks are so dense that a square millimeter will have from 25 to 50 cracks. The cracks allow galvanic corrosion cells to develop between the nickel coating and the chromium layer. When the corrosion cells develop, anodic and cathodic poles are created, and an electrical current is generated. By increasing the number of cracks, electrical current is spread out across the surface and the resulting galvanic corrosion is slowed significantly.

Another proprietary plating process operates on a similar principle by making the chromium layer very porous. Microscopic pores are created through the chromium layer by treating the underlying nickel plating. The pores act to spread out the galvanic corrosion cells similarly to the microscopic cracks.

Exterior uses of chromium-plated parts will require periodic maintenance, particularly if the reflective luster is desired. Interior parts, such as hardware, should be kept clean and free of water deposits and other foreign particles. Maintenance may be as minor as polishing and buffing to remove damaging deposits of salts. Frequent application of wax will keep chromium-plated parts appearing bright and reflective for years.

For steel parts, the chromium is applied after the steel has undergone copper and nickel plating. The softer copper gives a better layer to polish, and the nickel aids in brightening. Use a thick nickel plating under the chromium plating for additional corrosion protection. Plumbing hardware made from forged and cast brass commonly receives chromium plating. Door hardware, ornamental fixtures, and furniture are other typical chrome-coated fabrications.

Typically, chrome plating should be considered for relatively small metal fabrications or for selective ornamental applications. Car bumpers represent the upper limit of dimensions available. The finish matches that of stainless steel, with an almost imperceptible blue tint. Decorative chemical treatments can color chromium plating, giving very striking metallic deep blue and black hues. Certain additives can be introduced into the plating tank to develop black chrome. Other colors are also possible by developing interference films of chromium oxide.

LEAD COATINGS

Lead coatings are applied to base metal surfaces through the hot-dip method and electroplating. Tin is often added in small amounts to aid in bonding the lead to the base metal surface. Lead does not adhere well to steel, but will adhere to tin by alloying at the metal-to-metal interface.

When the content of tin increases to the 7% to 25% level, the coating is known as terne plate. *Terne* means "dull" and was originally used to distinguish the dull lead-tin coating from the brighter tin-plate coatings applied on iron in the late 1700s and early 1800s. Terne plate was developed in New York by Joseph Truman at about 1825. Early terne plate was tin-plated wrought iron dipped in molten lead.

Terne-coated steels have had more than two centuries of successful use in the building industry. They were originally provided in small "plates," 14 by 20 inches (356 by 508 mm) or 20 by 28 inches (508 by 711 mm), which were formed and soldered into large sections to make flat-seam roofs and guttering. Today, large sheets and coil stock are available in terne-coated steel.

Terne coating is provided in two weights, 20 lb. and 40 lb. This nomenclature is left over from years past, when stocks were warehoused in a particular format. It refers to the total weight of terne coating provided on both sides of 112 small plates. This particular quantity of plates, 112, refers to a "base box," or 436 square feet (42.45 square meters) of terne metal. This nomenclature is used regardless of whether sheets or coils are purchased. Of course, today it is wise to order the material with a nominal coating thickness or weight per unit area.

20 lb. Coating   0.047 lb/sf (0.229 kg/m^2)
40 lb. Coating   0.092 lb/sf (0.459 kg/m^2)

The 40 lb. coating is the recommended minimum coating for architectural use. The base metal used for terne coating is copper-bearing steel rolled to thin gauges. Standard thicknesses are indicated in Table 7-3.

TABLE 7-3. Standard Base Metal Thicknesses for Terne-Coated Steel		
GAUGE	INCHES	MM
24	0.024	0.61
26	0.018	0.46
28	0.015	0.38
30	0.012	0.31

Care and Maintenance

The soldered plates perform best when coated further with paints. Many of the original tin roofs were coated with a reddish brown paint known as "tinner's red." This was a red iron oxide paint applied liberally over the surface of the terne metal roofs. Today, a thick coating of zinc oxide primer or an epoxy primer followed by a finish coat of acrylic or urethane enamel will provide good long-lasting results. Deterioration of the thin terne-coated sheet metal usually occurs when moisture is allowed to collect on the underside of the metal. Corrosion occurring on the underside will eventually perforate the metal. It is recommended to paint the underside of the sheets with a thick prime coating.

With lead coatings, and the lead-tin coatings, porosity of the outer layer is a concern. The small pores in the surface coating create localized corrosion cells and fill up with rust products from the base metal, inhibiting further corrosion. The surface will discolor with streaks of rust as the underlying base metal corrodes.

Repair of rusting terne-coated surfaces requires cleaning the rust from the surface with wire brushes or steel wool. Try to avoid damaging regions where the terne coating is still in adequate shape. Apply a zinc-rich primer over the entire surface, and follow with a thick layer of all-purpose paint.

Do not use sand or other abrasive blasting on terne. The high-energy impact may perforate the thin steel sheet. Remove the portions of terne-coated steel that have totally deteriorated from rust. Check for regions where the terne plate appears to be thinning from rust on the underside. Remove these regions and replace all damaged underlayment. Resolder new terne-plated steel replacement segments where these bad sections have been removed. Do not cover corroding sections with new material, as this can cause acceleration of corrosion. Apply a thick coating of paint to the replacement material.

Joining and Forming

Terne-coated steel is readily soldered with 50-50 lead-tin solders. Other low-lead solders will also work well. Use stainless steel rivets and stainless steel fasteners to attach terne material to the base support. A thin terne sheet, regardless of the configuration to which it is formed, requires full support. Support is usually plywood or other wood material covered with felts and building paper.

Conventional forming and cutting equipment are used with this metal. The softness of the coating is the main concern as it can be easily damaged from working the metal. Scratches or scrapes to the coating will dramatically reduce its corrosion resistance.

Standard sheet metal construction details are utilized when working with this metal. Seams should be locked, using single or double folds. This minimizes exposure of the uncoated edge of the terne sheet.[3]

---

[3]Details indicated in the *SMACNA Architectural Manual*, 5th ed., (1993) indicate the recommended forming and attachment details for terne-coated metal.

Terne-Coated Stainless

A more recent development in terne-coated metals is coated stainless steel. Commonly known as terne-coated stainless, or TCS, this coated metal provides an initial reflective surface which quickly tarnishes to a medium-to-dark-gray. TCS was introduced in the 1960s as a premier architectural metal roofing material. The terne coating acts as an anode, a sacrificial coating, to the base stainless steel. Terne-coated stainless has shown exceptional results in both industrial and coastal exposures.

A thin terne coating gives some additional corrosion resistance to the base stainless steel sheet, but the main benefit is the dull silver-gray color. This color is long lasting, unlike those of terne-coated steels that require painting for survival.

Refer to Table 7-4 for a list of standard terne-coated stainless sheet thicknesses.

Terne Coating

A terne coating is applied by hot dipping in molten metal. It is applied to both sides of the stainless steel sheet to an average weight. For terne-coated stainless steel, this thickness is variable, ranging from 0.25 oz./sf to 0.55 oz./sf (76 to 168 g/m^2). A terne coating is specified under ASTM A308.

Various available coating thickness, as measured by the triple spot test, are specified as follows:

Coating Designation

LT01	no minimum	
LT25	0.25 ounces/sf	76 g/m^2
LT35	0.35 ounces/sf	107 g/m^2
LT40	0.40 ounces/sf	122 g/m^2
LT55	0.55 ounces/sf	168 g/m^2

The thickness of the coating on a stainless steel sheet varies slightly across the sheet. In general, the thickness is very slight, amounting to less than 0.75 mil (0.041 mm) per side.

Base Metal

Type S30400 Stainless Steel

Composition	18–20% Chromium
	8–12% Nickel

TABLE 7-4. Standard Thicknesses of TCS Sheet

GAUGE	INCHES	MM
26	0.018	0.46
28	0.015	0.38

Alternate

Type S31600 Stainless Steel

Composition   16–18% Chromium
                    10–14% Nickel

When these thin, reflective terne sheets are initially installed on a building, a high degree of "oil-canning," or surface distortion, is apparent. As the sheets weather, the reflectivity diminishes and so does the apparent distortion.

The thin-plated coating does not interfere with the formability of the stainless sheet. The base stainless steel is fully annealed and can be easily formed into many difficult configurations without damage to the film. Soldering and sheet metal construction are the same as for terne-coated steel.

Sheet Dimensions

The maximum sheet width available is 36 inches (915 mm). Lengths are available to 50 feet (15.24 meters).

Lead-Coated Copper

In the early 1900s the technique of applying lead to the surface of copper became commercially available. The Revere Copper and Brass Company introduced lead-coated copper to commercial use in 1930. The copper industry developed the process for hot dipping sheets of copper into a molten bath of lead with a small amount of tin, usually about 4%. The tin is needed to assist in the interface bond between the two metals. Lead will not adhere to other metals without some level of tin present.

The lead-coating process was developed in response to requests from the design community for a nonstaining, corrosion-resistant gray colored metal. Lead sheets would meet these criteria, but lead is very heavy and soft, and relatively expensive. Lead had extensive use as a roofing material in antiquity, but its use today as an architectural material is limited to gutter linings or intricate fabrications. Tin-coated steel and terne-coated steel were available, but these materials soon corroded if not painted. Stainless steel was just coming into prominence. Monel was the only corrosion-resistant gray metal available at an affordable cost.

By applying lead to copper, a gray sheet metal material is obtained. The lead coating reduces staining of adjoining materials while providing a lightweight construction material. The color of lead-coated copper is a soft, sometimes spangled, gray that weathers to a dark, even gray. As the lead coating oxidizes, a grayish-white, almost translucent, oxide will develop. Some of the oxide will run off onto adjacent materials, but the staining is much less than created by the green copper sulfate runoff characteristic of uncoated copper. Lead oxide is clear in color; however, some surface pollutants may cause the runoff to appear dark. Lead carbonate can form on the lead sheet by absorption of carbon dioxide from the atmosphere or exposure to organic acids. Lead carbonate appears as a white film or light-colored streaks on the darker lead surface.

A lead coating is available on one or both sides of a copper sheet. The coating is produced by one of two methods: dipping the copper sheet into a molten bath of lead or electrodeposition of the lead onto the copper sheet.

The first method, known as hot dipping, produces a rougher texture, similar to the heavily spangled appearance of hot-dipped galvanized steel. Sometimes the lead coating produces a melted, flowing appearance across the surface.

The hot dipping method removes some of the temper of the copper sheet. When the base copper sheet is of the standard cold-rolled temper, the resulting temper after the dip in molten lead is slightly reduced. Because of this reduction in temper, many applications use a heavier-gauge copper sheet. When very intricate or severe metal forming is required, the lower temper is desirable. In most applications, this reduced temper has little effect.

The forming properties of lead-coated copper are similar to those of uncoated sheet copper. The lead coating is very adherent to the copper surface, and forming operations should not craze or damage the coating. Sheared edges will show the red copper color only along the cross section. Shearing should not cause the lead to flake. If the lead coating peels or flakes from the surface, then a metallurgical bond did not occur. This could be the result of a contaminated lead bath or foreign matter on the copper sheet prior to hot dipping.

The electrodeposition process plates the copper sheet with lead. There is no reduction in temper of the copper in this process. The resulting texture is smooth and more uniform, owing to a consistent plating of the lead coating. The appearance is similar to that of terne-coated stainless steel. A metallurgical bond will not occur during plating. The corrosion-resistant characteristics of plated products are dependent on the thickness of the plated lead and will not exhibit the benefit of the duplex alloy layers that develop during hot dipping.

The process of applying lead to a copper sheet is specified in ASTM B101-83. The weight of the lead coating is specified as total pounds of lead applied to both sides of 100 square feet of copper sheeting. Two thicknesses of lead coating are designated in B101.

Class A Standard Coating

The class A coating is the standard commercial architectural coating.

Minimum	12 pounds of lead per side of 100 square feet
	5.4 kg of lead per side of 9.29 square meters
Maximum	15 pounds of lead per side of 100 square feet
	6.8 kg of lead per side of 9.29 square meters

This quantity of lead equates to approximately 0.002 inches (0.051 mm) to 0.003 inches (0.076 mm) of lead per side.

Class B Standard Coating

The class B coating is an industrial coating for use in heavily corrosive environments.

Minimum	20 pounds of lead per side of 100 square feet
	9.1 kg of lead per side of 9.29 square meters

Maximum 30 pounds of lead per side of 100 square feet
13.6 kg of lead per side of 9.29 square meters

Standard Dimensions

Sheet widths	24 inches	610 mm
	30 inches	762 mm
	36 inches	915 mm
Sheet lengths	120 inches	3,048 mm

Stocked gauges of the copper sheet are 16 and 20 ounces per square foot. Custom sizes and thickness are available from mill processors upon request.

Care and Maintenance

Corrosion failure in lead-coated copper can result from porosity in the lead coating or scratches through the lead coating to the copper sheet. Care should be taken to prevent the damage of the lead-coated sheet during fabrication and handling. Scratches and pits in the lead coating will develop corrosion cells at the copper surface. Corrosion in these areas is accelerated because of the added galvanic potential that develops between the lead and the copper. Particular care should be exercised in corrosive urban environments.

Lead-coated copper, allowed to weather naturally, has been shown to provide excellent service for more than 75 years. Vertical and severely sloped applications of lead-coated copper have shown superior corrosion resistance in industrial environments. Low slopes and horizontal surfaces will, after 30 years, begin to show a slight green color owing to erosion of the lead surface and the subsequent exposure of the base copper sheet.[4]

Abraded lead- and terne-coated surfaces can be repaired by applying 60-40 lead-tin solder with a large-surface copper soldering iron. The surface must be thoroughly cleaned of all oxides and foreign matter. This technique will work only on small surface areas.

Lead-coated copper can be contaminated with iron and steel particles. The particles can become imbedded into the lead-coated surface during forming and fabrication operations. Dies or rolls used to form the metal can be contaminated with small iron particles, which will imbed into the soft lead surface. The cutting of steel or iron near a lead coating can allow the hot steel particles to melt into the coating. When this occurs, the iron particles will quickly rust and streak the lead surface. Mild acids can be used to remove iron particles without harming the lead coating. Experiment with small sections to test the results. Use extreme caution when using acids. Follow closely all safety regulations provided with the acid, or check with professional cleaners for the correct procedures.

White streaks on lead-coated copper can result from formations of lead carbonate on the surface. A whitish coating can cover the entire surface in certain exposures, but lead carbonate is typically characterized by a filmy streak concentrated at ledges or joints. The carbonate can be removed by wiping the surface with trisodium phosphate. If this fails, consider treatment with equal

[4]H. B. Fishman, B. P. Darling, and J. R. Wooten. "Observations on Atmospheric Corrosion Made of Architectural Copper Work at Yale University." *Degradation of Metals in the Atmosphere*, ASTM STP965 (1988), pp. 96–114.

parts of ammonia hydroxide and isopropyl alcohol, followed by a thorough rinsing. Depending on the carbon dioxide level in the atmosphere, the white streaks will probably return. Keep in mind that small amounts of lead are removed as the surface is wiped. Follow the recommended steps for handling and working with lead, as mentioned later in this chapter.

A dark film will form on lead surfaces exposed to polluted atmospheres. The dark film is lead sulfate, which forms an adherent coating. To remove the finish, scrubbing will be necessary, along with the use of trisodium phosphate.

In general, lead-coated copper is a superior architectural metal in terms of performance and corrosion resistance. Reduction of serviceability is generally due to poor installation methods, design, and detailing, rather than to corrosion-induced deterioration.

Soldering and Joining

Lead-coated copper joints can be soldered. Excessive fluxing of the solder joint should be avoided. Rinse and neutralize the flux from the joint after the solder is applied. Use nonacid-core solders heavy in tin. Soldering lead-coated copper can develop lead-rich seals when solders of the 50-50 tin-lead ratios are employed. Some of the lead from the coating eludes the solder seal and makes a weak lead joint. Use 60-40 tin-lead solder to counteract this effect.

Use copper nails and copper rivets to join lead-coated copper. Exposed rivets or nail heads should have a layer of solder applied over them to create a color match and to seal the penetrations.

Use copper clips to attach hemmed lead-coated copper panels. Double and single locked seams should be used to join lead-coated copper panels. Forming techniques should follow the parameters for standard copper sheet metal construction. Always use a rigid substrate such as plywood for full support of the lead-coated copper; felts and building paper should cover the wood prior to its application.

Hazards in Working with Lead Coatings

Lead oxides that form on the surface of lead-coated and terne-coated metals are toxic if ingested. When working with lead- or lead-alloy-coated materials, wear gloves. Avoid touching the mouth; do not allow cigarette smoking, not because of flammability, but because of possible lead ingestion. Do not allow eating on lead surfaces.

Material Specifications

ASTM B101-83	Standard Specification for Lead-Coated CopperSheets for Architectural Uses
ASTM B370	Specification for Copper Sheet and Strip for Building Construction
ASTM E36	Method for Chemical Analysis of Brasses
ASTM B248	Specification for General Requirements for Wrought Copper and Copper Alloy, Plate, Sheet, Strip, and Rolled Bar
ASTM B601	Practice for Temper Designations for Copper and Copper Alloys—Wrought and Cast

NICKEL COATINGS

Nickel coatings are applied by electroplating or ion implantation. Nickel is often plated over a copper intermediate plate. The copper acts as a softer underlayment and allows better buffing and polishing. Chromium is added to nickel coatings when brightness is critical. On exterior parts, nickel coatings should be kept in the 0.8 mil to 1.5 mil range. For interior surfaces, the amount of nickel can be reduced to produce a coating thickness in the 0.3 to 0.5 mil range.

Nickel plated objects used architecturally are limited to small fabricated objects or selectively plated decorative surfaces. A nickel coating requires maintenance when used in industrial or urban environments. The surface will fog over as nickel sulfate develops from pollution products.

TIN COATINGS

Tin is a soft silvery-white metal. Too soft to use alone, tin is applied by hot dipping or electroplating onto other metals with better strength characteristics. Tin is also included in small quantities with other elements to develop enhanced alloys with particular properties. (See Chapter 2, Copper Alloys and, in this chapter, Terne and Pewter.)

Tin ore is never found in the pure state; however, deposits of the mineral cassiterite, also known as "tin stone," are dispersed throughout the world. These tin oxide ore deposits are rarely of any significant size because many sites—some of them quite ancient—have been depleted by early mining operations. There are no major domestic tin mining or production facilities, even though the United States is the largest consumer of tin. Most of the world's supply is concentrated in southeast Asia. China has 24% of the world's supply, Thailand has 15%, and Malaysia another 12%.

The price of tin is usually very stable, even with its limited supply. Pricing was regulated by the International Tin Council, a consortium of consumer countries and producing countries. When prices moved lower, the consortium bought tin to firm the price, then sell as prices rose. Apparently, stockpiles accumulated and the consortium was forced to sell, bringing tin prices down dramatically. The consortium has since disbanded.

History

Early man discovered that small amounts of tin could be added to molten copper to form the bronze alloys. Bronze was easier to cast than copper, and the resulting metal was harder and would hold a sharpened edge longer. Axes, awls, hammers, and weapons could be cast with this harder metal.

The advent of the Bronze Age, approximately 3000 B.C. to 1100 B.C., began a time of vibrant economic expansion for the region around Mesopotamia. This new metal allowed for rapid development of the tools necessary to the advancement of civilization. The use of tin as an alloying material signalled the beginning of the age of metals, which continues to this date. Archaeologists have uncovered tin mining and smelting operations that date back to 2870 B.C. in the area north of the Mediterranean city of Tarsus.[5]

During the eighteenth and nineteenth centuries, iron rectangles, approximately 12 inches square, were dipped into a bath of molten tin. The rectangles would then be nailed and lapped as roof shingles. Thomas Jefferson used hot-dipped tin-coated sheet iron shingles on Monticello, his Virginia estate, early in the nineteenth century. Jefferson had the shingles imported from Europe after he saw roofs in France, more than 100 years old, made from this material.[6] The tin-coated iron shingles lasted for 30 years, before wood shingles were applied over the tin. Recently, Monticello was restored and a new tin-coated stainless steel roof was applied. Stainless steel shingles replicating the original iron shingles were hot dipped into molten tin and nailed in layers, precisely replicating the early Jefferson design. Tin-coated stainless has the lustrous silver-white appearance characteristic of new tin. Eventually, it will become mottled in appearance as oxides develop on the surface.

In the late 1800s and early 1900s tin-coated steel products were in considerable use as architectural surfacing materials, particularly for roofs. Prior to 1890 most tin-coated steel and iron products were imported from Europe. By the start of the 1900s, the United States took over as the predominating tin-producing country. Products such as American Old Style, N&G Taylor Company Old Style Roofing Tin, U.S. Eagle, and Bassett Genuine Charcoal Iron Old Style Roofing Tin were in common use across the United States[7] in such prominent buildings as the White House and Independence Hall in Philadelphia. Tin plate was sold in boxes of 112 sheets. Sheet sizes were either 14 × 20 inches or 28 × 20 inches, and thicknesses of 30 gauge for roofing and 28 gauge for gutters and valleys. Rolls of seamed sheets were also available. Rolls of 14-inch, 20-inch, and 28-inch widths were available joined together with the standard double-lock seam, either soldered or unsoldered.

Alloys

The color of polished tin is a highly lustrous silver-white. The luster rapidly diminishes on exposure to the atmosphere, but further tarnishing is slow to develop. One major advantage of tin, as indicated by the universal use of the common tin can, is that the metal is nontoxic. Tin is suitable for contact with food.

The spotty surface on some organ pipes is the result of crystalline structure formations developed when a lead-tin alloy sheet used to make the pipes, cools. The interesting network of bubble-like crystals is generated by a differential shrinkage phenomena. The spotty texture is known for the superior

---

[5] John Noble Wilford. "Turkish Tin Begat the Bronze Age." *New York Times* (1994).

[6] Stephen D. Cramer, and M. Jeffery Baker. *Monticello Roof Restoration* (1993).

[7] The descriptive term "Old-Style" was often used to refer to the European tin-plated product, which used iron as the base metal. The U.S. product at the turn of the century used steel as the base metal, some of which was low grade.

Note: Quote from Sweets Catalog, 1906. "Every architect and roofer knows that the Steel Basis Plate as used in roofing today is not up to the standards of years ago, and we claim that we have, through modern processes, brought out a plate to furnish roofing which is identical with the Old Charcoal Iron Plate, as made before the advent of soft steel."

tone generating from the metal pipes. This phenomena is found to occur only on thin cast alloy sheets containing from 35% to 65% lead.

The corrosion-resistant characteristics of tin are comparable to those of zinc. Tin acts as a sacrificial metal when in contact with most other metals, similar to the way zinc offers galvanic protection to steel.

Tin is readily applied to clean, oxide-free metals by the hot dipping method. This technique of applying tin for corrosion protection to steel and iron has been in use for centuries. Tin can also be applied to aluminum, copper, and the copper alloys with little difficulty by the hot dipping method. When the base metal is immersed in the bath of molten tin, a metallurgical bond develops at the interface between the two metals. Tin forms a tight adherent bond, continuous over the surface of the steel. Coating weights range from 0.05 to 0.15 oz/sf (15 to 45 g/m^2).

The amounts of allowable impurities in tin used as a coating are designated under ASTM B339-67. For architectural uses, Grade A Standard is the designation for the pig tin grade. This specification calls for 99.8% purity. There are a total of seven classifications for various levels of purity. Lead is the most common impurity, which can cause the tin coating to appear spangled and less lustrous.

Tin Electroplate

Tin coatings over steel and copper can be applied by electrodeposition, also known as plating. *Electrolytic tin plate* is a common term describing the coating used on the majority of ferrous can stock. Tin oxides are nontoxic and act sacrificially to save the base steel material. When used in construction, tin-plated steels are found in pressed ceiling panels or small, inexpensive decorative trims. In regard to many of its past uses, tin coated metal has been supplanted by galvanized steel or aluminum.

A European product is available for roofing and flashing sheets, which utilizes a thin electroplate coating of tin on copper. The initial sheet is a very even gray color. As the plated tin weathers, a green patina can be seen forming on the surface.

Tin-plated copper is more prevalent in European countries than lead-coated copper. With the growing environmental concern about lead in the United States, tin-plated copper may develop into the material of choice for achieving a gray appearance.

Tin can be alloyed with zinc and nickel to form decorative coatings with a strong silvery luster. These alloys are applied by electrodeposition onto other metals. *Speculum* is the name given to a 58% copper and 42% tin alloy. This alloy is plated onto steel and polished to a bright silver appearance. At one time, this coating was used for mirrors because of its ability to resist tarnish in indoor environments. Speculum offers good resistance to sulfur compounds and other oxidizing agents. Exterior exposures will turn the surface gray in short order.

Tin-zinc electrodeposition coatings have been in use since the 1940s. With the enlightenment of pollution and chemical concerns associated with cyanide, the application of tin-zinc coatings were basically shelved for all but a few applications. The International Tin Research Institute developed a new method of applying the coating without cyanide electrolytic solutions. This coating, applied by the electrodeposition process, called **Stanzec**,[8] contains 20% to 30% zinc. Stanzec uses a noncyanide electrolyte in the plating bath. The coating develops a nonporous, bright, and reflective barrier over steel or other clean substrate. Because of the tin coating, a corrosive barrier is introduced. The zinc portion of the coating enables galvanic protection along regions where scratches breach the thin protective barrier. Stanzec is a matte finish with a white-silvery color. Stanzec can be polished to develop a reflective mirror finish but will require periodic cleaning to maintain the bright appearance.

Pewter

Pewter, an alloy of tin, is a rich gray metal from antiquity, well known to most consumers. The pewter tankard and tableware with the clean, consistent, flat gray color come to mind. Pewter, unlike any other metal, except perhaps lead, imparts the feeling of softness and weight. Perhaps this feeling is instinctively developed, inasmuch as this metal has been used since the days before the Roman Empire.

As an architectural metal, pewter has had little use. Ornamental uses have been limited to plaques and small statues. Products manufactured from this metal have always been considered valuable. An entire guild devoted to the working of pewter developed across Europe during the Middle Ages. This guild manufactured goblets and plates for the use of the nobility. Today, there is still a guild that manufactures pewter for ornamentation and plate wares.

Pewter had widespread use across Europe, particularly in England, during the fourteenth through the sixteenth centuries. The pewter alloy used during this period was 80% tin, 20% lead, and trace amounts of other metals. Pewter of this constitution was very soft and malleable. Shapes could be easily hammered and hand drawn from cast blocks of the metal. The alloy melted at low temperatures, approximately 450°F (232°C). The molten alloy flowed well into intricate molds, which could later be easily shaped and polished. Attempts to pour sheet pewter using this lead alloy were not practical. The leaded alloy lacked the ability to develop strength when rolled thin.

A new alloy known as Britannia metal was developed by reducing or eliminating the lead content and introducing antimony and copper. Britannia metal could be rolled into sheets. Lead, considered a toxic substance when digested, was removed from the alloy mix. This was wise, considering that pewter was still used extensively for eating utensils and drinking vessels.

From a purists viewpoint, modern day pewter is actually Britannia metal. It contains approximately 6% antimony and 2% copper with some trace amounts of other metals to produce a hard or a dark surface. The remaining element is tin. Because of the large proportion of tin, the polished surface is very bright and reflective.

There are no architectural uses of pewter, in part because of the limited knowledge of this metal, lack of available distribution, and the relative price

[8]S. J. Blunden. "Tin-Zinc Alloy Plating; A Non-Cyanide Alkaline Deposition Process." International Tin Research Institute; A. J. Killmeyer, Tin Information Center of North America

of the metal. Because of its soft, malleable nature, pewter would have limited use unless it is applied as a coating to other metals, such as copper or stainless steel.

As to corrosion resistance, this high-tin-content alloy would perform adequately. Initially bright surfaces would tarnish to a dark gray color on exterior exposures. In interior exposures, where touching and handling are restricted, little maintenance will be required. Tin is an expensive metal. At this writing, tin is almost three times as expensive as copper per unit weight.

Decorative patinas can be produced on a pewter surface. Chemical acceleration of the patina, through proprietary processes to modify the characteristics of the oxide layer, is available. Three basic coloration processes are used:

1. Immersion in heated alkaline solutions of polysulfides
2. Immersion in heated acid solutions of copper and arsenic
3. Development of oxide in phosphate solutions or caustic soda

Once the tone and color of the patina are achieved, the object is coated in wax or oil to resist further change in color.

Pewter-Coated Copper

Pewter-coated copper, or tin-coated copper, depending on the definition of pewter employed, was developed to produce a gray tone similar to that of lead-coated copper, but without the lead. Initially, the coating is reflective and has a bright silver-white luster. The initial luster can be reduced by pumicing the surface or by simple exposure to an outdoor environment for a short while. Like pure uncoated copper, tin exhibits interference colors as the thin stannic oxide layer forms on the surface. Eventually, a dense gray or blue-gray layer develops. As weathering occurs, a mottling of different tones appears over the surface. Sometimes a yellow tint is apparent. This is an oxide interference layer, which is more prevalent in copper-rich pewter coatings.

Pewter-coated copper is produced by dipping clean, oxide-free copper sheet into a molten bath of tin at 800°F (427°C). As the clean copper enters the molten bath, some of it goes into solution. A duplex interface of copper-tin alloy forms between the almost pure tin outer layer and the copper base sheet.[9]

When the copper sheet is removed from the molten bath, excess tin is removed by passing the sheet under an air knife. The sheet cools, leaving a shiny, slightly spangled tin surface. The copper base sheet loses some of its temper by immersion in the hot bath.

Forming and Joining

Pewter-coated copper, tin-plated copper, and tin-coated stainless steel can be easily formed and shaped with the use of conventional sheet metal equipment. Joints can be readily soldered with a low-lead solder to produce an appearance matching the base sheet.

[9]Edward J. Daniels. "The Hot-Tinning of Copper: The Attack on the Basis Metal and its Effects": pp. 199–205.

Care and Maintenance

Procedures for handling, transport, and storage of tin-coated metals are similar to those of other metals. They should be kept dry and separate from wood. Acid can leach from wood and stain a tin surface. A thin cottonseed oil coating can be applied to the surface as temporary protection against condensation and to reduce oxidation of the surface. Tin sheets, stacked on top of one another without protection, can be subjected to fretting corrosion. Fretting corrosion is a problem with hot-dipped coated metals. The coating has high and low spots, which can rub on adjacent sheets. Tin oxide on the surface of one sheet will act as an abrasive. As the sheets rub back and forth during transportation, the abrasive action may remove some of the protective coating. Such damage is visually superficial in most cases, but there is some reduction in the protection afforded by the tin coating.

Avoid contact between tin coatings and copper. The copper electrolyte will rapidly corrode the tin coating. Bituminous and asphalt compounds can also corrode a tin coating. Unfortunately, many old tin roofs, upon showing signs of deterioration, had asphalt compounds applied to the joints. These compounds only accelerated the deterioration of the metal.

On tin- or pewter-coated copper, tin acts as an anode. Accelerated corrosion will occur at breaches in the protective tin coating. Corrosion particles initially appear as black spots on the surface.

Pure tin coatings develop a fragile tin oxide, which is soluble to a degree. An early architectural use of the metal was tin-coated wrought iron. The shiny tin coating would last for about 20 years, depending on the exposure level. Once breached, the iron base metal rapidly corroded.

ZINC COATINGS

Zinc-coated steel is known commonly as galvanized steel. The method of galvanizing steel was introduced by the French chemist Malouin in 1742. Almost a century later, the British patented the process of hot dipping iron into molten zinc. Once the process was commercialized and the rolling of thin sheet steel became economical, galvanized steel sheet came into common architectural use. Decorative siding panels and elaborate cornices and guttering were affordable and practical, since this new process imparted a lasting effect to common iron.

One early entrepreneur, the Mesker Brothers Company of St. Louis, Missouri, developed and marketed kits for building facades. These kits were used by architects of buildings in many small midwestern towns. The kits included prefabricated galvanized cornices and wall panels anchored to wood structures. Many of these building facades are still in existence after almost 100 years of service.

Zinc coatings are applied to sheet steel through electrodeposition or hot dipping. For large fabricated steel parts, the hot dipping method is an economical way to apply a complete coating of zinc. Hot dipping large structural or fabricated parts can produce thicknesses of zinc in the 2 to 4 mil (0.05 to 0.10 mm) range. This method of applying zinc is known by the acronym HDGAF, for "hot dipped galvanized after fabrication." This method puts a continuous, unbroken zinc coating over the entire exposed surface of the metal. When using this method, it is necessary to allow for draining of the part to eliminate trapped molten zinc. Welded connections can be hot-dipped galvanized after they are cleaned and ground smooth. Hidden recesses and concealed cavities within a fabricated part can receive a coat of zinc using this method.

Hot dipping is also the preferred method of galvanizing sheet steel used in architectural applications. With sheet stock the zinc is applied to coils of metal in a continuous fashion. Hot dipping steel sheet or fabricated parts into molten zinc, either before or after fabrication, develops a metallurgical bond between the zinc layer and the base metal. This bond is very tight and adherent; it actually combines with the base metal.

Electrodeposition of zinc, however, does not develop a metallurgical bond between the zinc layers and the base metal. The thickness of the electrodeposited coating is not as great and therefore will not afford a superior corrosion-resistant barrier. Electrodeposition does produce a smoother surface, because of the even application of the zinc coating onto the steel. Electrodeposition of zinc should not be specified when the galvanized steel is to be exposed to corrosive environments, even if it is finish painted. Electrodeposition is not the preferred architectural coating process.

For sheet steels, the hot dipping method applies zinc to continuous coils of cold rolled steel sheet. The ribbon of metal is passed through molten baths of zinc at a rate of several hundred feet per minute. The amount of zinc applied to each side of the sheet is controlled by a set of wiping rolls. The coated sheet steel is allowed to cool, then recoiled for further operations such as leveling or paint finishing.

As the steel sheet is immersed in the molten metal, a thin transition alloy zone is created between the steel and the pure zinc surface. During the process, the steel reacts with the molten zinc, forming a series of zinc-iron alloy layers topped with a coating of zinc at the outermost surface. The zinc coating develops a tough metallurgical bond to the oxide-free steel, resisting mechanical removal by subsequent cold-working operations. The layers of zinc-iron alloy provided by the hot dipping method offer a twofold barrier of cathodic and mechanical resistance.

There are four distinctive layers developed during the metallurgic bonding of zinc to steel. The outer layer is nearly pure zinc. This layer provides cathodic protection to scratches through the coating and edges where the

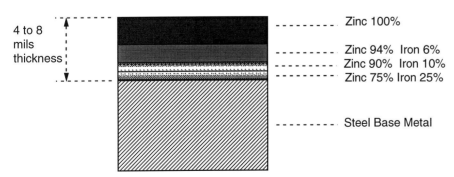

Figure 7-3. Cross section of zinc applied to steel.

galvanized steel sheet is cut. Moving inward toward the base metal, a layer approximately 94% zinc and 6% iron provides the second line of defense. This layer is denser and tougher than the outer layer of pure zinc. The next layer, 90% zinc and 10% iron, resists abrasion. This is the hardest layer of the zinc-iron alloy. The final layer, 75% zinc and 25% iron, is metallurgically bonded to the base steel sheet. By metallurgically bonding, this layer of zinc-iron alloy is electrochemically joined to the base steel. There is an exchange of electrons, which fuses the metal types together. The development of each of these layers depends on the length of time the steel part is immersed in molten zinc. Preheating the steel will aid in accelerating the growth of the layers.

Hot dipping provides excellent protection by covering all areas of the metal surface. Subsequent painting of the galvanized steel, using pretreatment processes to prepare the zinc outer layer for paint adhesion, results in a superior corrosion-resistant surface. Galvanizing under the paint will enhance the performance of the paint finish by sealing pores in the paint finish and by preventing undercutting corrosion.

Galvanized coatings have a characteristic spangled surface appearance. The spangles are the crystallization of the zinc grains on the surface of the metal. On hot-dipped galvanized steel, the spangles are larger and more variable in thickness and light reflectivity. When the spangles are allowed to crystallize without restriction, the coating is considered Regular Spangle and is specified with the prefix G (see Figure 7-4).

When the zinc coating is altered before crystallization occurs, a zinc-iron alloy layer develops, eliminating the outer pure-zinc layer. The surface is rough and gray in color and lacks the characteristic spangle. This rougher surface receives paint without special priming or preparation. This treatment or surface

Figure 7-4. Spangle effect of crystals cooling on steel. (*a*) Maximum spangle; (*b*) Minimum spangle.

is known as Galvannealed and is specified with the prefix A. The metal is heat treated or annealed after galvanizing to eliminate the zinc outer layer.

The weight of zinc coating on sheet steel is given in ounces per square foot of surface area on each side. The thickness of the zinc coating produced by the hot dipping method will vary slightly from side to side, as well as from edge to edge. The corrosion resistance of the zinc coating is directly proportional to its thickness. The thicker the coating, the better the corrosion protection afforded by the zinc. The coating designation for each prefix and its associated thickness are indicated in Table 7-5.

G90 is the commonly specified United States coating designation for galvanized steel sheets used in architectural applications. All major United States coil coated steel is specified with a G90 coating of zinc under the paint finish. This painted coil is used for roll forming metal siding panels and for press brake forming operations to create coping caps, gutters, and an assortment of other custom architectural metal shapes.

In the hot-dipped galvanized coatings, the spangle produced can also be affected by treatment of the zinc coating just prior to crystallization of the outer zinc layer. Unlike the dull alloyed coating designated by the prefix A, this treated galvanized coating has a reduced spangle, sometimes dull in appearance, characterized by tight, small crystals. This finish surface is called Minimized Spangle.

When a smooth surface is required, one that is free of the differential heights of the zinc crystals, a temper-rolled surface is specified. This smoothes the surface of the metal by eliminating the difference in crystal heights. This process tends to compress the alloy layers and has a small effect on ultimate corrosion resistance.

TABLE 7-5. Standard Galvanized Coating Designations

TYPE	COATING	MINIMUM (OZ/SF)*	MINIMUM KG/M²	FEDERAL SPECIFICATION QQ-S-775E
Regular	G235	2.35	0.72	Class a
Regular	G210	2.10	0.64	Class b
Regular	G185	1.85	0.56	
Regular	G165	1.65	0.50	Class c
Regular	G140	1.40	0.43	
Regular	G115	1.15	0.35	
Regular	**G90**	**0.90**	**0.28**	**Class d**
Regular	G60	0.60	0.18	Class e
Regular	G30	0.30	0.09	
Regular	G01	no minimum	no minimum	Class f
Alloyed	A60	0.60	0.18	
Alloyed	A40	0.40	0.12	
Alloyed	A25	0.25	0.08	
Alloyed	A01	no minimum	no minimum	

* The minimum oz/sf is based on the spot-test check specified under ASTM A90, which requires three checks of the coating across the sheet.

Quality Control

Hot-dipped galvanized steel should be inspected for surface irregularities and coating performance. Surface irregularities, such as coil breaks or cross breaks induced in the metal surface, should be rejected. Blemishes resulting from zinc strings, or coarse coatings in which thickness varies, are not harmful to the life of the sheet; however, the appearance may be objectionable. Forming operations on sheets with these blemishes will produce creases and deformations along the formed edge.

When forming 180-degree brakes, the galvanized coating should be adherent and not flake off. The thicker coatings above the G90 range can flake when undergoing severe forming operations. In some instances it may be better to galvanize the part after fabrication if a thicker coating is desired.

Galvanized steel is furnished from the supplier with a thin coating of oil to resist corrosion while it is stored. Storage stains resulting from moisture will show up as white, powdery rust. If the surface is to receive further painting, such stains will not pose a problem in corrosion resistance or paint adherence. If galvanized sheet is to be left uncoated, removal of a stain will remove some of the protective zinc. Coils or sheets that become wet should be dried immediately. Separate sheets to dry, and open coils to allow air to circulate.

Newly galvanized sheets or steel fabrications will not receive paint. Exposing the surface to the weather for a short period of time will create a coarse, grainy zinc oxide that will receive paint fairly well. Priming the galvanized surface is recommended to enable paint adhesion. Galvanized steel can also be specified with a phosphatized surface for paint adherence. This chemical process, performed at the mill, will remove the spangle and allow paint application. A primer is still recommended for improved paint adhesion.

Galvanized steels are available in five principal quality classifications as discussed in the following paragraphs.

Commercial Quality

Galvanized steel sheet is ordered to specification by the customer for a given application. Commercial Quality galvanized steel is a class of base steel that has variations in chemical composition. Additional processes to improve the galvanized surface appearance and control the temper of the base metal are possible. This quality classification is for metal fabrications intended to undergo light to moderate deformation.

Drawing Quality

Galvanized steel of Drawing Quality is specially processed for good drawing properties. There is no specification for the durability of the zinc coating. The zinc coating can flake or stretch during the process of drawing. View test samples to ensure the quality. Postgalvanizing may be necessary.

Drawing Quality Special Killed

The form of galvanized steel used for severe forming and drawing operations is Drawing Quality Special Killed. Again, the steel is specifically formulated for the operation desired. As with Drawing Quality, the zinc will probably flake or thin from the severe forming.

Structural Quality

This form of galvanized steel, as specified in ASTM A446, "Sheet Steel, Zinc Coated by the Hot Dipped Process, Structural Quality," has special mechanical properties developed for a required application. The mechanical properties are afforded by the type of steel. The galvanized coating is applied according to the various specifications for thickness.

Lock Forming Quality

Galvanized steel of Lock Forming Quality is for those applications in which the steel is to undergo operations that machine lock form the edges. This form of galvanized steel is recommended for most architectural applications. Machine lock forming operations require characteristics not available in the Commercial Quality classification. Lock Forming Quality is specified in ASTM A527, "Sheet Steel, Zinc Coated by Hot Dipped Process, Lock Forming Quality." For this classification, sheet and coil stock should carry the designation mark LFQ on the crating or packaging.

Corrosion Considerations

Zinc offers excellent corrosion protection for the base steel material. Zinc coatings, when applied by hot dipping, will provide cathodic protection at scratches and cut edges, as well as a tough barrier coating that resists mechanical abrasion.

Zinc coatings over the base steel, when breached, cathodically protect the exposed steel surface to a distance of up to 0.25 inches (6.35 mm). When exposed to corrosive environments, the zinc goes into solution and sacrifices itself over the steel. The more corrosive the environment, the better the protection afforded by the zinc over a larger area of exposed steel.

Paint alone, as a coating, offers only a barrier to corrosion attack. Once the barrier is breached, corrosion will continue, often flaking off paint as the iron oxide expands and undercuts the paint film. Galvanizing helps prevent this by sealing any pores in the paint and preventing underfilm deterioration.

Galvanized steel has been the material of choice for many applications. The galvanizing process is an economical means of extending the life of steel sheets and fabrications. Exposure tests of the standard G90 (1.25 oz/sf) galvanized coating have demonstrated the following expected life spans of the steel sheet:

11 years + for rural exposures
8 years for coastal exposures
4 years for industrial exposures

Adding a good paint finish will definitely extend the expected lifetime of a sheet metal surface.

Zinc coatings are sensitive to acid attack, particularly by sulfuric and hydrochloric acid. Galvanic corrosion will occur rapidly when zinc-coated steel comes in contact with copper electrolytes. Moisture draining from rooftop condensation equipment or roof drainage systems through copper piping will quickly corrode galvanized steel fabrications if allowed to come in contact with the zinc surface.

Sherardizing

Another method of coating zinc onto small parts of steel or iron is Sherardizing. This process requires the placement of a cleaned and degreased part into an air-free chamber containing zinc dust. The object is heated as it is tumbled in the chamber. The zinc dust coats the surface of the object with a thin layer. The surface is relatively smooth and free of the typical spangle from the hot dipping process. The nature of the process limits the size and shape of the part to be coated.

Maintenance and Repair

Galvanized steel sheets must be kept clean and dry during storage. Storage stains will develop on the surface of galvanized steel sheets if they are allowed to become wet. These storage stains can occur when sheets of galvanized steel or coils of galvanized steel are wetted in the spaces between the thin sheets or within the rolls of metal. The stain appears as a white, coarse compound on the surface of the metal. The damage is superficial unless the stain is severe, and can develop in as little as 24 hours. Remove the stain using household detergents. Do not use a wire brush.

If stains are severe, there are a couple of chemical treatments that may work to remove and brighten the surface. Use protective clothing and eye protection, as well as ventilation, when applying and working with hazardous chemicals.

1. Chromic Acid Solution Cleaning Treatment
   - Mix 48 ounces of chromium trioxide with a gallon of 5% diluted nitric acid solution.
   - Coat onto stain with a cotton cloth.
   - Rinse with hot water.
   - Repeat.
   - Dry.
2. Sodium Hydroxide Solution Treatment (This will brighten.)
   - Mix 32 ounces of caustic soda with a gallon of water.
   - Add 96 ounces of talc to develop a paste.
   - Apply with a nylon-bristle brush.
   - Rinse with clean water and dry.

On large galvanized members and on welded articles where the zinc coating is damaged or removed, cold galvanized applications will work. Good cold galvanized coatings are made of a minimum 95% zinc. The color will not match the original hot-dipped coating; however, the cold galvanized application will seal the surface and resist corrosion.

Variations on the Galvanized Theme

Previously discussed are the aluminized steel coating and the aluminum-zinc coating known as Galvalume, but there are other variations on zinc coatings over steel developed for industries and applications requiring the galvanic protection of zinc.

TABLE 7-6. Other Coatings Using Zinc as a Main Alloying Metal Coating

NAME	MANUFACTURER	
Zincrometal	Metal Coatings Intl.	One sided, chromate based with zinc top coat
Unikote	National Steel	One sided, electrolytic treated
Zincgrip O.S.	Armco	One sided
Galva-One	U.S. Steel	One sided electrogalvanized
Galvannealed steel		Treated sheet to enhance paint appearance
Paint tite	Inland Steel	Zinc iron coating
Zincgrip Ultrasmooth	Armco	Improved surface appearance
Galfan	International Lead Zinc Research Organization	95% zinc, 5% aluminum/mischmetal*

* *Mischmetal* is the German name for "mixed metal." This term is given to describe an alloy made up of cerium, lanthanum, didymium, and some of the other rare earth metals.

Most of these bear trade names, and many were developed for the automotive industry in light of their requirement for economical corrosion protection, paint adherence, and surface appearance. (See Table 7-6.)

The galvannealed steel sheet is galvanized sheet that has been thermally treated to induce an iron–zinc interface coating as discussed earlier. The Zincgrip and Paint tite products are variations of the galvannealed surface. The Galfan sheet is a relatively new treatment that shows superior characteristics. Adding the mischmetal to the aluminum–zinc binary alloy improves the surface smoothness and reduces the variations in zinc crystal (spangle) size. In addition, corrosion resistance and formability are improved over that of the standard galvanized. Accelerated exposures of the Galfan coating develop a light gray patina. The patina grows rapidly, but then stabilizes.

Soldering, Welding, and Joining

Galvanized steels can be soldered, brazed, and welded. Soldering will require the cleaning back of all oxide using a zinc chloride flux. Use standard sheet metal joints such as the single or double lock seams, then solder the surfaces together.

Zinc chloride fluxes will work well to prepare a joint for soldering. Dissolve a few strips of zinc or galvanized steel in a dilute mixture of hydrochloric acid (muriatic acid) until all effervescence ceases. The remaining fluid is zinc chloride. Coat the joint to remove oxides, and solder using 50-50 tin-lead solder or one of the 95-5 low-lead solders.

Galvanized steel can be welded easily with a spot welder. All the standard welding processes can be used to join galvanized steel. The zinc coating does not necessarily have to be removed before welding; however, on thicker parts the joint will be stronger if the zinc is removed. Zinc fumes created from the welding process will create a hazard. Use adequate ventilation when welding or soldering galvanized steels, as the zinc oxide dust is toxic when inhaled.

Refer to Table 7-7 for a listing of various welding processes used on galvanized steel.

TABLE 7-7. Welding Techniques Used on Galvanized Steel

	STANDARD WELDING PROCESSES
SMAW	Shielded Metal-Arc Welding
	Flux Cored Arc Welding
GMAW	Gas Metal-Arc Welding
GTAW	Gas Tungsten-Arc Welding
PAW	Plasma Arc Welding
	Carbon-Arc Welding and Brazing
	Gas Metal-Arc Spot Welding
	Gas Tungsten-Arc Spot Welding
	Resistance Spot Welding

ZIRCONIUM NITRIDE COATINGS— BRASS APPEARANCE

Zirconium nitride coatings provide a brasslike appearance. Typically used on hardware and other small fabrications, zirconium nitride provides a brass tone that resists tarnish. Brass and brass-plated coatings tarnish rapidly, taking on reddish color tones when lacquering is not applied. Clear coatings applied to brass reduce the metallic sheen.

Zirconium nitride is applied by the sputter plating process. In the sputter plating process, also known as sputter-ion plating, a thin film coating is applied over a base metal. The tight, wear-resistant coating provides an even layer over the metal surface. The color is slightly less reflective than polished brass, but it resists tarnishing and is very durable.

COPPER-STAINLESS STEEL-COPPER BONDED SHEET

When copper prices were high in the early 1970s, a material was developed that bonded thin stainless steel between two layers of copper; the stainless would offset the higher cost of the copper. The material did not fair so well in the marketplace. The added strength and slightly reduced cost did little to impact the solid copper sheet market. Besides, the scrap was not salvageable. All metal firms salvage their scrap materials regularly, and the sight of copper in the trash bin conflicted with their historical spirit. If and when scrap recovery improves, this product could have a very important place in the market. The additional strength of this sheet metal form would greatly enhance the use of copper.

This product is covered under ASTM B506-81, "Standard Specification for Copper-Clad Stainless Steel Sheet and Strip for Building Construction."

GOLD COATINGS

Introduction

Throughout history, gold, among all metals, has held the seat of prominence. Market pressures might move platinum, palladium, and a few other rare and "strategic" metals higher in value than gold, but just for the nonce. The average person will place a value on gold above that of all other metals. Throughout time, this yellow metal has always been held in high esteem.

Gold is the most noble of all metals. It will not react with common oxidizing agents to form carbonates or sulfates like other metals. Exposure to the atmosphere forms a thin oxide, but not much else will adhere to its surface. Gold is found throughout the earth, generally in a highly pure state. South Africa and Russia are two of the largest mining regions devoted to gold, whereas the United States mines much of its gold as a by-product of copper mining. The metal is so rare that all the gold mined from the beginning of time would form a cube approximately 50 meters per side.

Gold is also the most malleable metal. A gram of gold can be hammered and rolled to an incredibly thin foil, 8.4×10^{-5} mm thick, or stretched and pulled into a wire over 2 kilometers in length. Pure gold plate is soft and will wear easily if unalloyed. Because of its softness, gold is often alloyed with other metals, such as nickel.

The high cost of gold has limited its use as an architectural metal to few forms other than thin sheets, gold plating, or powder emulsions in paints. Used to enhance less elegant metals, applications of small sheets or leaves on ornamental features have had wide use in almost every major city on this planet.

Large domes of underlying copper or tin-coated steel can be easily coated with the metal gold through the use of techniques that date from the Roman era. Domes coated with gold leaf are often the most striking architectural structures visible on the rooftops of a city. They catch the eye because of the bright gold luster shining against the background of grays, browns, and dark greens of the cityscape. Gold, above all other metals, has an almost mystical effect on people. Used extensively as jewelry and money, gold will always be considered the ultimate decorative material.

Gold in leaf form, known as gilding, is easy to apply and relatively inexpensive, considering that it can produce a surface with an unsurpassed, elegant luster. Gilding, an ancient technology, was practiced by the Egyptians and Romans. The Romans gilded large objects, such as bronze columns and statues.

For large ornamental features, the ancients used a method known as *flame gilding*. Gold was mixed with mercury to form an amalgam, which was painted or sprayed onto a surface to form a thick coating. The surface was heated with a torch, evaporating the mercury and leaving behind a gold coating. Unfortunately, those early flame gilders died at an early age from mercury poisoning, as the mercury vaporized and was inhaled. The process today is outlawed.

The Gilding Process

Initial preparation of the underlying metal is crucial for the long-term performance of the gold leafing. A good corrosion-resistant surface below the gold leaf, such as aluminum or copper, free of oxides, performs best. Wrought iron, galvanized or coated with an epoxy primer, followed by a good thick coating of paint, will also suffice. A smooth surface is critical to develop a high luster in the gold leaf.

The gold is most lustrous when first applied. Any rubbing or "burnishing" of the surface actually reduces the luster by imparting small scratches to the gold surface. Sanding and polishing the surface of the base metal prior to application of the leaf adhesive is recommended, both for final luster and for preparing the surface by removing the oxides.

The adhesive used to adhere gold leafing is called *sizing*. Sizing is a contact cement applied to the base metal prior to application of the gilding. Sizing comes in various degrees of dwell time for tack. Sizing with a dwell time requiring hours to develop a tacky surface, will produce the most lustrous gold finish. There are other sizings available that reach tack levels for application of gold leaf in a very short time. Depending on the application and the environment in which the gold leafing is applied, different tack levels are desired. Exterior applied leafing requires a tack with a relatively short dwell time, otherwise dust and other foreign material may contaminate the sizing.

Applications performed in a shop can use the longer dwell times and produce a better luster. For a quick sizing, consider one that reaches tack level in approximately 1 to 3 hours. Humidity is more critical than air temperature. Slow sizings are available with tack times of 10 to 18 hours. The application and the desired technique will determine the sizing tack time.

There are excellent European sources of variable tacks for sizing; a number of French companies excel in the product. The United States has a few manufacturers of gold leaf sizing.

Once the sizing develops the required level of tackiness, lightly brush the gold leaves onto the surface. This is a technique perfected by the artisan. If the surface of the tack is too wet, it will appear through the gold sheets. As the tack approaches final drying, the difficulty of application increases, but the final quality also increases. A superior luster will occur when the tack is weakest.

Gold leaf is available in small squares or in rolls cut to width. The gold "leaves" are actually thin sheets of hammered gold applied to rouged tissue paper. The thickness of the gold leaf, which varies from one manufacturer to another, is generally described as approximately half that of rice paper, which would equate to a range of 0.1 to 0.2 mils (0.0025 to 0.0051 mm). When purchased, gold sheeting is usually provided in books of 25 sheets, $3^3/_8$ inches (86 mm) square. These are sold in boxes of 20 books, making a total of 500 small squares.

Boxes of gold leaf are available in different levels of quality, 14 karat, 16 karat, 18 karat, 22 karat, and 23 karat. The different karat levels have different color and quality characteristics. The cost of the boxes is related directly to the current market price of the metal.

The top qualities, 22 and 23 karat, are known as "deep" gold or XX Gold. This quality is used for all general surface leafing applications. Deep gold applications produce the rich golden tone typically associated with exterior gilding. The luster of this deep gold leaf can be unparalleled.

The 18-karat gold, known as Lemon Gold, is supplied only in the loose leaf form. Exterior use of Lemon Gold requires a good varnish coating and periodic maintenance. The other qualities of gold are not commonly used for surface gilding except as decorative treatments on glass. The 16-karat leaf is known as Pale Gold because of its pale yellow tint. The 14-karat gold leaf is known as White Gold because it is an alloy of 50% gold and 50% silver. This gold is silvery in color and will develop a slight tarnish.

After brushing the gold leaf onto the sizing, gently rub the surface with a soft cotton or velvet pad to remove any excess leaf material and eliminate creases in the thin gold film. Rubbing will put small scratches on the gold surface, reducing some of the luster but having no detrimental effect on the overall performance. Allow the sizing to dry. When lower-quality gold leaf is used, apply a clear protective coating to resist tarnishing. This leaf material is alloyed with other metals such as nickel and silver. The 23 k gold will not tarnish and does not need additional protective treatments. Therefore, it is recommended to use the better-quality gold leaf on exterior applications.

As with the sizing, the environment in which the gold leaf is applied can make the application of thin sheets difficult. Gold leaf adhered to tissue is available. The gold and the tissue are applied to the sizing, and the tissue is removed, leaving the gold in place. Application of gold in an exterior environment will require a windbreak or other precautions to prevent loss of material and the creation of creases in the thin leaf film. Moreover, for large surfaces the sizing should be applied in small regions, no more than can be covered with the leaf while the surface is still tacky.

Quality applications on large flat surfaces are tedious. Small fabrications or shaped fabrications with small workable surface areas pose reasonable challenges to the inexperienced.

Repair and Maintenance

A gold leaf surface showing deterioration is the result of corrosion of the underlying metal, not deterioration of the gold surface. As the underlying metal

TABLE 7-8. Comparison of Various Gold Leaf Materials

LEAF TYPE	NAME		
14 karat	White gold	Least expensive	Least durable
16 karat	Pale gold		
18 karat	Lemon gold		
22 karat	Deep gold		
23 karat	Deep gold	Expensive	Most durable

corrodes, small flakes of gold come off with the oxidizing base metal. The gold surface itself requires little maintenance.

Natural rainfall usually is sufficient to keep the surface clean. An occasional washing with mild detergents will help to keep the gold surface bright. In addition, periodic inspections of the surface can detect premature deterioration of the gold coating and enable the repair of damaged or breached coatings. Deterioration can result from abrasive, natural activity such as hail or from underfilm corrosion occurring on the base metal.

When deterioration is identified, it may be necessary to remove the gold leaf around the damage to enable cleaning and stabilizing of the base metal. The leaf should be scraped back to areas where the base metal is still intact. Clean the corrosion particles from the surface of the base metal with a stainless steel wire brush. Paint the base metal with a corrosion-inhibiting primer and reapply the gold leaf, using a quick-drying sizing. Lap the gold leaf over the unharmed leaf and lightly burnish the surface. Usually, the entire gold surface must be cleaned to remove all foreign particles and return the luster to match that of the newly replaced leaf.

Use soft cotton rags or other nonabrasive materials to clean the gold surface. Strong detergents and even mild acidic cleaners will not affect the gold surface. However, acids may find a way behind the gold and attack the underlying metal or sizing used to hold the gold in place. Therefore, proceed with caution if acids are used to remove adherent matter. Begin with a very small area and see how the surface reacts to the cleaning treatment before attacking the entire gilded surface.

Surface tarnish will develop on the lesser-quality gold alloys. A good commercial tarnish remover will work adequately. To prevent the return of tarnish, apply wax or lacquer to the coating. Both of these protective coatings will eventually wear out and require removal, cleaning, and reapplication.

Gold Plating Techniques

Selective or high-speed plating is an excellent technique for applying gold. Selective plating is an electrodeposition process available for use on large objects.

General Electrodeposition Process
In many electrodeposition techniques, the part to be coated must be dipped into a solution containing the plating metal. An electrical current is applied, running through the solution and to the workpiece to be coated. The workpiece is the cathode, and as current is applied the metal ions in solution migrate to the workpiece and deposit on the surface. The surface deposition is molecular, resulting in a very even, smooth coating of metal.

Selective Electrodeposition Process
A process was developed to allow selective deposition of one metal onto the surface of another metal. This process is not limited by the workpiece's having to be submersed in a bath. The selective plating process allows large objects—even the size of a house—to be electroplated, which

is a great advantage to complicated architectural fabrications and exterior ornamentation.

The technique uses a DC power source to feed a current through a solution containing the plating metal. A rectifier is used to supply current through two cables, one attached to a cathode and one attached to an anode. The cathode will be attached to the workpiece and the anode, usually made of graphite, is wrapped in cotton or a custom-shaped sponge.

The sponge or cotton anode is saturated in the solution containing the metal as an electrolyte. The saturated sponge or cotton wrap is passed over the surface of the workpiece, leaving a very thin deposit of electroplated metal. The solution is continually replenished, by dipping the cotton wrap or sponge into fresh solution. As the anode is passed over the surface a layer of metal is selectively "painted" onto the surface of the other metal. The metal is plated only at the area of contact of the anode to the cathode workpiece.

As the plating is applied, adjust the rectifier for proper current flow. If the current is too strong, the anode will heat up and evaporate the solution from the sponge.

Proper cleaning is essential. If the surface contains oil, grease, or other foreign particles, the plating will be deposited but will not adhere. Adhesion occurs on a microscopic level, one metal attaching, ion to ion, onto another metal. Therefore, it is critical to clean the workpiece effectively.

Resist and Other Decorative Techniques

Using the rectifier, it is possible to control the thickness of the plating or reverse the polarity and selectively remove the plating. There are resist films available for masking the workpiece surface. The films can be cut to develop decorative designs on the metal surface. The plating is applied within the boundaries of the film resist. The film is removed, and the plated design remains on the unplated background of the base metal.

It is critical to control the current when using resist films. High temperatures can remove portions of the film. When this occurs, the plating "bleeds" through to the regions being resisted. This requires removal of the plated metal.

Resist films can also be applied to a plated metal surface. Reversing the polarity, making the part an anode and the graphite wand a cathode, will remove the plated metal from the exposed surface. An interesting decorative technique is to remove a metal that was previously plated and then re-plate with another metal. Remove the resist, and a metal surface made of multiple metals remains. The sharp distinction between the change from one metal to the other is not detectable to the touch.

The resist coating used to perform selective plating is a fast-drying vinyl lacquer material. Good plating resists have adequate adhesion to withstand the heat, chemicals, and the different cycles of plating. They also must be easily removed by peeling from the surface in large sheets. If the resist comes off in small shards, the cost of cleanup will be astronomical. Therefore, it is important to use a resist that will peel away in sheets.

Plating solutions are available for the following metals:

Cadmium
Chromium
Copper
Lead
Nickel
Tin
Zinc
Gold
Platinum

Chemical Coloring

After the metal surface has received a plating of another metal, the resist can be left on and additional chemical processes can be applied to develop intriguing colors on the surface. The chemical coloring process develops oxide layers and patinas on the plated metal surface. Removal of the resist leaves a dramatic contrast in metallic color on the surface.

8

Paint Coatings on Metal

INTRODUCTION

Currently, the major use of metal as an architectural material is in the painted form. More than 60% of all metal used on commercial and residential structures is painted with an organic or inorganic coating. Organic coatings are those with a carbon-base complex molecular structure, and inorganic coatings are glass or ceramic based.

The requirements for both coating systems are that they provide a corrosion barrier and color. The corrosion-barrier requirement may entail merely a single, continuous, unbroken film inhibiting contact to the base metal or a dual coating, which involves a base film of sacrificial material covered with a top coat to provide color and further protection. Sometimes additional layers are applied to develop enhancements, such as a metallic appearance or a clear coat.

Perhaps the main reason to paint metal is to give color. There are innumerable pigments that can be further enhanced by variations in gloss and texture. Metallic coatings can be created by intermixing a layer of metal particles or mica flakes into the pigments or as an intermediate layer between pigment coatings and top coatings. Clear top coats can be used to produce depth in the coating, giving a rich, vibrant, almost pearlescent appearance on a metal surface.

Paint films can be baked on to produce a thin, continuous shell over the surface of a metal. The sleek, smooth surface of the metal is translated through to the surface of the paint. The paint becomes one with the metal surface, thus enhancing the look of hardness and durability.

Beyond just protecting the metal from oxidation and chemical attack, many paints also provide a level of abrasion resistance. Paint coatings can improve surface hardness, and thick coatings can provide a barrier to wear and abrasion. Many tests on the durability of paints include application of falling sand as an accelerated weathering evaluation. Other tests drop weights on the painted surface to determine whether the paint cracks or whether it bends with the deformation of the metal.

Textures can be developed with some forms of paints. Very thick films can be designed to "wrinkle" as they cure. This produces a unique skinlike texture on the paint finish. Because these wrinkle finishes are developed with very thick paint coatings, the durability is superior.

HISTORY

Among the early uses of paints was the application of varnishes to bronze statues. The Romans decorated their temples with bronze pillars adorned with various paints to add color and dimension.

Paints were later used on architectural metals as lead-rich or zinc-rich coatings over cast iron sheets, then as terne-coated steel sheets. Many a restoration project has uncovered the red-lead-painted terne roof. This primer coating also acted as a finish coating and provided superior corrosion resistance. The coatings were applied by brushing or rolling the paint over the surface.

The twentieth century was a time of new dimensions in painting metal. The development of aluminum and the advent of the automobile required a more specialized approach to paint finishing. Probably the single most important driving force in developing paints with various pigments was the automobile. Many people recall seeing the sleek black finish on a Model T Ford. Eventually, after the flood of black cars began to overrun the roads, the automotive industry sought to offer a variety of colors to the market. This demand developed new pigments and resins, and more efficient paint application systems were devised.

In 1938, Teflon was developed by the DuPont Company. Teflon was the name given to a very inert compound, called a fluorocarbon. Later, this incredible resin was modified to develop various fluorocarbon paint finishes, most notably the Kynar 500® coating.[1]

Continuous coil coating came into play during the 1940s. This method of applying paint to metal has greatly reduced the cost and enhanced the quality of paint finishes on metal. Coil-coated metal sheet now was available to the siding industry, in which large panels were roll formed into corrugated shapes. The dimensions of metal use expanded exponentially. Prefinished metals with hundreds of color choices were now available to the designer.

Soon after some of these first architectural projects were constructed, problems with color retention began to surface. This problem was also

[1]Kynar 500® is a registered trademark of Autochem North America Inc. There are a number of licensed Kynar manufacturers, which market their products under various trade names.

apparent in the automotive industry. Demand for better performance grew. A building's color chalking and fading, or worse yet, peeling, became a major factor in the development of quality paint finishes for metal.

APPLICATION TECHNIQUES

Paint coatings are applied to metal through two main techniques, spray coating and coil coating. Other techniques common in the coating industry, but not so common in architectural applications, are curtain coating, which applies paint by running a part or sheet below a falling "curtain" of paint, and dipping, which requires the immersion of the part or sheet in a bath of coating material.

Spray coating techniques apply the coating to metal on a per part basis. A part can be a fabricated assembly, or it can be a sheet of metal. Spray coating processes often use a conveyer system with the parts suspended and passed through a booth where the paint is applied by a spray nozzle. The spray can be wet or powder. The clean and chemically treated metal is suspended from a conveyer system or racked to receive a primer coat and a finish coat. Dust-free enclosures are necessary to keep dirt and other suspended particles from landing on the newly coated parts. Wet coatings can be either air dried or cured in an oven. Air-dried wet coatings can be applied to parts in the field, as well as in a shop. Precleaning and chemical treatments to the metal surface prior to paint application are critical.

Wet coatings can produce variable thicknesses across a part. The success of these coatings is highly dependent on the ability of the applicator. Variations of coating thickness can be problematic. Inconsistent appearance from one part to another, or across a large flat area, can be a major detriment to end quality results. Repainting to improve coverage can emphasize the variations if proper care is not exercised.

Powder coatings are applied to metal using a conveyer or metal suspension rack. These coatings require an electrostatic charge to be applied to the part. The powder is made up of small spherical paint resins, which receive a charge as they exit the spray gun nozzle. These particles adhere to the part, which has been given an opposite charge. As the powdered resins exit the gun, they cover the part evenly by adhering to regions on the part not yet covered. Powder that does not attach itself falls to the bottom of the paint chamber where it can be recycled and reused. An interesting and useful characteristic of powder coatings applied electrostatically is the way the small paint particles will wrap around a charged metal part. Paint powder that misses the part can come back and attach itself. Hidden recesses can be coated with powder attracted to the concealed charged surface (see Figure 8-1).

Once the part is covered with the small spherical particles of resin, it is passed into an oven where high temperature fuses the particles together into a continuous coating. The thickness of the coating is very controllable, and repairs can easily be made by grinding or sanding areas and recoating.

Powder coatings will not drip, sag, or run. The ease of application reduces the skill level required of the paint applicator and reduces waste through its recycling ability. These benefits, coupled with the absence of volatile organic compounds, decrease overall cost. Urethane, polyesters,

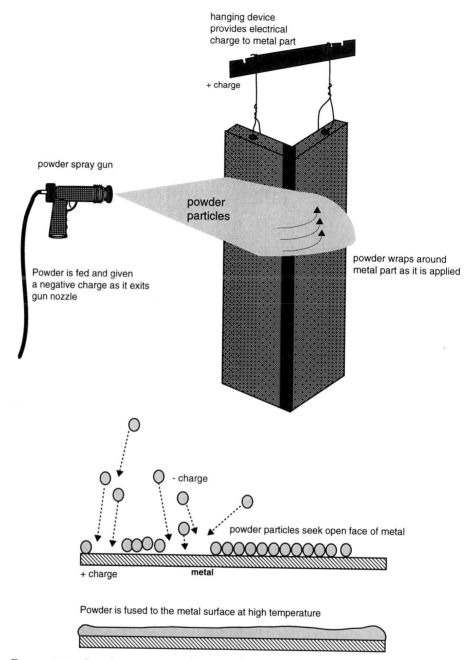

Figure 8-1. Powder-coating application of paint.

epoxies, and a limited number of fluorocarbon coatings can be applied as powder coatings. Because of their environment-friendly nature and reclaim system, powders are a growing choice for the future.

Coil coating is the technique of applying paint on a continuous ribbon of metal. A coil of metal is passed through a continuous processing system, which cleans and applies both the primer and the top coat by roller coating or under spray nozzles.

Coil coating techniques are the applications of choice for sheet metal. Coil coating is performed on thin sheet metal while it is still in coil form. The

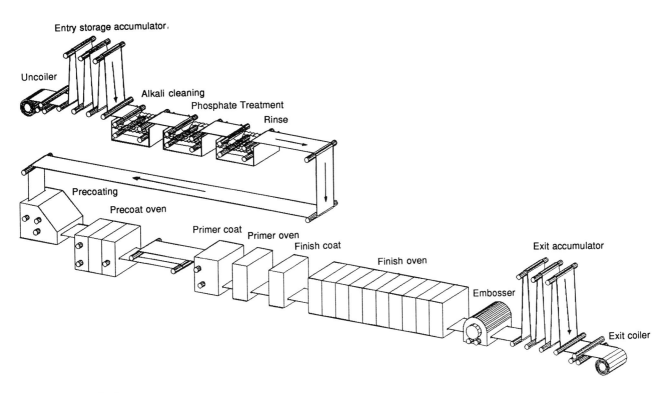

Figure 8-2. Coil-coating operation.

precoated metal is recoiled after application of the paint and later sheared to sheet sizes to undergo further fabrication, or roll formed into various shapes. The paint used in the coil coating process must be flexible and adherent enough to allow the metal to be postformed.

Coil applications use clean, chemically treated metal. Often the cleaning and chemical treatment occurs in line with the final paint application. The ribbon of metal passes through a point at which oxides and oils are removed by alkali cleaning processes or other cleaning processes. Next, it passes through a chemical treatment process where, depending on the metal used, phosphate or chromate solutions are applied to develop conversion coatings. The conversion coatings are necessary to ensure that the paint will adhere to the metal. Then a precoat primer is applied. The primer acts as the bridge coating between the final finish and the chemically treated metals. Primers can provide improved corrosion-resistant characteristics to the base metal or improved abrasion resistance. Typically, many final coatings will not adhere directly to the metal but will adhere tenaciously to the primer coating.

The primer usually passes through an oven in a curing process to remove solvents used in the primer application. This is critical; all solvents must be removed from the primer coat prior to the application of the finish coating. The primer coating is typically applied to both sides of the metal; this coating does not provide the pigment or finish color desired.

The finish coating, containing the desired color pigment, is next applied using a reverse roll that passes through a trough of paint and deposits the paint evenly across the surface. The finish coated metal is passed through a second oven to cure the paint, removing solvents from the coating. The coated metal ribbon is rewound as it exits the oven.

Intermediate techniques to improve quality are intermixed through the processing of the coil. Often, at the beginning end of the coil, a leveling system to flatten and stretcher level the metal is used. Stretcher leveling pulls the metal sheet from both ends until the yield strength of the metal is surpassed. This removes buckles and warps within the sheet. An accumulator, which maintains the speed at which the metal is passed through the process is also utilized. As different processes interrupt the timing of the sheet metal through the process, the accumulator expands and contracts to maintain the rate of the metal through the ovens.

Inspection systems are included throughout most processes to ensure that film thickness is consistent. These systems use nondestructive methods to verify that film thickness does not vary beyond certain standards across the sheet or along the length of the coil. Inspection systems can also incorporate visual comparisons with color chips for color matching.

Steps in the Coil Coating Process

1. Set up coil of metal.
2. Prepare metal.
   a. Clean the surface by degreasing in a hot alkaline bath.
   b. Dry the surface.
   c. Depending on the metal, use one of the following processes:
      chromate conversion (aluminum)
      iron phosphate (steel)
      oxide or phosphate treatment (galvanized)
      hot water rinse (tin)
3. Apply coating. This could involve first a primer coat, then (after cooling) a finish coat, with sometimes a third clear coating application.
4. Cure and cool.
5. Emboss or use other special processing treatment such as tension leveling.
6. Exit and recoil.

ENVIRONMENTAL CONCERNS

Lead-based paints and primers should not be used on architectural metal products. Eventual restoration may require removal of these lead paints, and this creates potential hazards resulting from dispersed airborne particles.

Today the primary hazard associated with paints used on metals is high-level volatile organic compounds, known by the acronym VOC. VOCs basically are solvents that are used to transport paint pigments onto metal surfaces. The paints are applied wet, then the solvents are evaporated, leaving the solid resin coating behind. These solvents are identified as a major cause of ozone depletion. As such, the EPA has moved to require industry to eliminate high VOC emissions. The current standards limit volatile organic compound emissions to 3.5 pounds per gallon of coating (0.42 kilograms per liter).

Many continuous coil coating operations have systems in place to control VOC emissions. Spray coating operations, on the other hand, are not as controllable. Because of the larger volume of space required, spray coating operators must consider alternatives to solvent-based coatings.

One alternative in the spray coating industry is the use of paints with high solid levels and reduced solvents, such as the powder coatings. These highly versatile coatings are electrostatically applied. Overspray is collected and often recycled. Very little of the volatile substances is released into the atmosphere.

Another alternative is water-based coatings, which use water as the carrier of the paint resin solids. These wet-applied finishes offer good flow characteristics with very low levels of solvents. Water-based coatings have been used for years as latex paints. The problem was in the long-term performance of these paints. Improved architectural qualities of chalk and fade resistance have developed from new acrylic resins using heavier, denser solids. Epoxies, with good durability, can also be applied with water-based coatings.

TYPES OF PAINTS USED ON METALS

There is a tremendous variety of paint and paint primers available for metal. Each type has certain advantages in relation to other paint formulations. However, architectural requirements impose major demands on the performance of paint finishes. Initial demands, such as color choice and gloss, limit some of the choices; often, they also limit the designer's options if long-term performance is a major end requirement.

There are at least two necessary components required for a successful paint application to metal: the base coat, or primer coat, and the top coat. Without a good primer coat, the top coat is worthless. Even the most expensive finish coating can peel or prematurely deteriorate if the primer is of poor quality or ineffectual.

PRETREATMENTS

Primers, in and of themselves, are the least effective means of protecting a base metal. Primers should be applied over surface pretreatments that enhance the corrosion resistance of the base metal and work to ensure proper paint adhesion to the surface. The chemical pretreatment of the metal surface is the most important factor in the long-term performance of a coating. Without an adequate pretreatment, you will not get the full service life out of the topcoat.

Proper pretreatment requires that the metal surface be as clean as possible, free of all dirt, oil, grease, excess oxides, moisture, fingerprints—everything. Consider preheating the metal, particularly aluminum, to ensure the elimination of moisture.

Cleaning can consist of acid treatments or alkali treatments depending on the metal and the severity of the cleaning requirements. Solvent degreasing is also used, but less frequently because of the damage some solvents pose to the environment. Mechanical cleaning by abrading the surface is also a technique used infrequently because of the surface texture it creates. Mechanical cleaning consists of shot blasting the surface or abrasive grinding.

Typically, steel and galvanized or zinc-alloy-coated steel or aluminum are the metals considered for paint coatings. Each of these requires a specific pretreatment to the metal surface to ensure paint adhesion and performance. Pretreatments enhance the corrosion resistance of the metal surface. The most commonly used on aluminum are chromate pretreatments. For steels and galvanized steels, phosphate pretreatments are common.

For aluminum surfaces, a conversion coating is necessary to convert the naturally forming aluminum oxide into a nonreactive, amorphous surface. Usually a chromate conversion coating is created after the aluminum surface is cleaned.

For aluminum, cleaning involves using either an etching solution, which takes advantage of aluminum's solubility at high pH, or a nonetching solution, which works as a detergent. Etching solutions often include sodium hydroxide, sometimes coupled with other alkali cleaners. These solutions will remove a small portion of the oxide film and leave the surface clean and porous. Nonetching solutions usually consist of metasilicate formulations that do not remove any aluminum surface but work to remove dirt and foreign particles.

After cleaning, the aluminum is subjected to an acidic chromate treatment, which develops chromium phosphate, chromium chromate, or zinc chromate. Each of these develops a barrier conversion coating that enhances final top coat adhesion and provides a level of additional corrosion resistance. The chromium chromate conversion coating offers good paint adhesion and excellent additional corrosion protection. This pretreatment leaves an irridescent gold color on the surface. The chromium phosphate pretreatment provides good paint adhesion but does not greatly enhance the corrosion protection. This pretreatment leaves a green color on the aluminum surface. Environmental concern with chromium waste products is developing into a major problem. No-rinse chromium pretreatments are being developed. These pretreatments reduce waste and provide good adhesion for paint applications. Additionally, nonchromium pretreatments are in development to elminate chromium altogether. These pretreatments, however, have not proven to perform as well as the chromium chromate treatments.

For steels and the zinc-coated steels, pretreatment is necessary to develop a surface that will allow paint adhesion. Paint will not stick to untreated galvanized steel. For steel, a conversion coating is necessary to develop improved corrosion resistance, particularly at areas where the paint may be undercut by corrosion, such as edges or scratches.

The typical cycle for pretreatment of steels and the galvanized steels is:

Clean, usually with some alkali formulation such as sodium hydroxide.
Water rinse.
Apply conversion coating.
Water rinse.
Seal.
De-ionized water rinse.

The conversion coating is an acidic bath containing primarily phosphoric acid along with some heavy metal salts used to impart various beneficial characterists to the surface of the metal. There are various levels of conversion coatings used on steels and galvanized steels. Ranked on overall corrosion protection from least to most favorable, the following list consists of common conversion coatings. This ranking also matches relative cost.

Five-stage iron phosphate
Zinc phosphate
Zinc-calcium phosphate
Zinc-nickel-fluoride phosphate
Zinc-nickel-manganese-fluoride phosphate

The phosphate crystals develop on the surface of the metal by a form of cathodic deposition. Areas not covered with the crystals attract the phosphate out of solution to form a barrier coating. Under a microscope, this barrier coating looks very rough as the crystals pile every which way. This roughening provides a better surface for paint adhesion and inhibits corrosion of the base metal surface.

After the conversion coating is developed, all excess solution is rinsed from the surface of the metal. The course, rough phosphate surface is sealed. The sealing of the surface is also very important to long-term performance. Sealing usually entails dipping the sheet in a chromium solution to develop a thin, complex layer of various chromium compounds. In an effort to eliminate or reduce toxic chromium from much of the coating industry, various organic coatings are being produced to develop bridge coatings between the phosphate surface and the finish top coating of paint.

PRIMERS

Primers are the barrier coatings used to protect the base metal. Only good quality primers should be considered for architectural metal uses. Primers should contain some additional corrosion-inhibiting compound such as zinc or chromium.

Primers use basically the same binders as the top coatings, with the exception that fluorocarbon coatings are never used as primers. Epoxies make excellent primers because of their hardness. Zinc dust can be added to epoxy primer mixes for additional protection.

COATING TYPES

There are two main families of paint coatings used on metals, inorganic and organic coatings. The organic coatings are the predominant types used on architectural metals. These are the coatings with a complex, carbon-base, molecular structure. Within this family, there are a number of coating groups that have various performance characteristics. Each group is described by its binder. A binder is a liquid or resinous material used to carry the pigment, as well as other additives, to the metal surface. The binder is the adhesive that

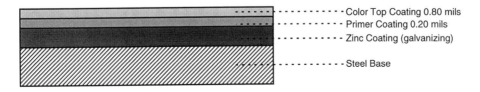

Figure 8-3. Cross section of primer and finish coatings on steel.

holds the coating together and produces the performance properties of the coating. Binders are also considered polymers. *Polymer* is the name given to giant chain molecules of a carbon base. Lacquers are polymers dissolved in various solvents. The solvent evaporates by forced heating in an oven or upon exposure to the atmosphere. When the solvent evaporates, it leaves behind the binder to develop a continuous coating of polymer. These solvents, during evaporation, expel large amounts of volatile organic compounds (VOCs) into the atmosphere. As mentioned, it is these volatile compounds that effect environmental damage if not collected during the drying or evaporation of the solvent.

The various binders and some of their major architectural performance characteristics are described in the following paragraphs.

Alkyds

The organic coatings, known as alkyds, or amine-alkyds, are inexpensive coatings, used on light-duty products, offering good durability. These coatings were common metal coatings in years past, but have been widely replaced by acrylics and urethanes. The alkyds can be formulated in many different ways to develop unique performance characteristics. They are essentially polyesters modified with a fatty acid derivative.

The alkyds require curing at moderately high temperatures to remove solvents such as benzene and naphthalene compounds, or they can be formulated into air dry-maintenance coatings. They can be used in postforming applications, but only in those of moderate levels. These coatings will age, lacking good color and gloss retention. Silicone-modified alkyds will develop better performance characteristics, and the addition of a clear top coating will improve color and gloss retention. Another benefit of the silicone-modified alkyds is that they can be formulated into air-dry coatings with good durability.

Vinyl derivatives to the alkyd coatings improve flexibility and forming characteristics. Vinyl alkyds are used where postforming operations are necessary but further sacrifice fade resistance. Alkyds are available in solvent-base, water-base or high-solid formulations. They are not available as powders.

Phenolics

The phenolics are organic resins that are formulated from phenol and formaldehyde. They are not typically used for architectural coatings. These organic resins offer good stain resistance and can provide adequate forming characteristics when coated over vinyl prime coats.

Acrylics

The acrylics are common architectural coatings. Acrylic coatings were first developed in the 1930s, but it was not until their formulations improved in the 1950s that they became popular as exterior finishes. There are many variations of acrylics used in both coil and spray applications. Thermoset acrylics are inexpensive coatings that offer excellent color and gloss choices. These coatings lack postformability, but have good durability. These are baked-on coatings, considered as baked enamels, that can be applied using both solvent-based or water-based carriers. Such coatings are used on exterior applications, but they will fade over time.

Silicone-modified acrylics show improved exterior performance. The silicone derivatives improve fade characteristics over the basic thermoset acrylics, but lose some of the already poor forming and flexibility characteristics.

Because these finishes are water based, they are finding wider usage as postforming coatings. Environmental pressures on the use of solvents have made these finishes relatively more economical in the overall application process. Acrylics are also available in high-solid paints and powder coatings.

Acrylic urethanes have been developed to provide further enhancements to this resin. This urethane derivative makes for excellent durability. Clean coatings made from this paint offer superior performance as coatings for brass and bronze work.

Epoxies

Organic coatings made of epoxy resins offer superior durability and impact resistance. They can also provide good formability and flexibility. These coatings are used as binders in powder-applied coatings and provide an excellent primer base for the more expensive exterior coatings. Epoxy coatings will deteriorate rapidly when used as finish top coatings on exterior metal applications and will yellow with exposure to ultraviolet radiation.

Epoxy coatings can develop very thick films with excellent adhesion to properly prepared surfaces. When epoxy coatings are combined with polyamide resins, zinc dust can be added to improve corrosion protection. Coupled with a finish coating of a fluoropolmer or polyester, epoxy primers deliver one of the more superior finishes available.

Interior applications of epoxy coatings can present various colors and textures. On interior metal surfaces, epoxy offers one of the most durable coatings possible on ornamental metal. Such surface coatings will resist abrasion and chemical attack. These coatings are very heat resistant.

Epoxies are available in conventional solvent-base coatings, as water-borne formulations, high-solid formulations, powder coatings, and two-component coatings.

Vinyls

Vinyl coatings are organic coatings, supplied as plastisols or organisols. The vinyls are solid resin particles dissolved by a plasticizer, which is basically an acid ester. The coating is applied in a thick solution containing the resin particles. The metal part being coated is then heated to fuse the solid particles into a continuous film. Patio furniture is often coated with these thick plastic coatings.

Vinyl plastisols do not use a solvent carrier, whereas the organisols do. These thick coatings provide excellent mar resistance and good corrosion and chemical resistance. They will not work well on large flat surfaces, because they tend to sag and are prone to tearing. The plastisols can be pigmented with almost any color available, but offer only moderate color retention and fade resistance.

Vinyls are available in conventional solvent formulations, as well as in water-borne, high-solid, and powder formulations.

Urethane Enamels

There are various types of urethane coatings used on architectural metal products. Acrylic-modified urethanes and polyurethanes are common architectural finishes sought for their wide range of colors and glosses. Often considered for their ability to provide the bright primary colors, urethanes can be applied wet as spray or coil coatings, or as powder coatings. Urethanes can be thermoset coatings, or they can be modified to cure at low temperatures. They can be coated over modified epoxy-base primers for better durability and hardness.

Urethane coatings provide excellent chemical resistance and good durability. Direct exposure to sunlight will chalk and fade urethanes to a minor degree in approximately 5 to 10 years. They can be recoated in the field using air-dry derivatives.

Urethane coatings are usually applied via two-part catalyzed systems. The parts are mixed together to activate certain polymer derivatives while inhibiting others. The shelf life of many of these coatings is limited.

Urethane coatings are also available with metallic flake additives. The high-gloss ability of urethanes make for very striking colors when metallic flakes are imbedded.

Polyesters

Polyester paints are organic coatings that possess good weatherability properties and are available in a wide variety of colors. They can be applied wet or as powders. Wet coatings can be applied to coil or fabricated parts. Polyester coatings are actually modified alkyds. They have good durability and when modified with silicone, they can perform well in exterior applications.

Silicone-modified polyester coatings were widely used during the 1960s and 1970s as coil coating paints used on sheet metal siding. The coatings will allow moderate forming into the various corrugated configurations. As the fluorocarbon coatings were developed and application techniques were enhanced, their superior weathering characteristics came to be in greater demand, eventually replacing polyester as the coating of choice for exterior metals.

Polyester coatings will not resist ultraviolet weathering as well as the fluorocarbons; however, they are available in medium to high glosses and as powder coatings. In Europe polyester coatings are widely used in high-quality architectural applications. The European polyesters are modified with triglycidal isocyanurate (TGIC) and applied with powder application equipment. These TGIC polyesters have shown superior performance in exterior exposures across Europe and Africa. There is environmental concern regarding the

TGIC compounds, and different formulations, using hydrooxyalkylamide as a less toxic modification, are now being considered.

The polyesters are also available in metallic colors. As with the urethanes, their high-gloss characteristics can enhance the metallic appearances.

Polyester coatings are available in solvent-base and high-solid formulations, powder coatings, and two-part liquid coatings. They are not currently available in water-borne formulations.

Pearlescent Additives

There are a number of pearlescent additives that can be used to enhance urethane and polyester coatings. Pearlescents are pigments that exhibit color through the optical effect of thin-film light interference. These finishes are top-of-the-line automotive paint additives used to enhance the use of a clear top coat. Such coatings are similar to metallics, with the exception that the pearlescent pigments act as optical filters that both reflect and transmit light.

Common pearlescent pigments are titanium-dioxide-coated mica and iron-oxide-coated mica. Light travels through the clear coating and reflects off the mica surface, part of the light passing through the mica surface to another surface, where the action is repeated. As the light is reflected, interference films develop, creating enhanced color. Interference is an effect that occurs when two or more light waves interact. The light waves travel by different paths back to the eye. If the waves are in phase, that is, of the same pattern and frequency, then they will combine to form a new wave with double the amplitude. When this happens segments of the light wave, or spectrum, are enhanced and thus show as red, blue, gold, or green interference colors. Interference does not destroy light, but only moves it about and separates it into its components.

Pearlescent finishes are very attractive. Their durability is dependent on the makeup of the basic coating. These are very high-gloss coatings and will show minor imperfections on the metal surface.

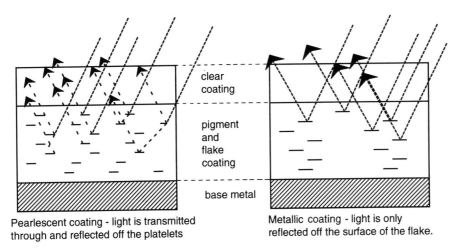

Pearlescent coating - light is transmitted through and reflected off the platelets

Metallic coating - light is only reflected off the surface of the flake.

Figure 8-4. Pearlescent paint finishes in relation to metallic finishes.

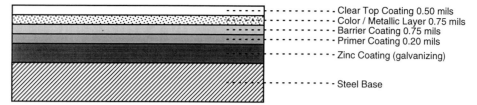

Figure 8-5. Cross section of primer, finish, and clear coatings used to generate metallics on fluorocarbon coatings.

Fluorocarbons

Fluorocarbon coatings are the premier exterior architectural coatings. The fluorocarbon resins, developed by the Pennwalt Corporation and known by the registered name of Kynar 500, are inert, nonreactive films. The Kynar 500 resin is used in formulating various paint coatings supplied by different paint suppliers. Some of the more common architectural trade names for these resins are:

Trademark	Producer
Duranar®	PPG Industries
Duranar XL®	PPG Industries
Hylar 5000®	Ausimont USA
Fluropon®	DeSota Inc.
Nebular®	Glidden Coatings and Resins
Optima®	Commonwealth Aluminum
Dubelar®	Dupont

The following formulation uses a fluoropolmer resin:

Lumiflon®	Asahi Glass Company

The fluorocarbon coatings do not readily adhere to metal and thus require proper pretreatment and priming. These resins use the strong bond between the fluoride and carbon atoms to develop very complex molecules. A continuous hard coating is developed on the metal surface by fusing the fluorocarbon resins into a continuous, unbroken film. The Kynar 500 resins are based on the complex polyvinylidene fluoride molecule (PVF2 or PVDF). They are formulated with various pigments and solvent carriers for wet spray or coil application. Most products use 70% Kynar 500 resins. This percentage within the formulation has shown superior performance under exterior application.

There are other fluorocarbon resins besides the PVF2. Among these is tetrafluoroethylene, also known as Teflon, which is often used to coat cooking utensils because of its high melting point and hardness. Teflon can also be used to coat metal substrates exposed to abrasion. Vinyl fluoride and chlorotrifluoroethylene are additional fluorocarbon coatings. Vinyl fluoride coatings are developed and used as architectural metal coatings similarly to the PVF2 resins.

The fluorocarbon coatings, which do not combine with oxygen, will resist water and oils. These coatings offer excellent fade and chalk resistance even in direct exposure to the sun. These finishes are currently available in low to medium glosses with colors limited to the earth tones; bright primary colors cannot be formulated. Metallic coatings and clear top coats are available to add further dimension to these coatings. The metallics utilize aluminum or mica flakes intermixed between color coatings and clear top coatings. Metallics require either three or four coats, rather than the basic two coats used in standard fluorocarbon finishes (see Figure 8-5).

In applying the metallics, a primer coat is first deposited, sometimes followed by a barrier coating of epoxy. Next, a color coating with the metallic flake intermixed is applied. Finally, a clear coating is distributed over the finish coat. Each of these coatings is oven baked to cure the solvent and form the continuous bond.

In the metallic coatings, there is a degree of directionality resulting from the alignment of the aluminum flake within the pigment coating. During the application of the metallic pigment, the small particles of metal or mica align themselves in the direction of the roll coating system. Applictions of spray-coated metallics have a reduced directionality. As the metal panel or sheet is viewed from different directions, particularly if one sheet is flipped in relation to another, a slight color difference is apparent.

The following standard metallic colors are available:

- Silver
- Bronze
- Copper
- Rose
- Blue

There are variations on these colors, depending on the manufacturer.

The fluorocarbon finishes have relatively high curing temperatures. When flat plate forms are finished, oven baking to the necessary cure temperatures can warp previously flat parts. There is an air-dry version of the Kynar coating for field finishing and touch-up. This finish has shown good color and fade retention properties, but long-term performance is dependent on in-field preparation and the variables of temperature and humidity that exist at the time of application.

The fluorocarbon coatings are superior to all other coatings in weatherability. Ultraviolet radiation and airborne pollution work very slowly on these inert films. If a fluorocarbon coating is applied correctly with a good, flexible primer and the appropriate pretreatment, the paint should provide more than 20 years of performance before minor defects begin to appear. These coatings are flexible and allow moderate forming to occur on prefinished sheets. Thin sheet metal, less than 0.063 inches (1.6 mm) can be brake press formed at a radius of approximately two times the thickness. Some small micro-checking of the surface may be apparent, particularly on aluminum. Micro-checking should not pose a problem if a good primer coating was applied and if the base metal has some form of galvanic protection, such as zinc in the case of steel. With aluminum, the micro-checking could be caused by the outer thin layer of aluminum oxide not being as flexible as the base metal. At these micro-checks, the paint should not come off. If it does begin to peel or flake, then an adhesion failure has occurred.

For thick aluminum sheet, such as 0.125 inches (3 mm) and heavier, consider V-cutting the back side. This will allow for a sharp corner and reduce the cracking at the paint surface.

Inorganic Coatings

The inorganic coatings are the vitreous coatings bonded to a metal substrate by fusion at very high temperatures. These coatings include the porcelain enamel coatings, or ceramic coatings, as some manufacturers call them. Basically, porcelain is a complex, glasslike substance composed of silica, borax, soda, and various metallic oxides. Such coatings are used on special low-carbon steels called enameling iron or enameling steel, aluminum alloys, copper, stainless steel, and a few other metals for jewelry purposes. The low-carbon enameling steels and the aluminum alloys make up the majority of metals with porcelain coatings used in architecture.

Porcelain coatings offer a wide range of colors and textures. The ability of porcelain enamel surfaces, basically glass in nature, to resist atmospheric chemical attack and ultraviolet radiation effects surpasses that of all other applied coatings—as long as the enamel is not breached. On the steels, porcelain protects effectively where the coverage is continuous and unbroken. If the coating is breached, there is no secondary zinc protection to fall back on. The steel will corrode severely and rapidly.

Porcelain enamel coatings are made of a material called *frit*. The frit is made of various components mixed thoroughly together to develop a variety of densities and characteristics. The components are melted in a smelter to temperatures of 2,500°F (1,371°C) and rolled thin. The thin sheet of molten hot glasslike material is passed through water-cooled rolls or deposited in water. The rapid shock resulting from the drastic temperature change shatters the substance into the tiny particles called frit. The frit is milled in water, with clay and coloring pigments added to create an emulsion called a *slip*.

For steel sheets, the slip is applied as a ground coat to both sides of the metal. The ground coat is of slightly different makeup than the top finish coat. The ground coat contains oxides that assist in adherence to the base metal and form a layer for the top coat to adhere to. Once fired, the ground coat is fused to the base metal, forming a permanent bond between glass and metal. The top coat contains the various pigments and additives to develop the texture and gloss. This coating is applied, and the whole mixture is fired a second time. Aluminum does not require a ground coat; the top coat forms the bond to the base material.

The slip is applied as a wet slurry film or as a finely ground powder using electrostatic adhesion. Once coated, the metal and film are fired at temperatures above 800°F (425°C) to remelt the frit and fuse the molten glass to the base metal. The temperature of most firings occurs in the range of 1,450°F to 1,550°F (788°C to 843°C) for steel and 950°F to 1,000°F (510°C to 538°C) for aluminum.

The metal can be fired and refired with decorative slips silk-screened over the top coat surface. The decorative surface can include signage or graphics of various colors. The density of these slips can be varied and applied over different sections of the top coat sheet. Once fired, these various densities will "float" to different levels, generating the illusion of depth.

The metal used is critical to the appearance of the porcelain surface. Porcelain is almost always applied to sheet material. The material must have undergone all necessary forming operations prior to enameling. All welding, piercing, and edge folding must have occurred prior to application of the slip and initial firing. Postforming will fracture the porcelain surface, rendering it useless as a protective coating. Preforming to tight dimensions or sharp corners will also fracture the surface by generating zones of high stress as the porcelain cools.

Thickness of the sheet is also highly critical if the now-formed panel is to retain its shape during firing. Typical thicknesses of panels are in the 0.075 inch (1.9 mm) to the 0.048 inch (1.2 mm) range. There are flat, unfolded porcelain panels available. These are laminated to a solid backing of some kind and require protection of the edge with an additive device, such as an aluminum extrusion.

With porcelain coatings, reds, as well as yellows, are difficult to achieve in consistent quality. This has to do with the clays involved. The red colors are thicker than other ceramic coatings. Generally, a base coating of white porcelain enamel is applied, then the red coating.

FORMABILITY ASPECTS

The ability of a coated metal surface to undergo postforming processes depends on the flexibility of both the top coating and the base coating. The distance of paint coating must stretch is another factor related to flexibility. Whether the formed edge is a gentle radius or a complete 180-degree fold makes a difference.

All paints designed for coil-coating applications usually are flexible. On rare occasions a primer coating may be too stiff or incorrectly cured to the point that it flakes upon postforming. Typically, the coil-coated paints are designed for maximum flexibility, because they are intended for a subsequent forming operation. In addition, coil-coated sheet is usually thinner. Thick sheets are not often coil coated. Typical thicknesses for architectural coil-coated sheet range from 0.0179 inches (0.45 mm) to 0.075 inches (1.9 mm).

Corrugated metal siding is formed from prepainted coil without any adverse effects to the coating. Sheet metal forms are often manufactured from prepainted sheets. Brake forming operations have little effect on the finish paint coating. Perforating, piercing, and some minor stamping operations can be performed on prepainted coil or sheet if the paint coating is flexible.

Spray coatings are usually applied to thicker sheet or prefabricated parts. Thick sheets, when formed after painting, often show small cracks in the coating along the bend line. The small cracks are indications that the paint coating could not withstand the strain of the forming operation and, thus, split. The paint should still adhere to the surface, however. If flakes of paint begin to release, this is a sign of poor adhesion either of the top coat to the primer or of the primer to the base metal. Sometimes the primer, rather than the top coat, is not capable of making the bend at the corners or the edges.

The same kind of cracks will occur, since the top coat is directly adhered to the prime coat.

Spray coating postfabricated parts is prone to producing variable thicknesses of coating. Close inspection of the coated surface for variations in gloss and color is necessary to ensure proper coverage. Often, signs of poor coating thickness will occur at folds or bends where access and direct spraying is more difficult.

The glasslike porcelain coatings are very hard and stiff. Postforming will not work. The surface will crack and flake with even simple shaping. These coatings require the parts to be completely prefabricated prior to any finishing operation.

CARE AND MAINTENANCE

Both organic and inorganic coatings require little in the way of maintenance. Generally, external applications of painted metal will self-clean from natural rains. If the need arises and soot has accumulated over the surface, consider first mild detergents and a water blast. If soot and stains remain, use a solvent only after testing a region to ensure that it does not dissolve some of the coating.

If sections of the coating have begun to flake away from the metal, then wire brush all loose paint or sand blast the region to remove all loose paint and apply a corrosion-resistant primer. For steel, use a zinc-rich primer or an epoxy primer with zinc dust. Apply a compatible paint top coating. A good latex acrylic will work adequately in mild atmospheres, and a urethane coating will work well in more corrosive environments.

Scratches through the paint film can be a problem when working with metal at a project site. Try to keep the metal covered with a protective PVC film with a low tack adhesive or with electrostatic adhesion. If the inevitable happens and the paint has been scratched, but not severely enough to require total refinishing, consider the following steps.

If the metal has an oven-baked finish, either coil coated or spray coated, any field-applied paint will require air drying. Air-dry paints will not exhibit the same performance as baked-on finishes, even air-dry Kynar paints. What

TABLE 8-1. Performance Comparisons of Various Paints Used on Metal

COATING TYPE	AS A PRIMER	CHALK AND FADE	GLOSS	DURABILITY	FLEXIBILITY	CORROSION RESISTANCE
Alkyds	Good	Fair	Full range	Good	Poor	Good
Phenolics	Good	Poor	Medium to low	Good	Poor	Good
Acrylics	Good	Fair	Full range	Good	Fair	Good
Epoxy	Excellent	Poor	Low to medium	Excellent	Fair	Excellent
Vinyls	Good	Poor	High	Excellent	Good	Excellent
Urethanes	—	Good	Full range	Excellent	Good	Excellent
Polyesters	Good	Good	Full range	Excellent	Good	Excellent
Fluorocarbons	—	Excellent	Low to medium	Excellent	Good	Excellent
Porcelains	—	Excellent	Full range	Excellent	Inflexible	Excellent

TABLE 8-2. Comparable Cost and Weatherability of Metal Paints

COATING TYPE	RELATIVE COST	AIR-DRY CAPABILITY	POWDER FORM	PROBABLE EXTERIOR SERVICE LIFE (ASSUMING GOOD PERFORMANCE)
Alkyds	Low	Yes	No	5 years
Phenolics	Low	Yes	No	5 years
Acrylics	Low	Yes	No	5 to 10 years
Epoxy	Moderate	Yes	Yes	Will yellow in short time
Vinyls	Moderate	No	Yes	5 to 10 years
Urethanes	High	Yes	Yes	10 years
Polyesters	High	Yes	Yes	10 to 15 years
Fluorocarbons	High	Yes*	Yes*	20 years+
Porcelains	High	No	No	40 years+

* Limited

will occur in a few years is variations in fading and chalking between the baked-on surface and the air-dry surface. The air-dry surface will show chalking and fading to a number of levels greater than the baked-on surface. If the original panels were dark in color, an air-dry touch-up paint may match initially, but after a few years will be lighter. Therefore, if the scratch is minor, perhaps it may be better to leave it on a dark surface. If the surface is light in color, place paint only in the scratch, wiping all excess paint from the surface.

The fluorocarbon coatings, when scratched, can be touched up with the air-dry versions or a good urethane or acrylic paint. Again, keep any touch-up to an absolute minimum. Excessive touch-up will ruin the appearance of the coating after a few years of exposure.

Refer to Tables 8-1 and 8-2 for comparisons of the performance, cost, and weatherability of the various coatings used on metal.

FIELD COATING OF METAL

Powder coatings are not typically considered for field applications. Powder coatings are available in thermoset coatings which require a catalyst to be present during the heating process. The requirement for a catalyst and the overall heating of the part to relatively high temperatures makes the thermoset coatings impractical, if not impossible, for most field applications.

Powders are also available in thermoplastics. Thermoplastics of the ethylene methacrylic acid polymer variety can be applied to metal without a catalyst. These powder coatings can be applied to a heated metal surface using a propane-fueled flame-spray nozzle. The powder enters the flame as it exits the gun. In the flame the powder melts and is sprayed on the preheated part. The resulting thermoplastic film is approximately 10 to 15 mils (0.25 to 0.38 mm) in thickness.

These acrylic thermoplastic coatings are very durable and chemical resistant. The thick build adds durability. The preheating of the metal is desirable to eliminate moisture and to aid in adhesion of the thermoplastic. The preheating is only to the range of 170°F to 190°F (77°C to 88°C). Other benefits to the thermoplastic powder coatings are they can be recycled and they use no volatile compounds as carriers. These coatings offer decent exterior exposure performance. Fading is the primary concern.

Air dry Kynars such as PPG's Duranar ADR have superior fade resistance but are difficult to apply. Used more frequently as repair coatings for Kynar coated metals, these coatings must be applied by persons trained and authorized by the paint manufacturer.

There are other paint coatings such as the urethane and acrylics previously mentioned which can be applied as air dry coatings. As with shop applications, the most important step is initial preparation. Scratching the surface, roughening up the surface and removal of oxides is critical for a successful application. For galvanized steel, leaving it exposed to the atmosphere for a few weeks or months will actually improve paint adhesion. Newly galvanized steel needs to be etched and primed before paint will adhere. Aluminum is similar. Roughening the surface will aid in paint adhesion, whereas anodized aluminum will receive paint well. Prepainted surfaces should be roughed and scoured to remove all loose paint. A primer should be used to aid in adhesion of the finish paint. If rust is apparent, complete removal and cleaning of the surface is necessary. Apply a rust-inhibiting primer such as zinc chromate or one of the rust inhibiting epoxy primers.

Precautions with Field Painting

The obvious precautions dealing with dust and debris embedding in the paint surface must be addressed for any successful field application of paint. Other than the flame applied thermoplastic acrylics, most field paint applications cure slowly. You cannot heat them to speed the drying of the paint. Consider the waterborne paints for field application. These apply easily and do not damage the environment as the water base evaporates.

Field touch-up of baked-on enamel finishes can result in major aesthetic drawbacks as the paint surface ages and weathers. The air-dry paint may match the coil coated or baked-on finish when initially applied, however, after the paint has been exposed to the weather, variable chalk rates occur and the touch up paint can show like the black spots on a Dalmatian. Limit the touch-up painting to the metal scratch itself. Do not spread the paint out or try to feather the appearance. With certain occurrences, long term appearances may best be achieved by not applying touch up paint.

Paints Which Develop Patinas

There are some paint systems which actually will mimic the weathering of the natural metals. These relatively expensive paints contain pure ground metals in an acrylic resin suspension.

The color tones achievable are:

patina green tone
blue patina tone
iron rust
black rust patina

mottled gold
mottled bronze
mottled silver or pewter tone

The metal particles suspended near or at the surface will weather both naturally or artificially, or they can be coated with a clear acrylic sealer to maintain a metallic appearance. As the coatings weather they simulate the natural weathering which occurs when the pure metals are exposed to the atmosphere. The surface tends to remain more mottled than the pure metal form. These coatings require priming of the base metal. Keep in mind a noble metal in a suspension is being applied to another metal, less noble. Therefore, extra care should be considered to reduce or eliminate the contact between the dissimilar metals.

COLOR MATCHING

Attempting to match colors can be madness. Paper paint chips are intended to show what the paint will look like on a metal. However, metal surfaces are not the same as paper, and the process of applying colors to paper is different from applying colors to metal. Require that all paint samples for color matching be on metal. If the metal is to be embossed, then specify that the paint samples are to be embossed.

If different sources are to supply products finished in like colors, then have them each prepare samples. The processes of formulating the paint, applying the paint, and baking the paint will produce enough variation from one manufacturer, or coating facility, to another that the final colors will not match precisely.

TROUBLESHOOTING

Problem	Solution
Dust in coating	Need a quicker-drying paint or a dust-free chamber. Generally, this occurs on air-dry paint applications.
Runs in coating	This problem occurs in wet coatings. Increase solid content or decrease solvents. Take better care in application.
Areas uncoated	If the paint is applied electrostatically, such as with a powder coating, the problem may be caused by a faraday effect. This will occur in tight corners or around perforated holes. The electric charge will repel the paint instead of attracting it to this area. Mask over the hole at the back side, or add a small strip of metal to temporarily modify the electromagnetic field causing the faraday effect.
Paint peels off in sheet	Primer failure. This can occur when the base metal is not correctly pretreated. The paint and primer must be removed, and the surface must

Problem	Solution
	be recoated. Verify that the correct primer is being used. Verify that the correct pretreatment is being performed.
Paint peels in shards	Top coat failure or primer not cured properly, preventing good adhesion to the top coat. The paint and primer must be removed, and the surface must be recoated. Verify that the correct primer is being used. Verify that the correct pretreatment is being performed.
Metal distortion shows through	Consider a low-gloss paint. Metal needs to be block sanded. High-gloss paints will show all variations in sanding and grinding.
Cracks along bends	Bending radius to tight. Increase radius.
"Orange peel" texture	Paint coating is too thick. When curing, the paint congeals slightly, creating variations in thickness. This can occur with porcelain coatings as well. It can also happen on porcelain when the slip is mixed incorrectly.
Porcelain flaking off aluminum extrusion	Incorrect aluminum alloy. Surface not rough, no "tooth."

TESTING AND PERFORMANCE CRITERIA

There are all sorts of testing criteria used to determine the performance characteristics of the various coatings. In the United States, the American Society of Testing Materials (ASTM) criteria provide methods of testing to validate certain coating performances. These criteria indicate what happens; they are not a pass-fail type of analysis. Among the more common tests are the following:

ASTM D-2247. This is a humidity test of 1,000 hours and 2,000 hours.

ASTM D-1735-62. This is a humidity test of 1,000 hours.

ASTM D-523-67. This is a gloss-determination test read at a 60° angle to the surface.

ASTM D-2244-68. This is a color uniformity test. Readings are taken both visually and with instruments.

ASTM D-1400-67. This is a dry film thickness test to determine the finish thickness on a metal sample.

ASTM D-3363-74. This is a hardness test. Results indicate the hardest pencil that will leave the film uncut. The test is indicative of a coating's resistance to scratching and marring.

ASTM D-2794. This is an impact test. It will indicate a coating's ability to withstand rapid deformation resulting from impact. A $9/16$-inch-diameter ball is dropped with a force of 160 pounds.

ASTM-1737. This is a bend test. A coated metal sample is folded 180 degrees over a 0.125-inch-diameter mandrel. The test gives an indication of the coating's flexibility and its ability to withstand forming processes.

ASTM D-968. This is the falling sand test. It records how many liters of falling sand can be applied before the primer is exposed, and then how many liters before the base metal is exposed. This is an abrasion-resistance test of the top coat and the primer coating.

ASTM B-117. This is the salt spray test. The coated metal sample is exposed to a minimum 1,000 hours of continuous corrosive spray. It indicates a coating's ability to withstand corrosive salt exposures.

The American Aluminum Manufacturers Association has developed a series of specifications for coated aluminum products. These indicate the level of performance that a particular exterior coating will achieve, or the minimum criteria an exterior coating must meet.

There are two basic specifications relating to a coating's exterior performance. AAMA 603.2 is for general light-duty paint finishes on aluminum. AAMA 605.2, "Voluntary Specification for High Performance Organic Coatings on Architectural Extrusions and Panels" contains the more rigorous criteria. The following list compares the two:

	AAMA 603.2	**AAMA 605.2**
Uses	Interior, light commercial	Exterior
Thickness of coating	0.8 mil	1.2 mil
Pretreatment required	None	Conversion coating
Primer coating	None	Primer required
Abrasion resistance	None	ASTM D-968, falling sand
Chemical resistance	Muriatic acid, mortar	Muriatic acid, nitric acid
Color retention	1 year Florida exposure	5 years Florida exposure
Chalk resistance	None	Maximum #8
Film adhesion	Dry and wet adhesion	Dry, wet, and boiling water
Erosion resistance	None	Less than 20% after 5 years Florida exposure

If a paint coating meets or exceeds the requirements set forth in AAMA 605.2, then it should perform well in most exterior exposures. The fluoropolymer coatings will usually meet or exceed these requirements, as will some of the modified polyesters. Most urethanes and acrylics will meet or exceed the AAMA 603.2 requirements. There are a few powder coatings developed that will meet the requirements of AAMA 605.2.

All coatings should be investigated for their performance in the environment in which they are to be exposed. Be certain that the sample coating is tested with the primer and the base metal of the finish product. Performance criteria will vary, depending on the pretreatment used and the primer coating.

General Material Specification

This appendix contains basic information for consideration by the specification writer. Various metals and the forms typical of the metal are presented. The specification lists the alloys generally considered for most building construction applications. Other alloys and the various tempers as listed in the text can also be considered.

The material specifications as provided by the American Society of Testing Methods, United States Federal Specifications, or British and German specifications, when available, are also listed for reference. Basic finishing requirements, as well as basic handling, storage, fabrication, and installation criteria are indicated as a guide. Refer to the full discussion of each metal in the text for additional limitations and requirements and apply where appropriate.

METAL: ALUMINUM

FORM: CASTING

Architectural Alloys

A03080
A03800
A04430
A05180
A07130

Applicable Specifications

American Society of Testing Methods
ASTM B 26 Specification for sand casting
ASTM B108 Specification for permanent mold casting

Federal Specifications
QQ-A-601E Sand casting
QQ-A-596 Permanent molt casting

Finish Criteria

Regardless of the finish specified, prior to ordering metal and as a requirement for approval, provide two (2) samples of adequate size and shape to demonstrate the surface texture with the specified finish treatment. Samples to be provided with protective coating similar to the coating intended as protection on the finish product.

For specific descriptions and limitations of the various finishes, refer to the full discussion of this metal in the text.

Initial Surface Treatment

The aluminum surface is to be given an initial finish by one of the following:

- mechanical finish treatment
- chemical cleaning in nonetching solution
- surface etching to produce a matte finish
- chemical brightening
- electropolishing
- primer application such as a chromate pretreatment

Reference the Aluminum Association's standard designation system for aluminum finishes.

This initial treatment will be performed by companies familiar and practiced with standard techniques and chemical processes.

After initial surface treatment, handle the metal surface with clean cotton gloves to avoid contamination. Do not allow oils or other contaminants to reach the surface of the treated metal, and do not store the treated metal for longer than 8 hours prior to final finish treatment.

Fabrication Concerns

Finish Surface Treatment

After the pretreatment, apply a finish coating or surface treatment of one of the following:

Anodic Coatings
    Architectural Class I Coatings
    Architectural Class II Coatings
Organic Coatings

Custom Finish Surfaces

Provide custom finish as specified. Finish must match closely to the sample in the architect's possession.

The various custom finishes available must start with one of the standard finishes, usually a mill finish. It is recommended that a standard mill finish be specified with the desired custom finish as an additive. In any case, the mill finish should be clean and free of scratches, oils, and other foreign matter before the custom finish is applied.

Fabrication Concerns

The aluminum casting should have minimum surface imperfections created during the casting process such as riser and feeder distortions. These should be ground down smooth or located in a non exposed region of the casting. Weld and grind smooth if necessary. Keep sandblasting to a minimum and care should be undertaken to eliminate undercutting from grinding operations.

During all fabrication procedures, protect the aluminum finish surface with a protective film. Keep the aluminum surface away from ferrous materials that may become imbedded into the surface.

Do not allow the aluminum surface to be scratched or dented during fabrication. Dents and scratches which cannot be removed will be cause for rejection.

Storage and Handling

Do not allow the aluminum surface to come in contact with iron or other matter that may become imbedded into the aluminum surface. In particular, do not allow wet lime, mortar, and green concrete to come in contact with the aluminum surface.

Do not allow the surface to become wet during storage. Store the metal in a dry place using crates with wood side rails as protection.

Installation Considerations

Use only compatible fasteners and clips to fix aluminum fabrications in position. Peel all protective films from the surface and wipe off all excess adhesive. Protect the aluminum surface from other activities occurring around the metal which may cause contamination of the passive layer. Such activities include, welding and grinding, stone cleaning (using acids), wet concrete work.

METAL: ALUMINUM

FORM: EXTRUSION

Architectural Alloys

A96061
A96063

Tempers to consider: T4, T5, T6

Applicable Specifications

American Society of Testing Methods
ASTM B244 Testing of anodized coating thicknesses
ASTM B137 Testing of coating weight and density for desired hardness and durability

ANSI
ANSI H35.1 Alloys and Tempers
ANSI H35.2-1984 American National Standard Dimensional Tolerances for Aluminum Mill Products

Other References:
The Aluminum Extrusion Manual produced by The Aluminum Association and the Aluminum Extrusion Council
AAMA 606.1 Voluntary Guide Specifications and Inspection Methods for Integral Color Anodic Finishes for Architectural Aluminum
AAMA 605.2 Voluntary Specification for High Performance Organic Coatings on Architectural Extrusions and Panels
AAMA 609.1 Voluntary Guide Specification for Cleaning and Maintenance of Architectural Anodized Aluminum

Finish Criteria

Regardless of the finish specified, prior to ordering metal and as a requirement for approval, provide two (2) samples approximately 305 mm in length with the specified finish. Samples to be provided with protective coating similar to the coating intended as protection on the finish product. In no cases will the metal be provided unprotected.

For specific descriptions and limitations of the various finishes, refer to the full discussion of this metal in the text.

Initial Surface Treatment

The aluminum surface is to be given an initial finish by one of the following:

* mechanical finish treatment
* chemical cleaning in nonetching solution
* surface etching to produce a matte finish
* chemical brightening
* electropolishing
* primer application such as a chromate pretreatment

Reference the Aluminum Association's standard designation system for aluminum finishes.

This initial treatment will be performed by companies familiar and practiced with standard techniques and chemical processes.

After initial surface treatment, handle the metal surface with clean cotton gloves to avoid contamination. Do not allow oils or other contaminants to reach the surface of the treated metal, and do not store the treated metal for longer than 8 hours prior to final finish treatment.

Finish Surface Treatment

After the pretreatment, apply a finish coating or surface treatment of one of the following:

> Anodic Coatings
>> Architectural Class I Coatings
>> Architectural Class II Coatings
> Organic Coatings

Custom Finish Surfaces

Provide custom finish as specified. Finish must match closely to the sample in the architect's possession.

The various custom finishes available must start with one of the standard finishes, usually a mill finish. It is recommended that a standard mill finish be specified with the desired custom finish as an additive. In any case, the mill finish should be clean and free of scratches, oils, and other foreign matter before the custom finish is applied.

Fabrication Concerns

Order the aluminum extrusion to arrive free of all alloy and mill stamps across the finish surface. Order architectural surface quality with a minimum of structural streaks, die lines, and surface imperfections. There should be an absence of all scratches, gouges, nicks, and other indentations.

During all fabrication procedures, protect the aluminum finish surface with a protective film. Keep the aluminum surface away from ferrous materials that may become imbedded into the surface.

Do not allow the aluminum surface to be scratched or dented during fabrication. Dents and scratches which cannot be removed will be cause for rejection.

Storage and Handling

Do not allow the aluminum surface to come in contact with iron or other matter that may become imbedded into the aluminum surface. In particular, do not allow wet lime, mortar, and green concrete to come in contact with the aluminum surface.

Do not allow the surface to become wet during storage. Store the metal in a dry place using crates with wood side rails as protection.

Installation Considerations

Use only compatible fasteners and clips to fix aluminum fabrications in position. Peel all protective films from the surface and wipe off all excess adhesive. Protect the aluminum surface from other activities occurring around the metal that may cause contamination of the passive layer. Such activities include welding and grinding, stone cleaning (using acids), wet concrete work.

METAL: ALUMINUM

FORM: WROUGHT

Architectural Alloys

A91100-H14
A93003-H14
A93004-H34
A93105-H14
A95005-H34
A95050-H34

Applicable Specifications

American Society of Testing Methods

ASTM B 202-92a Standard Specification for Aluminum and Aluminum Alloy Sheet and Plate

ASTM B244 Testing of anodized coating thicknesses

ASTM B137 Testing of coating weight and density for desired hardness and durability

Federal Specifications

QQ-A-250F-78 Aluminum and Aluminum Alloy Plate and Sheet

ANSI

ANSI H35.1 Alloys and Tempers

ANSI H35.2-1984 American National Standard Dimensional Tolerances for Aluminum Mill Products

Other References:

AAMA 606.1 Voluntary Guide Specifications and Inspection Methods for Integral Color Anodic Finishes for Architectural Aluminum

AAMA 605.2 Voluntary Specification for High Performance Organic Coatings on Architectural Extrusions and Panels

AAMA 609.1 Voluntary Guide Specification for Cleaning and Maintenance of Architectural Anodized Aluminum

Finish Criteria

Regardless of the finish specified, prior to ordering metal and as a requirement for approval, provide two (2) samples approximately 305 mm square of the specified finish. Samples to be provided with protective coating similar to the coating intended as protection on the finish product. In no cases will the metal be provided unprotected.

For specific descriptions and limitations of the various finishes, refer to the full discussion of this metal in the text.

Initial Surface Treatment

The aluminum surface is to be given an initial finish by one of the following:

- mechanical finish treatment
- chemical cleaning in nonetching solution
- surface etching to produce a matte finish
- chemical brightening
- electropolishing
- primer application such as a chromate pretreatment

Reference the Aluminum Association's standard designation system for aluminum finishes.

This initial treatment will be performed by companies familiar and practiced with standard techniques and chemical processes.

After initial surface treatment, handle the metal surface with clean cotton gloves to avoid contamination. Do not allow oils or other contaminants to reach the surface of the treated metal, and do not store the treated metal for longer than 8 hours prior to final finish treatment.

Finish Surface Treatment

After the pretreatment, apply a finish coating or surface treatment of one of the following:

Anodic Coatings
    Architectural Class I Coatings
    Architectural Class II Coatings
Organic Coatings

Custom Finish Surfaces

Provide custom finish as specified. Finish must match closely to the sample, in the architect's possession.

The various custom finishes available must start with one of the standard finishes, usually a mill finish. It is recommended that a standard mill finish be specified with the desired custom finish as an additive. In any case, the mill finish should be clean and free of scratches, oils, and other foreign matter before the custom finish is applied.

Fabrication Concerns

Order the aluminum sheet to arrive free of all alloy and mill stamps across the finish surface. Order architectural surface quality with a protective film layer applied at the mill.

During all fabrication procedures, protect the aluminum finish surface with a protective film. Keep the aluminum surface away from ferrous materials that may become imbedded into the surface.

Do not allow the aluminum surface to be scratched or dented during fabrication. Dents and scratches which cannot be removed will be cause for rejection.

Storage and Handling

Do not allow the aluminum surface to come in contact with iron or other matter that may become imbedded into the aluminum surface. In particular, do not allow wet lime, mortar, and green concrete to come in contact with the aluminum surface.

Do not allow the surface to become wet during storage. Store the metal in a dry place using crates with wood side rails as protection.

Installation Considerations

Use only compatible fasteners and clips to fix aluminum fabrications in position. Peel all protective films from the surface and wipe off all excess adhesive. Protect the aluminum surface from other activities occurring around the metal which may cause contamination of the passive layer. Such activities include,welding and grinding, stone cleaning (using acids), wet concrete work.

METAL: COPPER

FORM: WROUGHT

Architectural Alloys

Applicable Specifications

American Society of Testing Methods

ASTM B248 Specification for General Requirements for Wrought Copper and Copper Alloy Plate, Sheet, Strip and Rolled Bar

ASTM B 601 Practice for Temper Designations for Copper and Copper Alloys-Both Wrought and Cast

ASTM B 152 Standard Specification for Commercially Pure Copper Sheet and Strip

ASTM B 370-92 Standard Specification for Copper Sheet and Strip for Building Construction

Federal Specifications

QQ-C-390 Copper Alloy Castings

QQ-C-576 Commercially Pure Copper for Building Construction

Finish Criteria

Regardless of the finish specified, prior to ordering metal and as a requirement for approval, provide two (2) samples approximately 305 mm square with the specified finish. Samples to be provided with protective coating similar to the coating intended as protection on the finish product. Copper sheeting used for roofing can be formed and installed without surface protection in most cases. For interior uses and decorative copper features, it may be desired to coat with a protective film.

For specific descriptions and limitations of the various finishes, refer to the full discussion of this metal in the text.

Custom Finish Surfaces

Provide custom finish as specified. Finish must match closely to the sample in the architect's possession.

Fabrication Concerns

Copper wrought products should be ordered to arrive at the fabrication plant in dry wood crates with adequate edge protection. Copper should be delivered dry. If moisture reaches the copper surface, quickly dry the sheet, unpack, and inspect to insure that water has not traveled between sheets or coils of metal. Do not allow other metals to stack or rest on copper surfaces.

Optional requirement:
Order copper sheets with a benzotrianozole-saturated interleave paper. This tarnish-inhibiting interleave paper will provide excellent surface protection for the copper or copper alloy surface.

During all fabrication procedures, protect the copper surface from coming in contact with other metals that may dent or scratch the surface. On thin copper sheets used for roofing and flashing applications, minor scratches and surface tarnish is unavoidable. However, excessive dents or gouges will be rejected.

Storage and Handling

Store fabricated copper parts in clean wood crates. Avoid exposed nail heads that can dent and gouge the copper surface. Do not stack or store copper fabrications in areas where ferrous materials will come in contact with the metal surfaces. Do not allow moisture to collect on copper surfaces. Tilt stored copper fabrications exposed to weathering to allow surfaces to drain.

Installation Considerations

Use only copper nails or copper alloy fasteners to attach copper fabrications to final surfaces. Wipe fingerprints from the copper surfaces after installation. Separate copper surfaces from wet wood surfaces. Separate the copper surface from wood or concrete using 30# felts or other barrier films. During installation, handle the copper carefully. Dented panels and fabrications should be repaired or replaced to the satisfaction of the owner's representative.

METAL: COPPER ALLOYS—BRASSES AND NICKEL SILVERS

FORM: WROUGHT

Architectural Alloys

C21000 Gilding Metal
C22000 Commercial Bronze

C23000 Red Brass
C24000 Low Brass
C26000 Cartridge Brass
C27000 Yellow Brass
C28000 Muntz Brass
C75200 Nickel Silver

Applicable Specifications

American Society of Testing Methods

ASTM E 36 Method for Chemical Analysis of Brasses

ASTM B248 Specification for General Requirements for Wrought Copper and Copper Alloy Plate, Sheet, Strip and Rolled Bar

ASTM B 601 Practice for Temper Designations for Copper and Copper Alloys—Both Wrought and Cast

ASTM B 584 Standard Specification for Copper-Zinc-Lead Alloy Extruded Shapes

ASTM B 124 Standard Specification For Forging Brass Copper Alloys

ASTM B 283 Standard Specification For the Forging Process for Brass

ASTM B 103 Standard Specification for Phosphor Bronze Alloy C51100

ASTM B 100 Standard Specification for Phosphor Bronze Alloy C51100—*includes other alloys*

ASTM B 169 Standard Specification for Aluminum Bronze Alloys C60800 through C64210

ASTM B 171 Standard Specification for Aluminum Bronze Alloys C60800 through C64210

ASTM B 96 Standard Specification for Silicon Bronze Alloy C65500

ASTM B 122 Standard Specification for Nickel Silver Alloy C75200

Federal Specifications

QQ-C-390 Copper Alloy Castings

QQ-13-613 Copper alloys C26000 and C28000

QQ-b-750 Phosphor Bronze Alloy C 51100

QQ-C-450A-91 Copper-Aluminum Alloy (Aluminum Bronze C61400) Plate, Sheet, Strip and Bar

QQ-C-591 Silicon Bronze Alloy C65500

QQ-C-585 Nickel Silver Alloy C75200

Finish Criteria

Regardless of the finish specified, prior to ordering metal and as a requirement for approval, provide two (2) samples approximately 305 mm square with the specified finish. Samples to be provided with protective coating similar to the coating intended as protection on the finish product.

For specific descriptions and limitations of the various finishes, refer to the full discussion of this metal in the text.

Standard Finish Surfaces

Provide fabricated and finished product in one of the following mechanical finishes. Finish should match the sample submitted for prior approval.

Satin Finish
> fine texture
> medium texture
> course texture
> uniform texture
> hand rubbed nondirectional satin texture
> brushed texture

Specular (reflective)
> smooth specular
> specular
> buffed

Custom Finish Surfaces

Provide custom finish as specified. Finish must match closely to the sample in the architect's possession.

The various custom finishes available must start with one of the standard finishes, usually a mill finish. It is recommended that a standard mill finish be specified with the desired custom finish as an additive. In any case, the mill finish should be clean and free of scratches, oils, and other foreign matter before the custom finish is applied.

Clear Coatings

After the final finish is applied to the surface of the metal, clean all excess oxides and polishing compounds from the surface. Treat the surface with an oxide inhibitor such as benzotriazole and apply a clear protective coating as described herein.

Refer to the section of the text that describes the various clear coatings available on copper alloys. Specific limitations and precautions should be mentioned at this point within the specification.

Fabrication Concerns

The copper alloy sheet, plate, tube, or other wrought form must be received at the fabrication facility free of surface imperfections. Surface imperfections are oil and water stains, other foreign matter, and scratches, mars and dents. Any of these imperfections must be replaced or repaired. Finish fabricated products with indications of these imperfections will be rejected and require replacement. During all fabrication procedures, protect the copper alloy finish surface with a protective film or covering. Keep the metal surface away from ferrous materials that may become imbedded into the surface.

Storage and Handling

Do not allow the copper alloy surface to come in contact with iron or other matter that may become imbedded into the surface. In particular, do not allow wet lime, mortar, green concrete, acids, or heavy alkaline material to come in contact with the metal surface.

Do not allow the surface to become wet during storage. Store the metal in a dry place using crates with wood side rails as protection.

Installation Considerations

Use only brass fasteners and clips to fix the metal fabrications in position. Peel all protective films from the surface and wipe off all excess adhesive. Protect the metal surface from other activities occurring around the metal that may contaminate the surface. Such activities include, welding and grinding, stone cleaning (using acids), wet concrete work.

METAL: COPPER ALLOYS—SILICON BRONZE

FORM: WROUGHT

Architectural Alloys

C65500 Silicon Bronze

Applicable Specifications

American Society of Testing Methods
ASTM B248 Specification for General Requirements for Wrought Copper and Copper Alloy Plate, Sheet, Strip and Rolled Bar
ASTM B 601 Practice for Temper Designations for Copper and Copper Alloys—Both Wrought and Cast
ASTM B 96 Standard Specification for Silicon Bronze Alloy C65500

Federal Specifications
QQ-C-390 Copper Alloy Castings
QQ-C-591 Silicon Bronze Alloy C65500

Finish Criteria

Regardless of the finish specified, prior to ordering metal and as a requirement for approval, provide two (2) samples approximately 305 mm square with the specified finish. Samples to be provided with protective coating similar to the coating intended as protection on the finish product.

For specific descriptions and limitations of the various finishes, refer to the full discussion of this metal in the text.

Standard Finish Surfaces

Provide fabricated and finished product in one of the following mechanical finishes. Finish should match the sample submitted for prior approval.

Satin Finish
fine texture
medium texture
course texture
uniform texture
hand rubbed nondirectional satin texture
brushed texture

Specular (reflective)
    smooth specular
    specular
    buffed

Custom Finish Surfaces

Provide custom finish as specified. Finish must match closely to the sample in the architect's possession.

The various custom finishes available must start with one of the standard finishes, usually a mill finish. It is recommended that a standard mill finish be specified with the desired custom finish as an additive. In any case, the mill finish should be clean and free of scratches, oils, and other foreign matter before the custom finish is applied.

Fabrication Concerns

The silicon bronze sheet, plate, tube, or other wrought form must be received at the fabrication facility free of surface imperfections. Surface imperfections are oil and water stains, other foreign matter, and scratches, mars, and dents. Any of these imperfections must be replaced or repaired. Finish fabricated products with indications of these imperfections will be rejected and require replacement. During all fabrication procedures, protect the copper alloy finish surface with a protective film or covering. Keep the metal surface away from ferrous materials which may become imbedded into the surface.

Storage and Handling

Do not allow the silicon bronze surface to come in contact with iron or other matter that may become imbedded into the surface. In particular, do not allow wet lime, mortar, green concrete, acids, or heavy alkaline material to come in contact with the metal surface.

Do not allow the surface to become wet during storage. Store the metal in a dry place using crates with wood side rails as protection.

Installation Considerations

Use only silicon bronze fasteners and clips to fix the metal fabrications in position. Peel all protective films from the surface and wipe off all excess adhesive. Protect the metal surface from other activities occurring around the metal which may contaminate the surface. Such activities include welding and grinding, stone cleaning (using acids), wet concrete work.

METAL: STAINLESS STEEL

FORM: WROUGHT

Architectural Alloys

    S30200
    S30400

S30403 (low carbon)
S31600
S31603 (low carbon)

Applicable Specifications

American Society of Testing Methods

ASTM A 380 Standard Specification for the Cleaning and Descaling of Stainless Steel Parts, Equipment and Systems

ASTM B 117 Standard Specification for the Salt Spray Testing of Stainless Steel

ASTM B 254 Standard Specification for the Electroplating, Preparation and Processing

ASTM A 167-92B Standard Specification for Stainless Steel and Heat Resistant Chromium Steel Plate, Sheet, and Strip—covers the various alloy types

ASTM A 176-93a Standard Specification for Stainless Steel and Heat Resistant Chromium Steel Plate, Sheet and Strip—covers the various standard finishes

ASTM A 480/A and A480M-93a Standard Specification for General Requirements for Flat Rolled Stainless Steel and Heat Resistant Steel Plate, Sheet and Strip

Federal Specifications

QQ-S-766D-88 Steel, Stainless and Heat Resisting Alloys, Plate, Sheet and Strip for General Fabrication Purposes

QQ-P-35C Passivation Treatments for Corrosion Resistant Steel

Finish Criteria

Regardless of the finish specified, prior to ordering metal and as a requirement for approval, provide two (2) samples approximately 305 mm square of the specified finish. Samples to be provided with protective coating similar to the coating intended as protection on the finish product. In no cases will the metal be provided unprotected.

For specific descriptions and limitations of the various finishes, refer to the full discussion of this metal in the text.

Mill Finish Surfaces

Provide standard mill finishes as described by the National Association of Architectural Metal Manufacturers (NAAMM). The finishes are designated as one of the following:

No. 1
No. 2D
No. 2B
No. 2BA

Polished Finish Surfaces

Provide standard polished finishes as described by the National Association of Architectural Metal Manufacturers (NAAMM). The finishes are designated as one of the following:

No. 3
No. 4
No. 6
No. 7
No. 8

Custom Finish Surfaces

Provide custom finish as specified. Finish must match closely to the sample in the architect's possession.

The various custom finishes available must start with one of the standard finishes, usually a mill finish. It is recommended that a standard mill finish be specified with the desired custom finish as an additive. In any case, the mill finish should be clean and free of scratches, oils, and other foreign matter before the custom finish is applied.

Fabrication Concerns

Order the stainless steel to arrive free of all alloy and mill stamps across the finish surface. Order the stainless steel with a protective layer from the mill. All stainless steel surfaces to be provided with passive chromium oxide layer free of all surface free iron or surface pits created by subsequent finishing, welding, or grinding operations.

If the stainless steel is contaminated during fabrication by iron or other nonferrous material, clean and insure the surface has adequately developed passivity. Use appropriate passivation procedures described herein. Test the surface passivity to insure all free iron contamination has been removed.

During all fabrication procedures, protect the stainless steel finish surface with a protective film.

Do not allow the stainless steel surface to be scratched or dented during fabrication. Dents and scratches which cannot be removed will be cause for rejection.

Storage and Handling

Do not allow the stainless steel surface to come in contact with iron or other matter which may damage the passive film layer of chromium oxide. In particular do not allow salts, hydrochloric acid (muriatic acid, builders acid), grinding debris, or iron to come in contact with the stainless steel surface.

Store the metal in a dry place using crates with wood side rails as protection.

Installation Considerations

Use only compatible fasteners and clips to fix stainless steel fabrications in position. Peel all protective films from the surface and wipe off all excess adhesive. Protect the stainless surface from other activities occurring around the metal which may cause contamination of the passive layer. Such activities include welding and grinding, stone cleaning (using acids), wet concrete work.

METAL: LEAD

FORM: SHEET

Architectural Alloys

Hard lead alloys L52900 and L53000
4 to 6% antimony, balance lead

Applicable Specifications

American Society of Testing Methods
ASTM B29-79 Standard Specification for Pig Lead
ASTM B749 Standard Specification for Sheet Lead

British Standards
BS 1178 "Milled Lead Sheet and Strip for Building Purposes"

Finish Criteria

Prior to ordering metal and as a requirement for approval, provide two (2) samples approximately 305 mm square with the specified finish. Lead finish to be standard, mill rolled finish.

Fabrication Concerns

Lead to be unrolled into sheets, cut to the required shapes, and formed using hand equipment into the necessary configurations. This can occur in the field or at the shop.

Care should be established to avoid imbedding iron particles into the soft lead surface. The particles will rapidly rust on exterior exposures.

Storage and Handling

Keep the lead rolls on wood skids. Avoid allowing other materials to be stacked on lead. Do not allow lead to become contaminated from iron particles, particularly from particles created by grinding or cutting operations which, if hot, will imbed into the lead surface.

Installation Considerations

Use only copper bars and brass screws to attach lead to the substrate. Fold lead over all exposed fasteners and bars or "burn in " covers over the fastener heads. Do not allow finish lead surface to be contaminated with iron particles. If this occurs, clean particles from the surface and wipe all rust stains from the surface.

METAL: ZINC

FORM: ROLLED ZINC SHEET

Architectural Alloys

Alloy No. 700
99.80% Zinc minimum
0.7 to 0.9% copper
0.08 to 0.12% titanium

(Note: German Spec is 99.995% purity with balance copper and titanium)

Applicable Specifications
American Society of Testing Methods
ASTM B69 Zinc (Slab Zinc)

German Standards
DIN 1706 Electrolytic High Grade Zinc
DIN 17 770 Mechanical Properties of Zinc Sheet

British Standards
BS 6561A

Federal Specifications
QQ-Z-100A Zinc Sheet

Finish Criteria

Regardless of the finish specified, prior to ordering metal and as a requirement for approval, provide two (2) samples approximately 305 mm square with the specified finish. Samples to be provided with protective coating similar to the coating intended as protection on the finish product. Zinc sheeting used for roofing can be formed and installed without surface protection in most cases. For interior uses and decorative fabrications, it may be desired to coat with a protective film.

Standard finishes available

Mill finish—nonaged
Preweathered

Custom Finish Surfaces

Provide custom finish as specified. Finish must match closely to the sample in the architect's possession.

The various custom finishes available must start with one of the standard finishes, usually a mill finish. It is recommended that a standard mill finish be specified with the desired custom finish as an additive. In any case, the mill finish should be clean and free of scratches, oils, and other foreign matter before the custom finish is applied.

Fabrication Concerns

Order the zinc sheet to arrive free of all alloy and mill stamps across the finish surface. Order architectural surface quality with a protective film layer applied at the mill.

During all fabrication procedures, protect the zinc finish surface with a protective film. Keep the zinc surface away from ferrous materials that may become imbedded into the surface.

Do not allow the zinc surface to be scratched or dented during fabrication. Dents and scratches which cannot be removed will be cause for rejection.

Storage and Handling

Do not allow the zinc surface to come in contact with iron or other matter that may become imbedded into the aluminum surface. In particular, do not allow wet lime, mortar, and green concrete to come in contact with the zinc surface.

Do not allow the surface to become wet during storage. Store the metal in a dry place using crates with wood side rails as protection.

Installation Considerations

Use only compatible fasteners and clips to fix zinc fabrications in position. Peel all protective films from the surface and wipe off all excess adhesive. Protect the zinc surface from other activities occurring around the metal which may cause contamination of the passive layer. Such activities include welding and grinding, stone cleaning (using acids), wet concrete work.

Do not attempt forming operations in exposures below 10°C (50°F). Remove and replace all cracked and damaged zinc fabrications.

It is necessary to build in proper ventilation to reduce or eliminate condensation on the back side of thin zinc cladding. The underside of the zinc sheeting must be adequately drained to prevent the trapping of moisture. Protect the zinc from contact with other metals or woods which may attach to the protective oxide surface.

METAL: MONEL

FORM: WROUGHT

Architectural Alloys

Alloy 400
63% nickel (minimum)
28 to 34% copper

Applicable Specifications
American Standard Testing Methods
ASTM B 127 Standard Specification—covers plate, sheet and strip forms of Monel

ASTM B165 Standard Specification—covers tube and rod forms of Monel

ASTM B564 Standard Specification—covers bar and rod forms of Monel

Federal Specifications
QQ-N-281D-85 Class A—Nickel Copper Alloy Bar, Rod, Plate, Sheet, Strip, Wire, Forging and Structural and Special Shaped Sections

Finish Criteria

Regardless of the finish specified, prior to ordering metal and as a requirement for approval, provide two (2) samples approximately 305 mm square with the specified finish. Samples to be provided with protective coating similar to the coating intended as protection on the finish product.

For specific descriptions and limitations of the various finishes, refer to the full discussion of this metal in the text.

Provide the cold rolled finish surface from the mill or provide a mechanical finish as follows:

Satin finish similar to a No. 4 finish on stainless steel

Mirror specular finish produced by buffing, similar to a No. 7 or No. 8 finish produced on stainless steel.

Provide samples of the finish for approval prior to ordering metal.

Custom Finish Surfaces

Provide custom finish as specified. Finish must match closely to the sample in the architect's possession.

The various custom finishes available must start with one of the standard finishes, usually a mill finish. It is recommended that a standard mill finish be specified with the desired custom finish as an additive. In any case, the mill finish should be clean and free of scratches, oils, and other foreign matter before the custom finish is applied.

Fabrication Concerns

Order the Monel sheet to arrive free of all alloy and mill stamps across the finish surface. Order architectural surface quality with a protective film layer applied at the mill.

During all fabrication procedures, protect the Monel finish surface with a protective film. Keep the surface away from ferrous materials that may become imbedded into the surface.

Do not allow the Monel surface to be scratched or dented during fabrication. Dents and scratches which cannot be removed will be cause for rejection.

Storage and Handling

Do not allow the Monel surface to come in contact with iron or other matter which may become imbedded or scratch the surface. In particular, do not allow wet lime, mortar, and green concrete to come in contact with the Monel finish surface. Do not allow the surface to become wet during storage. Store the metal in a dry place using crates with wood side rails as protection.

Installation Considerations

Use only compatible fasteners and clips to fix Monel fabrications in position. Type S30000 stainless steel fasteners are acceptable. Brass fasteners can also be used to join Monel parts. Peel all protective films from the surface and wipe off all excess adhesive. Protect the Monel surface from other activities occurring around the metal that may cause contamination of the passive layer. Such activities include welding and grinding, stone cleaning (using acids), wet concrete work.

METAL: TITANIUM

FORM: WROUGHT

Architectural Alloys

Alloy No. Ti-35A (ASTM Grade 1)

Applicable Specifications
American Society of Testing Methods
ASTM B265-G1 Standard Specification for Commercially Pure Titanium
ASTM 358 Standard Specification for Titanium sheet
ASTM B 381 Standard Specification for Commercially Pure Titanium Strip
ASTM B 348 Standard Specification for Titanium plate
ASTM F67 Standard Specification for Titanium tubing

Finish Criteria

Regardless of the finish specified, prior to ordering metal and as a requirement for approval, provide two (2) samples approximately 305 mm square with the specified finish. Samples to be provided with protective coating similar to the coating intended as protection on the finish product.

For specific descriptions and limitations of the various finishes, refer to the full discussion of this metal in the text.
Specify one of the following mill finish surfaces:

Low matte surface finish
Medium matte surface finish

Submit samples for approval prior to ordering materials.

Custom Finish Surfaces
Provide custom finish as specified. Finish must match closely to the sample in the architect's possession.

The various custom finishes available must start with one of the standard finishes, usually a mill finish. It is recommended that a standard mill finish be specified with the desired custom finish as an additive. In any case, the mill finish should be clean and free of scratches, oils, and other foreign matter before the custom finish is applied.

Fabrication Concerns

Order the titanium sheet to arrive free of all alloy and mill stamps across the finish surface. Order architectural surface quality with a protective film layer applied at the mill or paper interleave.

During all fabrication procedures, protect the finish surface with a protective film. Keep the titanium surface away from other metals which may leave trace particles on the surface.

Do not allow the titanium surface to be scratched or dented during fabrication. Dents and scratches which cannot be removed will be cause for rejection.

Storage and Handling

Do not allow the titanium surface to come in contact with iron or other matter that may stain the surface. Do not allow the surface to become wet during storage. Store the metal in a dry place using crates with wood side rails as protection.

Installation Considerations

Use only type S30000 stainless fasteners and clips to fix titanium fabrications in position. Peel all protective films from the surface and wipe off all excess adhesive. Protect the titanium surface from other activities occurring around the metal which may cause contamination of the passive layer. Such activities include welding, grinding, and wet concrete work.

METAL: GALVANIZED STEEL

FORM: SHEET

Architectural Alloys

Hot Dipped Galvanized Steel Sheet
Lockformer quality
Commercial quality
G90 (standard architectural exposure zinc coating thickness)

Note: There are various zinc coating thicknesses available as well as a variety of coating applications such as electrodeposition. Other variations on the galvanizing technology are also available such as Galfan coatings and paintgrip modifications, that may have the desired finish surface.

Applicable Specifications
American Society of Testing Methods
ASTM A 525-91b Standard Specification for General Requirements for
    Steel Sheet, Zinc Coated by the Hot Dipping Process
ASTM A 526/A and A526M-90 Standard Specification for Steel Sheet,
    Zinc Coated by the Hot Dipping Process, Commercial Quality

ASTM A 527/A and A527M-90 Standard Specification for Steel Sheet, Zinc Coated by the Hot Dipping Process, Lock Forming Quality

ASTM A90 Weight of Coating on Zinc-Coated Iron or Steel Articles

ASTM A120 Pipe, Steel, Black and Hot Dipped Zinc Coated both Welded and Seamless for Ordinary Uses

ASTM A239 Locating the Thinnest Spot in a Zinc Coating in Iron or Steel Articles by the Preece Test

ASTM A361 Steel Sheet, Zinc Coated by the Hot Dip Process for Roofing and Siding

ASTM A 385 Providing High Quality Zinc Coatings

ASTM A780 Repair of Damaged Hot Dip Galvanized Coatings

Federal Specifications

QQ-S-775E-78 Steel Sheets, Strip, Carbon, Zinc Coated by the Hot Dipping Process

Finish Criteria

Galvanized steel, left exposed and uncoated, should have one of the surface finish appearances listed below. Submit two samples of the specified surface for approval, prior to acquiring metal.

Surface quality criteria:

Standard spangle (large spangles developed from hot dipping)
Minimized spangle (tight, small crystals)
Galvannealed surface (no spangles, zinc outer coating inhibited)

Fabrication Concerns

Order the galvanized sheet to arrive with a minimum of surface oil.

During all fabrication procedures, protect the galvanized finish surface by covering dies with plastic shielding such as "rhino hide" or other thick but pliable covering. Keep the zinc surface away from ferrous materials which may become imbedded.

Do not allow the galvanized surface to be scratched or dented during fabrication.

Storage and Handling

Do not allow the galvanized surface to come in contact with iron or other matter which may become imbedded into the zinc. In particular, do not allow moisture to come in contact with the galvanized coating while the material is being stored. Store the metal in a dry place using crates with wood side rails as protection.

Installation Considerations

Use only compatible fasteners and clips to fix galvanized fabrications in position. Protect the galvanized surface from other activities occurring around the metal that may cause staining or premature deterioration of the zinc coating.

METAL: ALUMINIZED STEEL

FORM: SHEET

Architectural Alloys

Aluminized steel Type 2

This is a steel sheet with a commercially pure aluminum coating applied to all sides. Nominal thickness of coating is 0.038 mm per side (0.0015 inches).

Applicable Specifications
American Society of Testing Methods
ASTM A463-88 Standard Specification for Steel Sheet, Cold Rolled, Aluminum Coated Type 1 and Type 2

Federal Specifications

Finish Criteria

Prior to ordering metal and as a requirement for approval, provide two (2) samples approximately 305 mm square with the specified finish. Samples to be provided with protective coating similar to the coating intended as protection on the finish product.

Fabrication Concerns

Order the aluminized steel sheet to arrive free of all alloy and mill stamps across the finish surface. Order architectural quality, type 2 with a protective film layer applied at the mill or paper interleave.

During all fabrication procedures, protect the aluminum finish surface with a protective film. Keep the aluminum surface away from ferrous materials which may become imbedded into the surface.

Do not allow the aluminum surface to be scratched or dented during fabrication. Dents and scratches which cannot be removed will be cause for rejection.

Storage and Handling

Do not allow the aluminized steel surface to come in contact with iron or other matter that may become imbedded into the soft aluminum surface-coating. In particular, do not allow wet lime, mortar, and green concrete to come in contact with the aluminum surface.

Do not allow the surface to become wet during storage. Store the metal in a dry place using crates with wood side rails as protection.

Installation Considerations

Use only compatible fasteners and clips to fix aluminized steel fabrications in position. Peel all protective films from the surface and wipe off all excess adhesive. Protect the surface from other activities occurring around the metal that may cause contamination of the aluminum surface layer. Such activities include welding and grinding, stone cleaning (using acids), wet concrete work. Do not allow water to settle on the aluminum surface.

METAL: GALVALUME

FORM: SHEET

Architectural Alloys

Aluminum–zinc coating over steel
AZ 165 (AZ 55) coating classification

This coating classification provides nominally 165 grams of aluminum–zinc per square meter of surface (0.55 ounces per square foot of surface).

Applicable Specifications
American Society of Testing Methods
ASTM B 792 Galvalume

Finish Criteria

Regardless of the finish specified, prior to ordering metal and as a requirement for approval, provide two (2) samples approximately 305 mm square with the specified finish. Samples to be provided with protective coating similar to the coating intended as protection on the finish product.
Finish surface to be provided in one of the spangle levels:

regular spangle
extra smooth spangle

Fabrication Concerns

Order the galvalume sheet to arrive free of all alloy and mill stamps across the finish surface. Keep the surface away from ferrous materials which may become imbedded into the surface.
Do not allow the aluminum–zinc coating to be scratched or dented during fabrication. Dents and major scratches which cannot be removed will be cause for rejection.

Storage and Handling

Do not allow the aluminum–zinc surface to come in contact with iron or other matter that may become imbedded. In particular, do not allow wet lime, mortar, and green concrete to come in contact with the surface.
Do not allow the surface to become wet during storage. Store the metal in a dry place using crates with wood side rails as protection.

Installation Considerations

Use only compatible fasteners and clips to fix galvalume fabrications in position. Peel all protective films from the surface and wipe off all excess adhesive. Protect the surface from other activities occurring around the metal which may cause contamination of the aluminum–zinc protective layer. Such activities include welding and grinding, stone cleaning (using acids), wet concrete work.

B

Comparative Metal Thicknesses— Sheet Metal

THICKNESS		STEELS	ALUMINUM	LEAD	COPPER	COPPER ALLOYS	TITANIUM*	MONEL
Inches	mm	Gauge	Nominal Thickness	Pounds per sq ft	Ounces per sq ft	Nominal Thickness	Nominal Thickness	Nominal Thickness
0.188	4.78		0.188			0.188		0.188
0.172	4.37	8						
0.156	3.97	9		10				
0.141	3.57	10						0.141
0.125	3.18	11	0.125	8		0.125		0.125
0.109	2.78	12			80	0.109		0.109
0.100	2.54		0.100					
0.097	2.47				72			
0.094	2.38	13		6				
0.090	2.29		0.090			0.090		
0.086	2.19				64			
0.080	2.03		0.080			0.080		
0.078	1.98	14		5				0.078
0.075	1.91				56			
0.070	1.79	15	0.070			0.070		
0.065	1.64				48			
0.063	1.59	16	0.063	4		0.063		0.063
0.056	1.43	17						
0.055	1.40			3.5				
0.054	1.37				40			
0.050	1.27	18	0.050			0.050		0.050
0.049	1.23				36			
0.047	1.19			3		0.047		
0.044	1.11	19						
0.043	1.09				32			
0.040	1.02		0.040					
0.039	0.99			2.5		0.039		
0.038	0.95	20			28			0.038
0.034	0.87	21						
0.032	0.81		0.032		24	0.032	0.032	
0.031	0.80	22		2			0.031	0.031
0.028	0.71	23					0.028	
0.027	0.69				20		0.027	
0.025	0.64	24	0.025				0.025	0.025
0.024	0.61			1.5			0.024	
0.022	0.56	25			16		0.022	
0.020	0.51		0.020				0.020	0.020
0.019	0.48	26					0.019	0.019
0.016	0.41			1	12		0.016	
0.015	0.38						0.015	
0.010	0.25		0.010	1	8			0.010

*For titanium, other thicknesses in excess of what is indicated are possible. Check manufacturers.

For a particular metal type there are mill tolerances established which allow for variations in thicknesses. These tolerances are necessary and are a factor in the cold reduction process used to create sheet metal. The table below gives typical mill variations for aluminum alloys.

Example: Mill Tolerances for 3003 and 5005 alloys of aluminum.

NOMINAL THICKNESS (inches)	RANGE OF TOLERANCE (inches)		NOMINAL THICKNESS (mm)	RANGE OF TOLERANCE (mm)	
	Minimum	Maximum		Minimum	Maximum
0.188	0.1810	0.1990	4.78	4.60	5.05
0.125	0.1195	0.1305	3.18	3.04	3.31
0.100	0.0945	0.1055	2.54	2.40	2.68
0.090	0.0855	0.0945	2.29	2.17	2.40
0.080	0.0755	0.0845	2.03	1.92	2.15
0.063	0.0595	0.0665	1.60	1.51	1.69
0.040	0.0368	0.0438	1.02	0.93	1.11
0.032	0.0295	0.0345	0.81	0.75	0.88
0.025	0.0230	0.0270	0.64	0.58	0.69
0.020	0.0180	0.0220	0.51	0.46	0.56

Useful Conversion Factors

Useful Conversion Factors

TO CONVERT	INTO	MULTIPLY BY
mils	μm	25.38
mils	mm	0.0254
mils	cm	0.00254
mils	m	2.54×10^{-5}
inches	μm	2.54×10^{4}
inches	mm	25.4
inches	cm	2.54
inches	m	0.0254
feet	mm	304.8
feet	cm	30.48
feet	m	0.3048
in^2	mm^2	645.2
ft^2	m^2	0.0929
in^3	cm^3	16.3872
in^3	m^3	1.639×10^{-5}
ft^3	m^3	0.02832
yd^3	m^3	0.7646
psi	Pa	6.895×10^{3}
ksi	MPa	6.895
lb	kg	4.536×10^{-1}
ounces	kg	0.0283
oz/ft^2	kg/m^2	3.0515×10^{-1}
ft-lbs	joules	1.356
BTU	joules	1.054×10^{3}
calories	joules	4.184
°F	°C	$T_C = (T_F - 32)/1.8$
°C	°F	$T_F = 1.8 T_C + 32$
°C	°K	$T_K = T_C + 273$
μm	mils	0.0394
μm	inches	3.937×10^{-5}
mm	mil	39.37
mm	inches	0.0394
cm	inches	0.3937
m	inches	39.37
mm	feet	0.00328
cm	feet	0.0328
m	feet	3.28
mm^2	in^2	0.00155
m^2	ft^2	10.758
cm^3	in^3	0.061
m^3	ft^3	35.31
m^3	yd^3	13.069
Pa	psi	1.45×10^{-4}
MPa	ksi	0.145
kg	lbs	2.2046
kg	ounces	35.274

Glossary

abrasive belt
A flexible belt coated with hard substances and used for continuous grinding and polishing operations.

abrasive blast
Abrasive material applied at high pressure through a compressed air apparatus.

age hardening
A change in properties over time and at ambient temperature experienced by some alloys. As the metal rests, it becomes harder, usually accompanied by brittleness.

alloy
This very old term, dating back to the Roman times, refers to the combination of two or more metals melted together into a mixture, usually for some specific characteristics.

alpha-beta phase
The alpha-beta phase is the zone of transition from the alpha structure to a beta structure. At this point, both crystal structures are apparent. These copper alloys are used for hot forming, forging, and stamping.

alpha phase
The alpha phase relates to the crystal structure copper and its low zinc content alloys exhibit. The alpha phase refers to a face-centered cubic structure. Properties include good ductility and ability to cold work.

amorphous
Noncrystalline. Relates to those materials and forms which are not made of crystals.

angel hair
Very fine grit lines apparent on the polished surface of a metal.

annealing
A heat treatment performed on a metal. The metal is heated to a suitable temperature, held at that temperature, then slowly cooled. Usually used to improve machinability and cold working.

anode
Typically the positive pole of an electrolytic cell or system. The anode is the electrode where oxidation occurs. Anodizing, for instance, makes the aluminum sheet the anode to develop aluminum oxide on the surface.

anodic coating
The oxide coating developed on an anodic electrode. *See anode.*

anodizing
The process which uses the polarity of an electrochemical reaction. A metal is the anode that undergoes an oxidation process.

artificial aging
An accelerated aging process as compared to natural aging. Performed by heating a metal above room temperature.

ASTM
Abbreviation for the American Society of Testing Materials, the organization that sets standards for testing, evaluating, and manufacture of materials.

austenitic
The term used to describe the iron carbon alloy crystal featuring a face-centered cubic structure with carbon interstitial. Based on the material known as austenite contained within high carbon steel.

benzotriazole
A tarnish-inhibiting compound. Used over copper and copper alloys to resist surface oxidation. Also works well as an underfilm corrosion inhibitor for lacquer finishes.

billet
The initial form of a metal generally intended for forging or extrusion processes.

brake forming
A cold working process performed on sheet or plate forms of metal using a set of dies applied under pressure.

brass
The name given to a series of copper-zinc alloys. Usually associated with a golden yellow color. However, some brass alloys are referred to as bronzes, such as commercial bronze and architectural bronze.

brazing
A welding process using a filler metal applied at high temperature between two closely spaced metals. The process is similar to soldering except the filler is usually an alloy variation of the joining metals.

brazing alloy
A nonferrous filler metal used in brazing operations.

bright anneal
Annealing operation in a controlled environment to eliminate surface discoloration.

brinnell hardness
A hardness reading obtained from taking a measurement of the depth of surface indentation in relation to the size of the indentation.

bronze
A term used to describe alloys of copper and tin or copper and silicon. The term used to be specific to copper-tin alloys; however, it has grown to encompass those copper alloys of dark brown tones.

buffing
The term given to describe a method of polishing using rouges and large cotton pads to produce a reflective surface.

burnishing
A polishing method using metal-to-metal contact by rapidly passing coil or sheet over a polished metal roll. The effect is a shining of the surface. Cold rolled copper sheet is burnished on cold, unlubricated steel rolls.

carbon precipitation
The condition that develops when steel, including stainless steel, reaches high temperatures. Carbon comes out of solution and forms carbides on the surface of the steel. This particular problem develops when stainless steel is welded. Carbon precipitates to the surface and combines with chromium to develop chromium carbides. This formation weakens the stainless steel in terms of corrosion resistance by steeling chromium.

carbon steel
The description given to steels containing varying quantities of silicon, carbon, manganese, and copper. Other elements are present in very minor amounts.

castability
A relative measure of a metal's or metal alloy's ability to be cast.

cast iron
The family of cast ferrous alloys containing at least 2% carbon.

cathode
The electrode opposite the anode. Electrons enter at the cathode end causing a reduction condition, whereas the anode end undergoes an oxidation condition.

caustic dip
An alkaline solution, often containing caustic soda (sodium hydroxide). Frequently used as a metal surface etchant and degreaser.

cementation
The fusing of a metallic powder to a metal surface by pressure. Sometimes heat is added. Also refers to the fusing of metal powders together using high pressure.

ceramic
Crystalline compounds of both metal and nonmetals. A very large family of materials falls into this category. In the context of metals, we usually consider only the porcelain enamels when referring to ceramics.

chatter
A mark or surface imperfection created by unstable tooling. Surface lines created by vibration of polishing belts on the surface of sheet material.

chromate treatment
Surface treatment of metal in a chromium solution. Generally desired to create a conversion coating for pretreatment prior to finish painting.

coining
The compression of a metal within a closed die. The surfaces are confined and an imprint is induced in the metal surface from the die. Designs such as imprinting on a coin surface or the one-sided imprinting on a sheet of metal. Similar to embossing, except the imprint occurs on one surface.

cold forging
The repeated hammering of metal into a die at room temperature.

cold rolled sheet
After hot rolling, the cold sheet is reduced between steel rolls. The resulting product has better surface characteristics such as smoothness and uniformity. Grain direction and stress are induced in the sheet.

cold working
Forming and shaping metal at room temperature. A condition where strain hardening develops as the metal undergoes the cold working operation.

conversion coating
A film on the surface of a metal made by developing complex compounds by electrical or chemical action. The film is a product of the base metal and oxides of the base metal with or without other complex compounds, such as chromium chromates or phosphates.

corrosion
The electrochemical deterioration of a metal caused by atmospheric or induced chemical reactants.

crazing
A fracturing of a metal surface. Characterized by small microcracks.

creep
The movement of a metal over time while under stress. Metals such as lead and zinc are prone to creep conditions because of their low elastic levels.

crevice corrosion
A principle of corrosion created by the concentration of corrosion products in a crevice or seam.

dead soft
Temper produced by full annealing. Characterized by reduced hardness and tensile strength to the lowest levels obtainable.

deep drawing
The forming of parts to an extreme depth without substantial necking of the metal.

degreasing
The removal of oils, grease, and dirt by chemical or vapor action.

density
The mass per unit volume. Usually in grams per cubic centimeter.

descaling
The removal of scale from a surface. Scale is the heavy oxide that develops on some metals during hot forming and casting operations.

dezincification
The selective removal of zinc from brass alloys subjected to corrosion cell development. The zinc comes out of solution and is redeposited on the surface of the metal.

dry sand mold
Sand mold used for casting. The sand mold is dried at temperatures above 100°C to remove moisture.

ductility
Malleability. The ability of a metal to deform without fracturing or cracking.

elasticity
The property of a metal to return to its original form after the stress which deformed the metal is removed.

electrodeposition
Depositing a metal onto a charged surface or electrode by passing an electrical current through a solution. Also known as electrolytic deposition.

electroless plating
Electroless plating is the process in which one metal is immersed in a solution containing another metal in ion form. The metal in solution is plated on the other by means of a chemical reduction at the surface.

electrolyte
A solution that will conduct an electric current. Generally, rainwater carrying dissolved salts picked up from the atmosphere is an electrolyte.

electrolytic copper
Copper refined by electrolytic deposition. Copper is removed from a solution where its ions are dissolved and redeposited onto a cathode of copper.

electroplating
A metal is immersed in a solution containing the ions of another metal. A charge is imparted into the metal being immersed causing it to become a cathode. The ions come out of solution and deposit onto the surface of the immersed metal.

eletropolishing
A metal is immersed into an electrolyte and a charge is applied making the metal an anode. The high points of the metal are dissolved into the solution, smoothing the surface of the metal, thus producing a polished effect.

embossing
Passing a metal between two complementary dies and imparting a pattern onto the surface by raising the pattern.

etching
Chemical or electrochemical treatment to the surface of a metal to remove a portion of the surface along with foreign contaminants and oxides.

expansion coefficient
The rate of expansion of a unit area of metal related to a change in temperature.

extrusion
The product of a process involving high pressure and temperature in order to push a billet of metal through a cross section cut in a die. The resulting metal obtains the cross section closely resembling the cutout in the die.

fatigue

A condition developing from repeated or cyclical stresses below the tensile strength of the material. This condition leads to eventual rupture of the material.

ferritic

The name given to describe the alpha iron-carbon alloy form. This form is magnetic, has a face-centered cubic crystal, and is stable. Refers to the ferrite form of iron carbon alloys.

fire refined copper

Copper refined by a furnace process only. Also referred to as fire refined tough pitch copper.

fish eyes

A malady associated with clear lacquers and paints. Created by a contaminant on the surface of the metal prior to finishing. When the lacquer is applied, it is repelled by this contaminant and forms a half-moon shape commonly called a "fish eye."

fluidity

The ability of a metal to be poured into a casting. The higher the fluidity, the easier detail can be achieved in the casting process.

fluorocarbon

A complex compound of carbon in which the hydrogen atom is replaced with fluorine. Very durable and moisture resistant, this molecule forms a very tough, nonreactive film.

fluoropolymer

A linear polymer in which some of the hydrogen atoms are replaced with fluorine. These plastic complex molecules have excellent heat and corrosion resistance. An additional characteristic of fluoropolymers is a high molecular weight.

flux

An etching compound used to prepare a metal surface by removing oxides and scale compounds.

free cutting

The ability of a metal to come off in small shards rather than long strips when cut or machined.

frit

The material used to develop porcelain enamel coatings. Frit is created by rapidly cooling a mixture of various clays and other compounds which have been fused together at high temperature. The fused compound, while hot, is cooled rapidly, fracturing into many small particles.

gage
The thickness or diameter of a sheet of wire, which varies arbitrarily from ferrous to nonferrous or from sheet to wire. Also, a visual measuring device used as a comparative measurement.

galling
A condition whereby surface friction develops between two surfaces to the point where the high spots drag or deform.

galvanic corrosion
Corrosion condition created from two dissimilar metals or between two dissimilar areas on a metal. An electrical current develops between the metals and the more reactive metal decays in relation to the more noble metal.

galvanize
The coating of a metal with zinc.

galvanizing
The process of coating a metal with zinc. A transition zone is developed between the zinc and the base metal when the hot dip galvanized method is used. A simple plating process occurs during the electrogalvanizing process.

galvannealed
The zinc coating produced when galvanized except the metal is kept hot for a longer period of time to develop a zinc-iron alloy coating.

GMAW
Stands for"Gas Metal-Arc Welding.' Also known as MIG welding. This is an arc-welding process wherein coalescence is produced by heating with an arc between a continuous consumed filler metal electrode and the metal part. Shielding of the weld is by an externally applied gas mixture.

grain
An individual crystal in a metal or alloy.

green sand
A sand mixture used for casting in which the sand is moist.

ground coat
The ground coat is the initial porcelain enamel coating applied over steel prior to the finish coating. The ground coat contains oxides which assist in adherence to the base metal and form a layer for the top coat to adhere to. Once fired, the ground coat is fused to the base metal forming a permanent bond between glass and metal.

GTAW
Stands for "Gas Tungsten Arc Welding." Also known as TIG welding. This arc-welding process involves coalescence of the metal part by heating with an arc between a tungsten electrode and the work. Shielding is performed by

a gas mixture of argon and helium. Filler metal may or may not be used in this process.

hardening
Increasing the hardness of a metal by heating or cooling.

hardness
The descriptive term relating to the surface of the metal and its ability to resist impact.

heat strengthened
A general term referring to the controlled heating and cooling of a metal. By heating and rapidly cooling, some metals will harden and increase in yield strength.

heat tinting
Coloration process on a metal surface by generation of an oxide layer. Usually color is obtained through light interference.

heat treated
See heat strengthened.

highlighting
The selective polishing of portions of the metal surface. Usually performed on the raised surface created by embossing or on areas of ridges and folds to enhance the appearance.

hot dipping
Immersing one metal into a molten bath of another metal to create a thin coating.

hot forging
Repeated hammering of metal into a die at an elevated temperature, near plasticity.

hydrogen embrittlement
A condition of low ductility caused by the absorption of hydrogen.

Incralac
A clear acrylic lacquer coating developed by the International Cooper Research Association. This lacquer coating contains benzotriazole for underfilm tarnish and corrosion resistance.

interference
A phenomenon whereby a wavelength of light undergoes changes induced by the encounter with another wavelength.

intergranular corrosion
Corrosion occurring at the boundaries of grains. This is a galvanic corrosion condition whereby a polarity is developed between grain boundaries.

investment casting
Also known as the "lost wax" method of casting. In this process, a wax mold is created. From this, a ceramic shell is applied to the wax mold. The wax is then removed by steam or other heating method to leave the ceramic shell. The metal to be cast is applied into the ceramic shell.

killed steel
Adding a strong deoxidizing agent such as silicon or aluminum in order to remove oxygen and create a steel that does not have a reaction between carbon and oxygen during initial solidification.

knurling
The opposite of embossing. The design is imprinted into the metal by passing between dies which imprint into the surface.

lead burning
Colloquial term used to describe welding of lead surfaces.

lost wax process
Investment casting in which wax is used and melted away during the casting process.

malleable
Used to describe a metal's ductility

martensitic
Refers to the austenite frozen by quenching. This traps carbon in a body-centered cubic crystal structure.

matte finish
A satin or low reflective finish. Characterized by the lack of gloss and reflectivity.

metallic
A term used to describe the deep reflective nature seen in polished metal surfaces.

metallic bond
The principle bond between metal atoms.

metal spraying
Coating a metal or other object with molten metal sprayed onto the surface. Also known as thermal spraying.

MIG welding
Acronym stands for "Metal Inert Gas" welding operations. Represents a shielding gas operation while metal is fed into the weld zone. See GMAW.

mill
The term used to describe the plant where metal is created into sheet, coil, bar, extrusion, wire, plate, or cast billet form.

mill finish
The finish on the metal surface initially provided at the mill.

mill scale
The heavy oxides on the surface of a metal when it is first formed at the mill.

minimized spangle
A coating of zinc applied by the hot dipping method with very small grain size development. It provides a smoother surface.

mischmetal
A naturally occurring mixture of rare earth elements. One of the constituents of the metal coating Galfan.

modulus of elasticity
A measurement of the elasticity of a material. It is a ratio of stress to strain.

native metal
A naturally occurring high-purity form of metal.

noble metal
A relative term describing the higher electrical potential of one metal in relation to another metal.

oil canning
A metaphorical term describing the visible distortion of a flat metal surface. Also described as the instability of flat plate diaphragms.

orange peel
An irregularity characterized by a rough, pebble-like surface created when a coarse-grained metal form is stressed beyond its elastic limit. Can occur when deep drawing.
Also a characteristic of paint and ceramic coatings applied too thick.

ore
The mineral form of a metal that is mined for refinement.

passivation
The creation of a metal surface that is less reactive to external effects.

patina
Describes the surface oxide, carbonate, sulfate, or chloride that develops on the surface of a metal.

PAW
Acronym stands for "Plasma Arc Welding." This process involves coalescence by heating with a constricted arc between an electrode and the metal form. Shielding is obtained from the ionized gas sometimes supplemented with a shielding gas mixture exiting from an orifice surrounding the electrode.

peening
The mechanical working and finishing of a metal surface by application of metal shot or small hammers.

perforating
The piercing of metal and the creation of holes or slots.

periodic table
The classification of the elements in accordance with the periodic system. This system arranges the elements by atomic numbers so that those with similar properties are grouped together.

permanent mold
A metal, graphite, composite, or ceramic mold created for repeated castings.

pewter
A tin alloy. Usually alloyed with small amounts of copper. In the past, it had lead alloyed to produce a darker color.

pickling
The removal of surface oxides (scale) by chemical action. Usually acids such as hydrochloric and hydrofluoric acid.

pig
An initial metal casting used for later recasting.

pig iron
High carbon iron made from the initial reduction of iron ore in a blast furnace.

pigment
Various organic or inorganic compounds mixed into paints and held in suspension to provide color and gloss characteristics.

pitting
Describes the result of a corrosion process that dissolves metal in a very small localized region. The term also is used to describe a surface malady found on hot rolled sheet metal. Characteristics are small indentations found on the metal surface.

plasma arc cutting
A metal cutting process that utilizes an electrical arc in a restricted orifice and a jet of hot ionized gas. The electrical arc melts the metal in a very localized region and the jet of gas blows the metal away.

plasma spraying
A thermal spraying process involving the melting of one metal and spraying onto another surface. Molten metal is propelled against the surface by a jet of hot gas exiting from a plasma torch.

plasticity
The ability of a metal to deform without rupture and without elastic return.

plating
The application of a thin, adherent layer of one metal onto another.

polarity
The development of positive and negative charges at opposing ends of a cell.

porcelain
Inorganic, vitreous coatings bonded to a metal substrate by fusion at very high temperature.

powder metallurgy
The science of producing metal powders and using these powders to produce shaped metal forms.

precipitation hardening
Hardening in metal created by the precipitation of a particular alloying constituent from a saturated solid state.

press
A device used to form, punch, or pierce metal by inducing a ram perpendicular to a fixed platform.

press brake
A device used to form, punch, or pierce metal by lowering a ram onto a die. The characteristic of this device is an open frame and a linear bed in which the die is confined. A complementary die is held in the ram.

pretreatment
The process of treating a metal surface for subsequent finishing processes. Pretreatments can consist of simple cleaning and surface etching, or they can involve the generation of complex conversion coatings.

punching
Similar to press forming except the metal is generally pierced by the die. The hold created matches the configuration of the die doing the piercing.

quenching
Rapid cooling of a heated metal.

recrystallization
Formation of a new grain structure; usually accomplished by heating (annealing).

resin
A group of substances obtained as gums from trees or manufactured synthetically. These are soluble in organic solvents and are used in varnishes and paints.

resist
A film applied to the surface of the metal to prevent electrical conductivity or to protect against abrasive action. Also used to prevent selective contact with etching solutions.

riser
A localized reservoir connected to section of a casting to act as a feed for additional molten metal into the casting as it cools or shrinks.

Rockwell hardness
An indentation hardness test based on depth of dent created by a specific penetration tool into a flat specimen under certain fixed conditions.

roller leveling
Leveling of sheet metal by passing through a device having a series of small, staggered rolls that produce bending in first one direction, then in the opposite direction.

roll forming
The linear deformation of sheet metal in a series of steel rolls. Each roll produces a portion of the shaping as the metal moves down the series.

rosin paper
Paper saturated with a powdered form of the common resin produced from several varieties of pine trees. It has been alkali treated to neutralize the acidity.

sacrificial metal
A metal that is coupled with another metal, usually by coating the other metal, in which the coating metal is more active in certain environments than the base metal. The metal will then undergo sacrificial action in the environment, saving the base metal from corrosive attack.

sandblasting
High-pressure application of clean dry sand to impart course surface finish or to remove surface oxides and scale.

satin finish
A matte finish or semi-reflective finish with low reflectivity. Usually associated with small linear polishing lines and a diffuse reflectivity.

scale
Heavy, tenacious oxide on the surface of metal generated by the hot initial casting.

selective plating
The plating of exposed regions of a base metal with another metal. The nonexposed regions are masked with a resist material capable of resisting the electrical and chemical action of the plating tool and plating solution.

Sendzimir mill
A cluster of rolls with small rolls applied against the metal and massive backup rolls providing pressure to the small rolls.

shear
(1) The perpendicular applied force necessary to overcome the bonds holding a metal together. (2) A device used to cut metal. Involves the sliding motion of a blade applied at pressure and at right angles to the metal surface.

shear strength
The stress required to produce fracture in a metal's cross section. The forces are applied perpendicular to the planes of the metal surface.

sherardizing
A method of galvanizing. Used on small parts by applying a zinc dust onto a heated part.

shot blasting
See abrasive blasting. Difference is shot blasting refers to the use of steel shot in lieu of various blast media.

shot peening
See peening.

silver soldering
Denotes brazing of nonferrous metals using a silver base alloy filler material.

sintering
The bonding of adjacent surfaces of metals in a mass of metal particles by atomic attraction. The metals are heated to temperatures below the melting point of any one constituent.

slip
The slurry coating applied to a metal surface before firing. The frit is milled in water with clay and coloring pigments are added to create an emulsion called a slip.

SMAW
Acronym stands for "Shielded Metal-Arc Welding." This arc welding process joins the metal by heating with an electric arc between a covered metal electrode and the work. Shielding is produced as the covering of the metal electrode decomposes. Filler metal is the consumed electrode.

smelting
Thermal process used to refine ore. The ore is heated in a furnace and the natural impurities are removed to further purify the metal.

smut
A undesirable product left on the surface of metal after undergoing plating, etching, or pickling. The product is a chemical reaction with impurities within the metal.

soldering
A process that joints metal together by heating two clean surfaces and applying a filler metal in the molten state. Molten metal is applied to the clean surfaces by capillary action.

solution heat treatment
The heating of an alloy to a temperature and then sustaining that temperature for a period of time necessary to cause one or more alloying constituents to enter back into solid solution. Followed by rapid quenching to bind these constituents into solution.

specific gravity
The ratio of the density of a material to that of water.

spinning
The forming of a seamless metal part by forcing a mandrel into a spinning sheet held over a die. The metal, while spinning, conforms to the die creating a circular or cylindrical configuration.

sponge
A porous form of metal that is the result of the decomposition of a compound without fusion. Forms of iron and titanium are manufactured this way.

spot welding
The welding by electrical resistance of two overlapping surfaces of metal.

sputtering
The application of a thin film of metal onto the surface of another metal, glass, or other substance by atomizing in a hot gas jet.

stamping
A general term referring to almost all press operations.

statuary finish
A conversion coating generated on copper alloys by reactions with various acidic solutions to produce degrees of light bronze to black tones.

steel
A generic term referring to iron and carbon base alloys.

strain
The measurement of the change in dimension when a metal undergoes stress.

strain hardening
An increase in hardness and strength due to plastic deformation caused by cold working the metal.

stress
Force per unit area.

stress-corrosion cracking
Failure of a metal part due to exposure to the combined influences of pollution and stress.

stretcher leveling
The equalization of stresses in a metal sheet by pulling the sheet from opposing ends.

superplasticity
A characteristic of certain metals to undergo extreme plastic deformation without thinning or rupturing.

tack welding
A small, localized weld used to restrain and align a part for further welding.

tarnish
A thin oxide layer that develops on a metal surface. The oxide causes dulling and discoloration of the metal surface.

temper
The hardness and strength produced by mechanical working and thermal treatment.

temper rolling
A light rolling operation on steel and stainless steel to improve the surface appearance. The temper of the sheet is increased only slightly.

tensile strength
The ratio of maximum load to unit cross section. Also known as ultimate strength.

terne
A term given to describe a lead-tin alloy coating.

thermal spraying
See plasma spraying.

TIG welding
An acronym representing "Tungsten Inert Gas." *See* GTAW.

tinning
The immersion of a metal in a molten bath of lead-tin alloy. Typically associated with the coating applied to edges rather than the entire sheet surface.

toughness
The ability of a metal to absorb impact and deformation without rupturing.

tough pitch copper
See fire refined copper.

transverse seam
The seam on a roof or wall panel running perpendicular to the length of the panel.

triple spot test
A test performed on galvanized coatings to establish the thickness of the coating. Three samples are taken from various regions of a sheet of metal. The thickness, or rather the weight per unit area, of the zinc coating is established and the average is taken and compared to a chart to establish the ASTM coating rating.

ultimate strength
See tensile strength. The maximum stress a metal can withstand before rupture.

vacuum deposition
The deposition of a thin metal coating on a cool metal surface while in a vacuum.

vapor degreasing
The removal of foreign particles such as oils, dirt, and grease by suspending the metal part in the vapors occurring over a boiling solvent. The vapors are denser than air and thus the metal can realistically be immersed in them.

vapor deposition
The deposition of a metal onto another heated metal surface by reducing a compound at temperatures below the melting point of the base metal. The reduction process vaporizes in the heated atmosphere and redeposits on the metal surface.

Vickers hardness test
A test performed on metals in which a diamond pyramid indenter is applied at various loads. Readings are taken from the indentor. The Vickers hardness test enables the testing of various materials from soft lead to hardened steels.

welding
The fusion of at least two metal segments by application of high temperature or high pressure, or both. Filler metals may be used in the process. *See* TIG and MIG welding.

white bronze
A term used by some to describe the copper alloy Nickel Silver. Other names for Nickel Silver are German Silver and Dairy Bronze.

work hardening
See strain hardening.

wrought
The term given to describe the form of metal, both ferrous and nonferrous, making up the categories sheet, strip, plate, wire, bar, and rod. It is distinguished from cast forms and extruded forms.

wrought iron
A high iron silicate form of iron. More malleable than most other forms of iron.

yield strength
The stress at which a metal initially deviates from a consistency in the relation of stress and strain. A 0.02% offset in the stress-strain ratio is usually considered for most metals.

Bibliography

General

American Society of Metals. *ASM Metals Reference Book,* 2nd ed. Metals Park, OH: American Society for Metals, 1986.

Battelle Columbus Laboratories. *Corrosion of Metals in the Atmosphere,* MCIC-74-23. Columbus, OH: Metals and Ceramics Information Center, 1974, pp. 20–67.

Brown, D. *Metallurgy Basics.* New York: Van Nostrand Reinhold Company, Inc., 1983.

Cubberly, et al. *Tool and Manufacturing Engineers Handbook,* Desk Edition. Dearborn, MI: Society of Manufacturing Engineers, 1989.

Delmonte, J. *Origins of Materials and Processes.* Lancaster, PA: Technomic Publishing Company, Inc., 1985.

Dieter, G. *Mechanical Metallurgy.* New York: McGraw-Hill Book Company, 1986.

Gayle, M., Look, D., and Waite, J. *Metals in America's Historic Buildings.* Washington, D.C.: U.S. Department of the Interior, 1992.

Godard, H.P. *The Corrosion of Light Metals.* New York: John Wiley and Sons, Inc., 1967.

Horne, T. "Corrosion." *AOPA Pilot* (May 1982), pp. 49–55.

Ross, R. *Metallic Materials Specification Handbook.* New York: E. & F.N. Spon Ltd., 1980.

SMACNA. *Architectural Sheet Metal Manual,* 5th ed. Chantilly, VA: SMACNA, 1993.

Smith, C.S. *A History of Metallography.* Cambridge, MA: The MIT Press, 1988.

Uhlig, H. *The Corrosion Handbook.* New York: John Wiley and Sons, Inc., 1948.

Uhlig, H. and Revie, R.W. *Corrosion and Corrosion Control.* New York: John Wiley and Sons, Inc., 1985.

Venetsky, S. *From the Camp Fire to the Plasma.* Moscow: Mir Publishers, 1989.

Venetsky, S. *Tales About Metals.* Moscow: Mir Publishers, 1981.

Wick, et al. *Tool and Manufacturing Engineers Handbook*, Volume III, Materials, Finishing and Coating. Dearborn, MI: Society of Manufacturing Engineers, 1985.

Wilkes. *The Encyclopedia of Architectural Design, Engineering and Construction*, Volume 4. New York: John Wiley and Sons, Inc., 1989.

Aluminum

AAMA. *Aluminum Curtain Wall Design Guide Manual.* Chicago: Architectural Aluminum Manufacturers Association, 1979.

Aluminum Association. *Aluminum Construction Manual*, Section A. New York: The Aluminum Association, 1966.

Aluminum Association. *The Aluminum Extrusion Manual.* Washington, D.C.: The Aluminum Association, 1987.

Cassidy, V. "Anodizing: Two-Step Becomes Three-Step." *Modern Metals* (October 1991), pp. 22–25.

Jarvis, E. and Jarvis, H. *Facts for Foundrymen*, 6th ed., Niagara Falls: The Niagara Falls Smelting and Refining Division, 1946.

Mahn, R. "Why Electrolytic Coloring of Aluminum?" *Plating and Surface Finishing* (October 1986), pp. 32–35.

NAAMM. *Metal Finishes Manual*, Finishes for Aluminum, AMP 501-88. Chicago: National Association of Architectural Metal Manufacturers, 1988.

Peters, John. *Aluminum in Modern Architecture*, Volume I. Louisville: The Reynolds Metal Company, 1956.

Weidlinger, P. *Aluminum in Modern Architecture*, Volume II. Louisville: The Reynolds Metal Company, 1956.

Copper and Copper Alloys

Anaconda American Brass. *Architectural Metals.* Waterbury, CT: Anaconda American Brass Company, 1960.

Anaconda American Brass. *Copper Metals by Anaconda:* Their Properties and Applications. Waterbury, CT: Anaconda American Brass Company, 1961.

Copper Development Association. *Copper Brass Bronze Design Handbook*, Architectural Applications. Greenwich, CT: Copper Development Association, 405/7R.

Copper Development Association. *Copper Brass Bronze Design Handbook*, Sheet Copper Applications. Greenwich, CT: Copper Development Association, 401/OR.

Copper Development Association. *Standards Handbook*, Cast Copper and Copper Alloy Products. New York: Copper Development Association, 1978.

Copper Development Association. *Standards Handbook*, Copper Brass Bronze Wrought Products. Greenwich, CT: Copper Development Association, 1985.

Copper Development Association. *Standards Handbook*, Wrought and Cast Products, Specification Index. Greenwich, CT: Copper Development Association, 1988.

Copper Development Association. *Standards Handbook*, Wrought Copper and Copper Alloy Mill Products. New York: Copper Development Association, 1983.

Crosbie, M.J. "Restoring a Lady's Tarnished Beauty," *AIA Journal* (July 1984), pp. 43–53.

Fishman, et al. *Observations on Atmospheric Corrosion Made of Copper Work at Yale.* Philadelphia: American Society of Testing Materials, 1988, pp. 96–114.

Gowen, J. *Modern Applications of Sheet Copper in Building Construction.* New York: Copper and Brass Research Association, 1955.

Hughes, R. and Rowe, M. *The Colouring, Bronzing and Patination of Metals.* New York: Watson-Guptill Publications, 1991.

Leidheiser, H. *The Corrosion of Copper, Tin, and Their Alloys.* New York: John Wiley and Sons, Inc., 1971, pp. 3–27.

NAAMM. *Metal Finishes Manual,* Finishes for the Copper Alloys, AMP 502-88. Chicago: National Association of Architectural Metal Manufacturers, 1988.

Revere Copper and Brass, Inc. *Copper and Common Sense.* New York: Revere Copper and Brass, Inc., 1982.

Metallic Coatings

Britton, S.C. *Tin Versus Corrosion,* Publication No. 510. Middlesex, England: International Tin Research Institute, 1975.

International Tin Research Institute. *Pewter,* Publication No. 494. Middlesex, England: International Tin Research Institute, 1975.

International Tin Research Institute. *Working with Pewter,* Publication No. 566. Middlesex, England: International Tin Research Institute, 1975.

Safranek. *The Properties of Electrodeposited Metals and Alloys.* Orlando, FL: American Electroplaters and Surface Finishers Society, 1975, pp. 445–447.

Painted Metals

Freeman, D. *Phosphating and Metal Pre-Treatment.* New York: Industrial Press, Inc., 1986.

Mock, J. *Introduction to Prefinished Metals.* Lancaster, PA: Technomic Publishing Company, Inc., 1983.

NAAMM. *Metal Finishes Manual,* Applied Coatings, AMP 505-88. Chicago: National Association of Architectural Metal Manufacturers, 1988.

Stainless Steel

Ailor, W. *Atmospheric Corrosion.* New York: John Wiley and Sons, Inc., 1982.

Allegheny Ludlum Steel Corp. *Stainless Steel Handbook.* Pittsburgh, PA: Allegheny Ludlum Steel Corp., 1959.

Baker, E.A. and Lee, T.S. *Long-Term Atmospheric Corrosion Behavior of Various Grades of Stainless Steel.* Philadelphia: American Society for Testing and Materials, 1988, pp. 52–67.

Burstein, T. "Revealing Corrosion Pits." *Nature.* Vol. 350 (March 1991), pp. 188–190.

Kane, R. *Super Stainless Steels Resist Hostile Environments, Advanced Materials & Processes.* Materials Park, OH: American Society of Metals, (July 1993), pp. 16–20.

Kearns, J.R., Johnson, M.J., Pavik, P.J. *The Corrosion of Stainless Steels in the Atmosphere.* Philadelphia: American Society for Testing and Materials, 1988, pp. 35–51.

Lula, R.A. *Stainless Steel.* Metals Park, OH: American Society of Metals, 1986.

NAAMM. *Metal Finishes Manual,* Finishes for Stainless Steel, AMP 503-88. Chicago: National Association of Architectural Metal Manufacturers, 1988.

Nickel Development Institute. *Answers for Architects.* Toronto: Nickel Development Institute, January 1988.

Nickel Development Institute. *Technical Manual for the Design and Construction of Roofs of Stainless Steel Sheet.* Toronto: Nickel Development Institute, April 1989.

Smits, B.A. *Architecture—A Demanding Market for Stainless Steel.* Toronto: Nickel Development Institute, August 1986.

Specialty Steel Industry. *Designer Handbook,* Design Guidelines for the Selection and Use of Stainless Steel. Washington, D.C.: Specialty Steel Industry of the United States, December 1993.

Specialty Steel Industry. *Designer Handbook,* Stainless Steel Architectural Facts. Washington, D.C.: Specialty Steel Industry of the United States, December 1993.

Specialty Steel Industry. *Designer Handbook,* Stainless Steel Fasteners. Washington, D.C.: Specialty Steel Industry of the United States, December 1993.

Stainless Steel Producers. *Cleaning and Descaling Stainless Steels.* Washington, D.C.: American Iron and Steel Institute, 1982.

Steel

American Iron and Steel Institute. *Steel Products Manual,* Sheet Steel. Washington, D.C.: American Iron and Steel Institute, October 1979.

NAAMM. *Metal Finishes Manual,* Finishes for Carbon Steel and Iron, AMP 504-88. Chicago: National Association of Architectural Metal Manufacturers, 1988.

Zinc

COMINCO. *An Architect's Guide to Specifying and Using Zinc Sheet.* Cominco LTD/Product Technology Center, January 1993.

Rheinzink. *Architecture with Rheinzink—Roofing and Wall Cladding.* Datteln, Germany: Rheinzink GMBH, January 1988.

Index

Photography Credits

Dan White, White and Associates
Timothy Hursley, The Arkansas Office

Craig Sands, Kansas City Star Photographer
Don F. Wong, Don F. Wong Photography
Rheinzink Canada LTD., Rheinzink Canada LTD.

Robbins Bell Kreher Arch., Robbins Bell Kreher Arch.
Thane Brethour, Thane Brethour Photography
James L. Stanfield, National Geographic Society
Japan Titanium Society, Japan Titanium Society
KPK Stainless, KPK Stainless
Bruce Matthews, Matthews Communication
Doug McCurry, Doug McCurry
Rafael Architects, Rafael Architects
Jordan Mozer, Jordan Mozer and Associates
Mackey Mitchell Architects, Mackey Mitchell Architects
Nick Merrick, Hedrich Blessing
Mike Christianer, Shaughnessy Fickel Scott Architects
Rick Darby, Rick Darby
Frank O. Gehry Associates, Frank O. Gehry Associates
Timet Industries, Timet Industries
Vicki Eudaly, Vicki Eudaly
Wolfgang Hoyt, Wolfgang Hoyt
Andrew Wood, Andrew Wood

KCPL Building; black and white photos
American Heritage Museum;
    Harold Washington Library
Bartle Hall Expansion with helicopter
cover photo—Weisman Art Museum
Carmel Convent; curved zinc roofing in
    Europe; Stamped shingles
University of Tampa
RLDS Temple Complex
Hagia Sophia
Nigata City Aquarium; Suma Aualife Park
dB Soft Building
Bartle Hall Expansion—overall view
Grady Hospital
Missouri Public Utilities
Cyprus Club
Custom Planter
Sears Tower
Johnson County Detention Center
RLDS Temple Complex
Barcelona Fish Sculpture
Private residence along the coast
AMA Headquarters
Liberty Science Center
Felix Restaurant